MIX
Papier aus verantwortungsvollen Quellen
Paper from responsible sources
FSC® C105338

Cliff Orori Mosiori
Walter Kamande Njoroge

High Speed Semiconductor Physics

Theoretical Approaches and Device Physics

Anchor Academic
Publishing

Mosiori, Cliff Orori, Njoroge, Walter Kamande: High Speed Semiconductor Physics.
Theoretical Approaches and Device Physics, Hamburg, Anchor Academic Publishing
2015

Buch-ISBN: 978-3-95489-432-1
PDF-eBook-ISBN: 978-3-95489-932-6
Druck/Herstellung: Anchor Academic Publishing, Hamburg, 2015

Bibliografische Information der Deutschen Nationalbibliothek:
Die Deutsche Nationalbibliothek verzeichnet diese Publikation in der Deutschen
Nationalbibliografie; detaillierte bibliografische Daten sind im Internet über
http://dnb.d-nb.de abrufbar.

Bibliographical Information of the German National Library:
The German National Library lists this publication in the German National Bibliography.
Detailed bibliographic data can be found at: http://dnb.d-nb.de

All rights reserved. This publication may not be reproduced, stored in a retrieval system
or transmitted, in any form or by any means, electronic, mechanical, photocopying,
recording or otherwise, without the prior permission of the publishers.

Das Werk einschließlich aller seiner Teile ist urheberrechtlich geschützt. Jede Verwertung
außerhalb der Grenzen des Urheberrechtsgesetzes ist ohne Zustimmung des Verlages
unzulässig und strafbar. Dies gilt insbesondere für Vervielfältigungen, Übersetzungen,
Mikroverfilmungen und die Einspeicherung und Bearbeitung in elektronischen Systemen.

Die Wiedergabe von Gebrauchsnamen, Handelsnamen, Warenbezeichnungen usw. in
diesem Werk berechtigt auch ohne besondere Kennzeichnung nicht zu der Annahme,
dass solche Namen im Sinne der Warenzeichen- und Markenschutz-Gesetzgebung als frei
zu betrachten wären und daher von jedermann benutzt werden dürften.

Die Informationen in diesem Werk wurden mit Sorgfalt erarbeitet. Dennoch können
Fehler nicht vollständig ausgeschlossen werden und die Diplomica Verlag GmbH, die
Autoren oder Übersetzer übernehmen keine juristische Verantwortung oder irgendeine
Haftung für evtl. verbliebene fehlerhafte Angaben und deren Folgen.

Alle Rechte vorbehalten

© Anchor Academic Publishing, Imprint der Diplomica Verlag GmbH
Hermannstal 119k, 22119 Hamburg
http://www.diplomica-verlag.de, Hamburg 2015
Printed in Germany

ABOUT THE AUTHOR

Cliff Orori Mosiori -PhD (Physics), M.Sc Physics - (Electronics & Instrumentation); B. Ed (Sc.) – Physics /Chemistry - Kenyatta University , Kenya
cliffmosiori@gmail.com; mosiori@rvist.ac.ke, mosiori.cliff@ku.ac.ke

Cliff Orori Mosiori is a Solid State Physics lecturer at Kenyatta University and a Physics Masinde Muliro University of Science and Technology, a visiting Electronic Engineering lecturer at Rift Valley Institute of Science and Technology. He is also a serving solid state research scientist at Kenyatta University. He is an Editor to peer reviewed journals with the following international Journal publishers: *International Journal of Advanced Research in Physical Science* -(**IJARPS**); *Academic and Scientific Publishing* - (**ASP**); *Research Journal of Engineering and Technology* - (**RJET**) and *American Journal of Materials Engineering and Technology-* (**AJMET**). He has published the following books in Physics: *Inorganic Ternary Thin Films – Analysis of Optical Properties*- Anchor Academic Publishers; *Physics of Thermal Phenomena* – Introduction to Thermal Physics; *Digital Electronics*; *Optical Coating* – Coating Properties & UV Protection; *Thin Film Device Physics for Solar Cell Applications* & *Characterization of $Cd_xZn_{1-x}S$ and PbS Thin Films for Photovoltaic Applications*- Electrical and Optical Properties all with Lambert Academic Publishing.

PREFACE

Solid state physics is a sub-genre of condensed matter physics. Some learners consider it a very boring and tedious subject in Physics and even some call it a "squalid state". Once it is properly introduced, solid state physics can be a very favorite subject of all in physics because it is full of variety, excitement and deep ideas. In fact, condensed matter physics is by far the largest single sub-field of physics. In fact, new condensed matter areas now under investigations include *high speed semiconductor physics* on which this book attempts to discuss. It has proved to be extremely important in technology in that it is based on very fast performing electronic device including photonics and device physics in high speed electronic gadgets, superconductivity and point-group symmetries for high speed processing devices. New books on the use of high speed semiconductors in devices have been either directed to the practitioner by emphasizing the state of the art or to the university student by using major simplifications in a treatment that achieves analytical and closed mathematical format solutions because it is difficulty to describe numerical techniques and their validity in the restricted space of technical publications or the limited time in the lecture halls. Recent and modern devices, with their small dimensions require an understanding of the physics of reduced dimensions, the use of statistical methods, and the use of one-, two-, and three-dimensional analytical and numerical analysis techniques. These techniques bring alternate approximations and simplifications. This book is an attempt to find a common ground for graduate students and lecturers teaching this subject.

This subject is not oversimplified in this textbook specifically graduate students interested in learning more about high speed semiconductor devices and their theoretical analysis as small dimensional devices but also made very appealing to graduate students who have already achieved a good understanding of the principles of semiconductors and related devices like transistors that are used in fast switching. Both one-dimensional classical approximations and numerical procedures for better accuracy have been incorporated in this book in order to clarify semiconductor behaviuor consistency. Students will find very unusual mathematical features or approaches. They are requested to carefully recognize their origin either from their underlying nature or the mathematical approximation of each model used. Most chapters discussed here have been designed to emphasize the concepts of semiconductor behavior to complement arguments or derivations made in other chapters.

Topics covered in this book are built in standard solid state physics references available in most online libraries or in other books. Studies on the complexity high speed semiconductor physics arises from condensed solid state matter. We believe that the content covered in this book gives a deep coverage in some topics or sections and gloss coverage on others. These topics are likely to differ a great deal from what is deemed important elsewhere in other books or literature. We know that there are many extremely good books on solid-state physics and condensed matter physics but very few books are restricted to high speed semiconductor physics though there are many good resources online.

Chapter one covers the general semiconductor qualities that make high speed semiconductor devices effect and includes the theory of crystals, diffusion and its mechanisms, while chapter two covers solid state materials, material processing for high speed semiconductor devices and an introduction to quantum theory for materials in relation to density of states of the radiation for a black body and its radiation properties. Chapter three discuss high speed semiconductor energy band theory, energy bands in general solid semiconductor materials, the Debye model, the Einstein model the Debye model and semiconductor transport carriers in 3D semiconductors while chapter four discuss effect of external force on current flow based on the concept of holes valence band, and lattice scattering in high speed devices.

Chapter five briefly describes solid state thermoelectric fundamentals, thermoelectric material and thermoelectric theory of solids in lattice and phonons while chapter six scattering in high field effect in semiconductors in inter-valley electron scattering and the associated Fermi Dirac statistics and Maxwell-Boltzmann approximation on their carrier concentration variation with energy in extrinsic doping chapter seven covers p-n junction diodes, varactor diode, *pin* diode Schottky diode and their transient response of diode in multi-valley semiconductors. Chapter eight discusses high speed metal semiconductor field effect transistors.

(*Author*)

Dr. Cliff Orori Mosiori
Department of Physics
Kenyatta University,
P. O. Box 43844 – 00100
Nairobi

Edited by:
Dr. Walter Kamande Njoroge
Department of Physics
Kenyatta University,
P. O. Box 43844 – 00100
Nairobi

DEDICATION

I wholeheartedly and most sincerely dedicate this book to my loving wife **Alice Mwango Nyamwange,** the firstborn daughter to Mzee Johnson Nyamwange Nyakenyanya who without fail gave me constant encouragement and the much love I desperately needed as I was writing this book. I specially dedicate this book to all my loving children; **James Ogoti** Orori, (now in Form two at Michinda High School – Elburgon), **John Okongo** Orori (a standard seven pupil at Workers Primary School – Nakuru), **Nehemiah Mienya** Orori (in standard two at Workers Primary school) and **Joyce Kerubo** Orori – (at Mary Hill Academy- Nakuru), a daughter in whom I see the love for my late mother replicated. A special dedication to my loving father, Mr. **Mosiori Maigo Ontiri** and my late mother, Mrs. **Ebisibah Nyaboke Mosiori,** without forgetting my brothers; James Twara Mosiori, Josphat Ombwera Mosiori, Cedric Nyanguka Mosiori, Elikana Moenga Mosiori, Robert Omosa Mosiori, Peter Maigo Mosiori and sisters; Jane Nyanchama Mosiori, Everlin Mosiori and Mary Nyamunsi Mosiori, who gave me a lot of support. To my brother Mr. **Cedric Nyanguka Mosiori** who inspired me without knowing it during his High School days at Kakamega Technical School in the mid of 1985 when he made several attempts to publish a book in the form of a Novel but never succeeded because he could not afford the cost of publishing a book in Kenya. I salute you all my Teachers from high school (Kisii High School), University lecturers, Kenyatta University and Academic Advisors, Dr. Walter Kamande Njoroge, Chairman, Department of Physics and Prof John Okumu, DVC- Academic, Kenyatta University, Kenya. I will not forget Kenyatta University that shaped my life. May God bless the support I received from Rift Valley Institute of Science and Technology lectures who among them; Mr. David Maru, Mr. Sangiriaki, Mr. Chege, Mrs. Mary Maina, Mr. Mutai, the Rotich's and the chief workshop technician, Mr. Brian and all the others. To KITI SDA church that shaped my spiritual and moral life, I dedicate this book to you too. I dedicate it to you all that know me.

My parents
Mr. Mosiori Maigo Ontiri & Mrs. Ebisibah Nyaboke Mosiori
*They gave me the **only thing** they did **not have**; Education*

My Academic Advisers
Dr. *Walter Kamande Njoroge & Prof John Okumu - Kenyatta University*
They gave me a Golden heart to work, learn and research in the Scientific World.

Alice Mwango Nyamwange

She is my brilliant and beautiful wife,
Without whom I would have been nothing.
She gave me Three Handsome Sons;
Eldest Ogoti, Okon'go and Mienya,
And a beautiful daughter, Joyce Kerubo
In place of My Mother and Her Mother too
Who she named after her humble but late mother
I admire her Brilliance, Beauty and above all, her Confidence
She always Comforts, Consoles and Encourages me.
*Cares for me and **NEVER** Complains nor Interferes with my work,*
*She Asks Nothing, Endures all for **ME** to make me feel **Full***
*And writes my **Heart** with Dedications, Encouragements*
And lots of Love Songs but above all
Without Erasing its Loving Long lasting Effects
Oh Dear Alice…. !!!
From the bottom of my Ears and Heart,
*I love **YOU***
Waooh !!!!...!!!

TABLE OF CONTENTS

ABOUT THE AUTHOR 5

PREFACE 7

DEDICATION 11

CHAPTER ONE
CRYSTALS IN SEMICONDUCTOR MATERIALS 17
    Crystals 17
    Strength of Crystals 19
    Plastic Behaviour 20
    Shear Strength Crystals 21
    Dislocation 23
    Etching 29
    Diffusion 35
    Colour centers 46

CHAPTER TWO
THEORIES FOR MATERIALS PROCESSING 69
    Material Technologies 69
    Material processing 70
    Quantum Theory 76
    Black body 81
    Bohr's Atomic Theory 87
    Schrödinger Equation 92

CHAPTER THREE
ENERGY BAND THEORY 107
    Energy Bands Theory in Solids 107
    Bond and Structures 108
    Energy band structure Theory 109
    The p–n junction 119
    Semiconductors Models for Heat Capacity 123
    The Debye Model 123
    The Einstein Model 124
    The Debye Model 125
    Semiconductor transport carriers 131

3D semiconductors Crystal structures 134

CHAPTER FOUR
BRILLOUN ZONES 137

    Bloch Theorem 139

    Eigen Value Equation... 143

    Bloch parameter, k 144

    Effect of External Force 145

    Current Flow in Crystals 148

    Concept of Holes... 149

    Valence Band 150

    Conduction Band... 151

    Mobility... 152

    Lattice Vibrations.. 152

    Actual vibration... 158

    Lattice Scattering for Mobility.. 160

CHAPTER FIVE
SOLID STATE THERMOELECTRIC FUNDAMENTALS 167

    Thermoelectric Materials 167

    The Absolute Scale or Thermodynamic Kelvin, Temperature Scale.................. 167

    Temperature-Dependent Effects 172

    Thermal Equilibrium and the Zeroth Law of Thermodynamics 174

    Thermodynamics Systems and Processes 177

    Thermodynamic Equilibrium of state variables 183

    Thermodynamic States and State Variables... 185

    Thermodynamic Processes for Pure Substances 189

    Electrical Equilibrium 194

    The thermal equation of state for a solid.. 201

    The Equation of State for an Ideal Gas 201

    The Equation of State for a Van Der Waal gas 206

    Equations of State for Other Systems 208

    System Parameters and Partial Derivatives.. 209

    The Partial Derivatives of State Variables 211

    Thermodynamics... 215

    The Thompson Coefficient... 216

Thermoelectric Theory of Solids 217
Lattice and Phonons 217
Phonon Dispersion 219

CHAPTER SIX
CARRIER SCATTERING 221
Mobility 222
Field effect due to Scattering 223
Electron Scattering 224
Carrie Density 225
Effect of Fermi Dirac statistics in scattering 230
Distribution of Electron due to scattering 231
Maxwell-Boltzmann Approximation in carrier scattering 233
Extrinsic Doping and Carrier Scattering 236
Generation Recombination process 241

CHAPTER SEVEN
SOLID STATE CONTINUITY EQUATIONS 247
Einstein and Continuity Equations 247
Semiconductor Diodes 252
P-N junction Diodes 252
Varactor Diodes 270
The P-I-N Diodes 275
Schottky diodes 277
Multi-Valley semiconductor Diode Theory 284
Diode Diffusion Theory 284

CHAPTER EIGHT
HIGH SPEED FIELD EFFECT TRANSISTORS 287
FET operation 287
Mobility in FETs 292
Hetero-junction in FETs 295
FET Current transport 299
BJT Models 301
The BJT Current Model 303
Hetero-junction Bipolar Transistor (HBT) 304
Hetero-junction FET 311

CHAPTER NINE
SWITCHING IN HIGH SPEED BIPOLAR JUNCTIONS 321

- Semiconductors and junctions 321
- Drift currents in semiconductors 322
- Metal-semiconductor junctions 327
- Semiconductor-semiconductor junctions 329
- p-n diode in high speed semiconductor devices 330
- Band diagram of unbiased high speed junction 331
- Long p-n diodes 336
- Minority carrier variations 340
- Switching delays in high speed pn-diodes 348

CHAPTER TEN
HIGH SPEED SWITCHING IN BIPOLAR TRANSISTORS 359

- BJTs Principle of operation 359
- Currents in a BJT 365
- High Speed Switching of the BJT 373
- Switching cycle 373
- BJT Schottky diode clamp 382
- Equivalent BJT circuit 382

APPENDIX 385

CHAPTER ONE
CRYSTALS IN SEMICONDUCTOR MATERIALS

Crystals

In ideal crystals atoms were arranged in a regular way but real crystals differ from that of ideal ones. Real crystals always have certain defects or imperfections and the arrangement of atoms in any volume of the crystal is imperfectly irregular. Real crystals always contain defects in abundance because of the uncontrolled conditions under which they are formed. These defects also affect the colour of a crystal making them their crystals valuable as gems. Even crystal prepared in advanced laboratories also contains defects. The characteristics of behavior of a material depend upon the material, type of defect and other properties. Properties like density and elastic constants are mostly proportional to the concentration of defects. Small defect concentrations will have a very small effect. There are other properties like the colour of an insulating crystal or the conductivity of a semiconductor crystal that are more sensitive to the presence of small number of defects.

The term defect carries with it the connotation of undesirable qualities. Defects are responsible for many of the important properties of materials. Material science involves the study and engineering of defects so that solids will have desired properties for specific applications. Therefore defect free crystals are very few. An ideal silicon crystal is of little use in modern electronics. The use of silicon in electronic devices is dependent upon the introduced small concentrations of chemical impurities such as phosphorus and arsenic which give it desired properties. Some properties of materials such as stiffness, mechanical strength, ductility, density and electrical conductivity which are termed as structure-insensitive in solid state physics because they are not affected by the presence of defects in crystals as shown in the figure below.

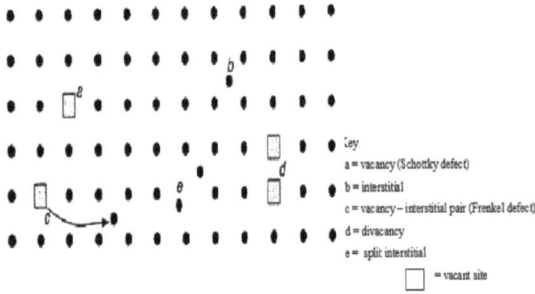

Properties like hysteresis and dielectric strength condition in semiconductors are termed as structure sensitive. They greatly affect minor changes in crystal structure and this are introduced due to defects or imperfections. Therefore, crystalline defects are classified the based on their geometry as follows:

    (i) Point imperfections
    (ii) Line imperfections
    (iii) Surface and grain boundary imperfections
    (iv) Volume imperfections

At dimensional level, the point defects are close to those found in interatomic space. Linear defects have a length of several orders of magnitude greater than their width while surface defects have a small depth with their width and length being several orders larger. Finally the volume defects which are regarded as pores and cracks have substantial dimensions in all measurements. All these defects contribute to the change of solid state properties of materials used in high speed electronic devices. It is traditional to make a distinction between the properties of the *lattice* and the properties of the *electrons*. The term lattice refers to the positions of the atoms themselves. Atoms are not stationary but they are considered to move slightly compared to the distances between the atoms. They vibrate about their average position but they do not move throughout the crystal. Most of the electrons are considered to be localized and thus they are always associated with the same particular atom. Localized electrons do not carry any current even when a force is applied. This makes them to be considered as part of the lattice. Although some of the electrons are essentially free to move throughout the solid, these free electrons determine the ability of a material to carry an electrical current.

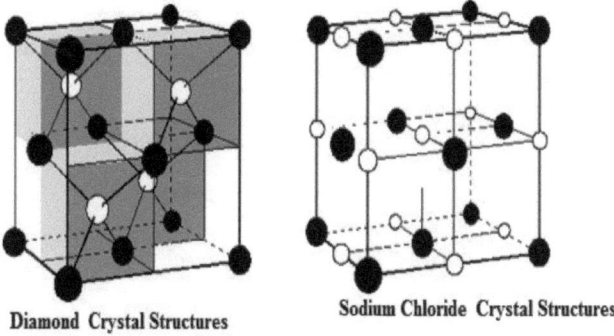

Diamond Crystal Structures   Sodium Chloride Crystal Structures

Silicon and Germanium forms the diamond crystal structure and all the atoms are identical while the PbTe and PbSe occur in the sodium chloride crystal structure. It is convenient to speak of the properties of the lattice and the properties of the electrons, but this division of properties is somewhat artificial. Sometimes it is important to remember that the lattice influences the behavior of the electrons and the electrons influence the behavior of the lattice. At the minimum, it is important to treat both systems on an equal footing for a reliable description of thermoelectric effects.

Strength of Crystals

Mechanical strength of a crystal depends on the amount of dislocations available. Plastic deformation in metals is caused by the formation of slip bands in which one portion of the material is sheared with respect to the other. Since metals are crystalline it is evident slip represent the shearing of one portion of a crystal with respect to the other upon a rational crystal plane. Some of the deformation operations correspond to slip while others correspond to a dislocation. The configuration below in; (a) shows the cylinder as originally cut (b) and (c) correspond to edge dislocations while (d) corresponds to screw dislocation.

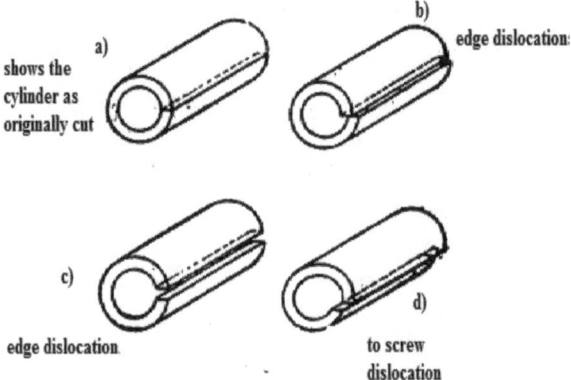

a) shows the cylinder as originally cut
b) edge dislocation:
c) edge dislocation
d) to screw dislocation

The intensity of X-ray beams reflected from actual crystals is about 20 times greater than that expected from a perfect crystal and therefore in a perfect crystal intensity is very low because of long absorption paths encountered by multiple internal reflections. This causes the width of the reflected beam from an actual crystal to be about longer than that obtained from a perfect crystal. This is because the actual crystal is small, roughly equiaxed crystallites and about 10^{-4} to 10^{-5} cm in diameter. This causes a slight disorientation with respect to each other resulting into the boundaries amorphous material. The disorientation explains the width of the beam. This is the Mosaic Block theory in which the size of the crystallites limits the absorption path and increases the intensity. These boundaries are actually arrays of dislocation lines.

A crystal can be deformed by simply applying stresses on it. If this stress is very large in the order of about 10^6 -10^7 dynes per cm^2 then a small deformation occurs and undergoes plastic deformation and this requires postulating a new type of defect called dislocations. Poorly prepared crystal are infested with dislocations and defects which interfere with each other's motion and as the crystals are purified and improved, dislocation largely move out of the crystal creating vacancies and interstitials. At low thermal equilibrium concentrations and the unimpeded motion makes it possible for the crystal to deform and the crystal is very soft.

Plastic Behaviour

Plastic deformation occurs when a crystal slides over another crystal with respect to the other and result in a slight increase in the length of the crystal ABCD under the effect of a tension FF applied to it as shown in figure below.

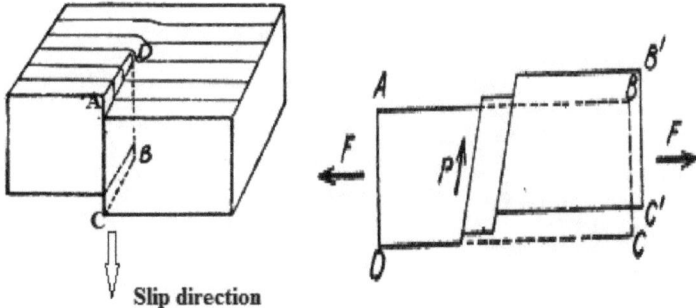

Slip direction

The process of sliding is called slip and the direction and place in which the sliding takes place are called the slip direction and slip plane respectively. The outer surface of the single crystal is deformed and a slip band is formed which can be observed by means of an optical microscope.

Shear Strength Crystals

The model given in the figure below which shows a cross-section through two adjacent atomic planes separated by a distance d is used to calculate the theoretical shear strength of a perfect crystal. In the figure, the full line circles indicate equilibrium positions of the atoms without any external force and τ indicate shear stress in the direction shown. Consider a case where all the atoms in the upper plane are displaced by an amount x from the original positions as shown by the dotted circles. Using the figure above, the shear stress τ is plotted as a function relative displacement of the planes from their equilibrium positions. The periodic behavior of τ is found to become zero for $x = 0$, a/2, a, ... etc. where **a** is the distance between the atoms in the direction of the shear. It is assumed that this periodic function is given by;

$$\tau = \tau_c \sin \frac{2\pi x}{a}$$

where the amplitude τ_c is the critical shear stress. To calculate such that for x << a, we have;

$$\tau = \tau_c \frac{2\pi x}{a}$$

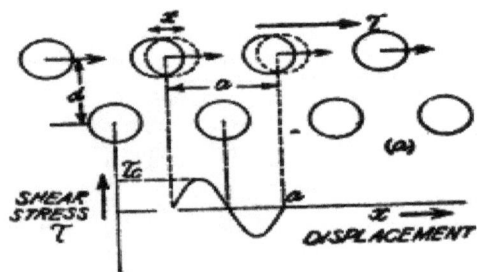

From the definition of shear modulus the force required to shear the two planes of atoms, we have;

$$G = \frac{Stress}{Strain} = \frac{\tau}{y} = \frac{\tau}{x/d} = \tau_c \frac{2\pi x}{a}$$

From the above formula G is the shear module such that

$$y = \frac{x}{d} = \frac{\tau}{G}$$

is the elastic strain to give;

$$\therefore \tau = G\left(\frac{x}{d}\right)$$

When the two equations are compared, we have;

$$\tau_c \left(\frac{2\pi x}{a}\right) = G\left(\frac{x}{d}\right) \quad or \quad \tau_c \frac{G}{2x}\left(\frac{a}{b}\right)$$

or

$$\tau_c \sim \frac{G}{2x} \sim \frac{G}{6}, \; if \; a \sim d$$

Based on this expression, the maximum critical stress above which the crystal becomes unstable is about one sixth of the shear modulus. In a cubic crystal, G c_{44} = 10^{11} dynes per cm for a shear in the <100> direction.

Dislocation

The Edge Dislocation

A dislocation is a region of a crystal in which the atoms are not arranged in the perfect crystal lattice. There are two extreme types of dislocations which are the edge type and the screw type. Any type of dislocation is a more complicated defect than any of the point defects. It is also noted that any particular dislocation is a mixture of these two types. Edge dislocation is the simplest as shown in figure below.

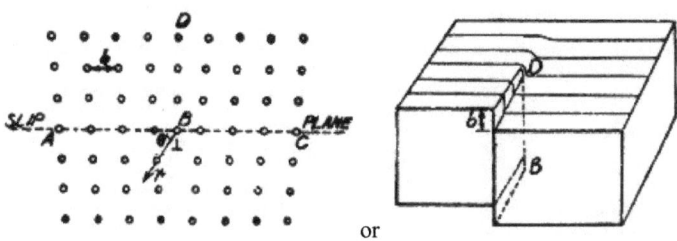

or

The part of the crystal above the slip plane at ABC has one more plane of atoms DB than the part below it. The line normal to the paper at B is called the dislocation line and the symbol at B is used to indicate the dislocation. All dislocations have intense near the dislocation line where the atoms do not have the correct number of neighbours and this region is called the core of dislocation. A few atom distances away from the center, the distortion is very small and the crystal is almost perfect locally. Dislocations are produced when the crystal solidifies from the melt. Plastic deformation of cold crystals also produces dislocations. Dislocations are of importance in determining the strength of ductile metals. These dislocations can be experimentally observed by many techniques. Electron microscopes can be used to study dislocations in their specimens of the order of a few angstroms which may transmit 100 kV electrons.

The Screw Dislocation

This type of dislocation called Burger's dislocation and was introduced by Burger in 1939. Consider a sharp cut made through a perfect crystal and the crystal on one side of the cut be moved down by one atomic spacing relative to the other as shown below.

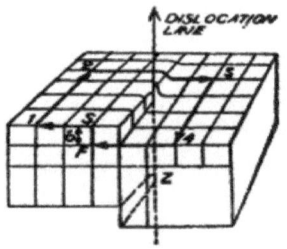

A line BD of distortion exists along the edge of the cut called the screw dislocation. The pitch of the screw may be left-handed or right-handed and one or more atom distances per rotation. The distortion is very little in regions away from the screw dislocation of while atoms near the center are in regions of high distortion so much so that the local symmetry in the crystal is completely destroyed.

Motion of a Dislocation

Many dislocations move just like the point defects lattices. However they are more constrained in motion because a dislocation must always be a continuous line. Dislocation move by a climbing or by a slipping or by a gliding as illustrated in the figure below. When the upper half is pushed sideways by an amount **b**, as shown then under shear motion tends to move the upper surface of the specimen to the right. Edge dislocations for which the extra half plane DB lies above the slip plane are called positive and if it is below the slip plane it is called negative edge dislocation. When the extra half plane BD reaches the right hand side of the block, the upper half of the block has completed the slip or glide by an amount b. Climb of a dislocation corresponds to its motion up or down from the slip plane. If the dislocation absorbs additional atoms from the crystal, it moves downward by substituting these atoms below B in the lattice. If the dislocation absorbs vacancies it moves up as the atoms are removed one by one from above B from the lattice sites.

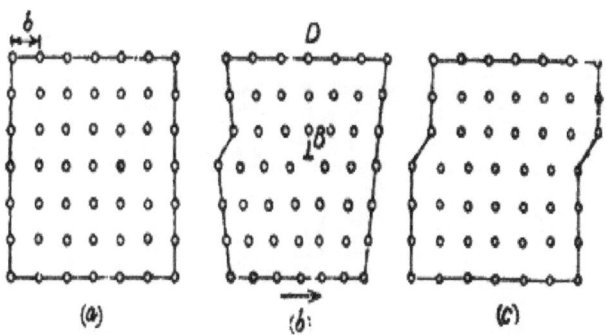

(a)   (b)   (c)

Dislocation Stress field

The Burger's Vector

The Burger's vector b above denotes the actually dislocation-displacement vector. Thus a dislocation can be described by a closed loop surrounding the dislocation line and this loop is called the Burger's circuit. Burger's circuit is formed by the undisturbed region surrounding a dislocation in steps. These steps are integral multiples of a lattice translation. They are theoretically completed by going an equal number of translations in a positive sense and negative sense in a plane normal to the dislocation line. Such a loop must close upon itself by an amount called a Burger's vector given by;

$$s = n_a a + n_b b + n_c c$$

where n_a, n_b, n_c are equal to integers or zero and a, b, c are the three primitive lattice translations.

The Burger's circuit demonstrated by *S-1-2-3-4-F* in the figure above show a screw dislocation. Starting at some lattice point S, the loop fails to close on itself by one unit translation parallel to the dislocation line and this is the Burger's vector pointing in a direction parallel to the screw dislocation.

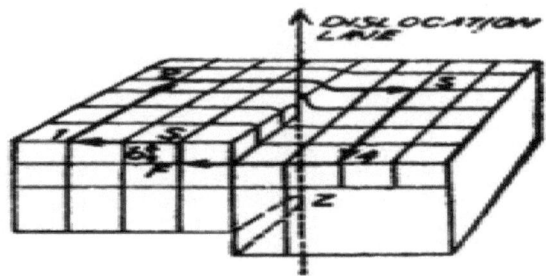

If the loop is continued, it will describe a spiral path around the Burger's dislocation just like the thread of a screw. Thus, b is a vector giving both the magnitude and the vector of the dislocation and must be in multiple of the lattice spacing so that an extra plane of atoms could be inserted to produce a dislocation. The dilatation Δ at a point near an edge dislocation can now be described by;

$$\Delta = \frac{\Delta V}{V} = \frac{b}{r}\sin\theta$$

where **b** is the Burger's vector.

This Burger's vector measures the strength of the distortion caused by the dislocation, in which r is the radial distance from the point to the dislocation line and θ is the angle between the radius vector and the slip plane. The atoms on a sheared lattice in screw dislocation being are displaced from their original positions according to the equation of a spiral ramp given by;

$$u_z = \frac{b}{2\pi}\theta$$

where the z-axis lies along the dislocation and u, is the displacement in that direction, θ is measured from one axis perpendicular to the dislocation. Thus when θ increases by 2π the splacement increases by a quantity b, the Burger's vector, which measures the strength of the dislocation. The Burger's vector of a screw dislocation is parallel to the dislocation line while that of an edge dislocation.

Stress Fields around Dislocations

The core of dislocations is a region within a few lattice constants of the centre of dislocation. The regions outside the core are stable regions and the strains in these regions are elastic strains. Consider a cylindrical shell of a material surrounding an axial screw dislocation.

Assume the radius of the shell be **r** and the thickness *dr*, then the circumference of the shell is $2\pi r$ and let the it shear by an amount b, so that the shear strain is given by;

$$e = \frac{b}{2\pi r}$$

This will result into a corresponding shear stress in the good region given by;

$$\tau = G.e. = \frac{G.b}{2\pi r}$$

where G is the shear modulus or modulus of rigidity of the material

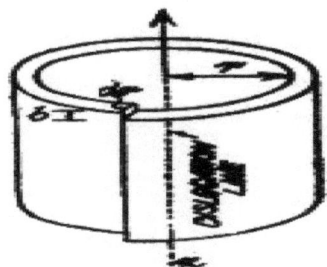

Considering the dimensions shown above, a distribution of forces is exerted over the surface of the cut for producing a displacement b result into work done by the forces to give an amount of energy E_s of the screw dislocation as

$$E_S = \int F.b.dA$$

The force, F is the average force per unit area at a point on the surface during the displacement and the integral extends over the surface area of the cut and therefore this the average force is obtained as;

$$F = <\tau> = (1/2)\tau$$

is half the final value when the displacement is b i.e.,

$$\therefore F = \frac{Gb}{4\pi r}$$

Substituting this force in the energy expression, we get;

$$E_S = \int \frac{Gb^2}{4\pi r} dA.$$

Since the expression dA = dz sr as per the dimensions used, then for a dislocation of length l, we have ;

$$E_S = \int_0^l \int_{r_0}^{R} \frac{Gb^2}{4\pi r} dA$$

$$= \frac{Gb^2}{4\pi} l \log \frac{R}{r_0}$$

The total elastic energy per unit length of a screw dislocation is thus given by;

$$E_S = \frac{Gb^2}{4\pi} \log \frac{R}{r_0}$$

where R and r_0 the proper upper and lower limits of r. This energy is found to depends upon the values taken for R and r_0 suitably and when it is equal to about the Burger's vector b or equal to one or two lattice constants and the value of R is not more than the size of the crystal.

Stress field of an edge dislocation

Consider the cross-section of a cylindrical material of radius R whose axis is along the z-axis and in which a cut has been in the plane y = 0, which becomes the slip plane. The portion above the cut is now slipped to the left by an amount b, the Burger's vector along the x-axis so that the new position assumes the shape shown dotted in figure below. The calculation of the stress field is done on the assumption that the medium is isotropic having a shear modulus G and Poisson's ratio υ. The positive edge dislocation is produced along the z-axis and if we assume that σ_{rr} is the radial tensile stress, along the radius r and $\sigma_{\theta\theta}$ is the circumferential tensile stress acting in a plane perpendicular to r. τr_θ will denote the shear stress acting in a radial direction.

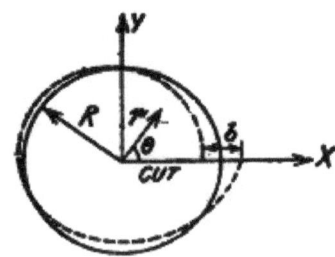

In isotropic elastic continuum, σ_{rr} and $\sigma_{\theta\theta}$ compression or tension acts in a plane perpendicular to r. It can be shown that the constants of proportionality in the stress vary as G and b of the edge dislocation in terms of r and θ are given;

$$\sigma_{rr} = \sigma_{\theta\theta} = \frac{Gb}{4\pi(1-v)} \cdot \frac{\sin\theta}{r}$$

and

$$\tau_{r\theta} = \frac{Gb}{2\pi(1-v)} \cdot \frac{\cos\theta}{r}$$

where the positive values of σ are for tension and negative values for compression. Above the slip plane σ_{rr} is negative giving a compression while below the slip plane corresponds to a tensile stress. It may be noted that for r = 0, the stresses become infinite and so a small cylindrical region of radius r_o around the dislocation must be excluded. The value of r_o, is calculated by letting r_o = b, so that the magnitude of the strain is in the order of $1/2\pi(1-v)\approx 1/4$ so that the energy of formation of an edge dislocation of unit length for edge dislocation will be given by;

$$E_e = \int \frac{1}{2} stress \, x \, strain = \frac{1}{2}\int_{r_0}^{R} \frac{Gb^2}{2\pi r(1-v)} dr = \frac{Gb^2}{4\pi r(1-v)} \log\frac{R}{r_0}$$

The core energy of edge dislocation should be added to the elastic strain energy. For a screw dislocation in the Z-direction in a cylindrical material, the stress field is given by a shear stress, according to the expression;

$$\tau z_\theta = \tau_\theta z = \frac{Gb}{2\pi r}$$

Etching

Types of Etching

In order to form a functional Micro-Electro-Mechanical Systems structure on a substrate, it is necessary to etch thin films deposited and also the substrate itself. There are two classes of etching processes.

Wet etching

This is the simplest etching technology and only requires a container with a liquid solution that will dissolve the material in question. One must find a mask that at least etches much slower than the material to be patterned. Some single crystal materials such as silicon exhibit anisotropic etching in certain chemicals. Anisotropic etching is in contrast to isotropic etching means different etch rates in different directions in the material. This is a simple technology that gives good results when the right combination of etchant and mask material to suit your application and it also works very well for etching thin films on substrates, and can also be used to etch the substrate itself. Anisotropic processes allow the etching to stop on certain crystal planes in the substrate, but still results in a loss of space, since these planes cannot be vertical to the surface when etching holes or cavities.

Dry etching

The dry etching technology can split in three separate classes called reactive ion etching (RIE), sputter etching, and vapor phase etching.

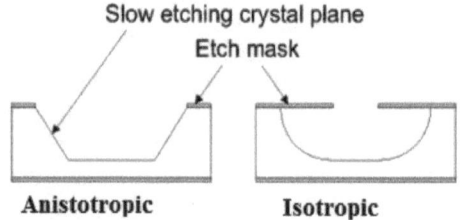

Types of wet etching

Reactive ion etching

In RIE, the substrate is placed inside a reactor in which several gases are introduced and a plasma material is struck in the gas mixture using an RF power source, breaking the gas molecules into ions. The ions are accelerated towards, and react at, the surface of the material being etched, forming another gaseous material. This is known as the chemical part of reactive ion etching. There is also a physical part which is similar in nature to the sputtering deposition process. A schematic of a typical reactive ion etching system is shown in the figure below.

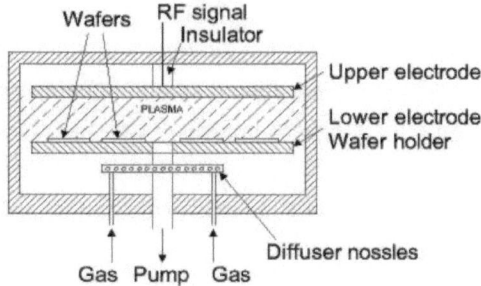

Parallel-plate reacctive ion exchange system

If the ions have high enough energy, they can knock atoms out of the material to be etched without a chemical reaction. It is a very complex task to develop dry etches processes that balance chemical and physical etching, since there are many parameters to adjust. By changing the balance it is possible to influence the anisotropy of the etching, since the chemical part is isotropic and the physical part highly anisotropic the combination can form sidewalls that have shapes from rounded to vertical.

Sputter etching

Sputter etching is essentially RIE without reactive ions. The systems used are very similar in principle to sputtering deposition systems. The big difference is that substrate is now subjected to the ion bombardment instead of the material target used in sputter deposition.

Vapor phase etching

Vapor phase etching is another dry etching method, which can be done with simpler equipment than what RIE requires. In this process the wafer to be etched is placed inside a chamber, in which one or more gases are introduced. The material to be etched is dissolved at the surface in a chemical reaction with the gas molecules. The two most common vapor phase etching technologies are silicon dioxide etching using hydrogen fluoride (HF) and silicon etching using xenon di-flouride (XeF_2), both of which are isotropic in nature. Usually, care must be taken in the design of a vapor phase process to not have bi-products form in the chemical reaction that condense on the surface and interfere with the etching process.

Point Defect in Crystals

The point imperfections occur due to the imperfect packing of atoms during crystallization. Therefore they can be regarded as lattice errors at isolated lattice points. They also occur due to vibrations of atoms when the material is exposed to very high temperatures. Point imperfections are completely local in effect and sometimes they are regarded as a vacant lattice site and that is why point defects are always present in crystals. The presence of point defects results in a decrease in free energy and therefore one can compute the number of defects at equilibrium concentration at a certain temperature as,

$$n = N \exp[-E_d / kT]$$

where, **n** - number of imperfections, **N** - number of atomic sites per mole, **k** - Boltzmann constant, E_d - free energy required to form the defect and **T** - absolute temperature. Theoretically, the value of E is of the order of 1 eV. This is when it is calculated on $k = 8.62 \times 10^{-5}$ eV /K, at T = 1000 K, $n/N = \exp[-1/(8.62 \times 10^{-5} \times 1000)] \approx 10^{-5}$.

Vacancies

Vacancy is the simplest point defect known and refers to an empty or an unoccupied site of a crystal lattice. It can also refer to a missing atom or vacant atomic site as shown in figure below.

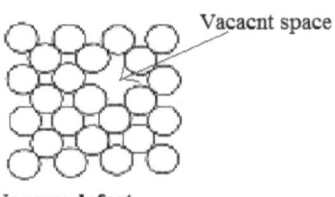

Vacany defect

Vacancy defects may also arise from imperfect packing during original crystallization. When the thermal energy due to vibration is increased, there is always an increased probability that individual atoms will jump out of their positions of lowest energy. They can result also from thermal vibrations of the atoms at higher temperatures where each temperature has a corresponding equilibrium concentration of vacancies and interstitial atoms. An interstitial atom is an atom transferred from a site into an interstitial position. For most crystals the vacancy defect at any thermal energy change is of the order of I eV per vacancy as thermal vibrations of atoms increases with the rise in temperature.

Single or two or more vacancies may condense forming a di-vacancy or tri-vacancy. Where such a vacancy exists, the atoms surrounding a vacancy tend to be closer together, thereby distorting the lattice planes. At thermal equilibrium, the vacancies exist in a certain proportion in a crystal and at higher temperatures; these vacancies have a higher concentration and tend to move from one site to another more frequently. Vacancies accelerate all processes associated with displacements of atoms that include diffusion, powder sintering and scattering.

Interstitial Imperfections

Interstitial imperfection occurs when an extra atom is lodged within the crystal structure of a closed packed structure having low atomic packing factor. This vacant space is known as interstitial position or voids. An extra atom can enter the interstitial space or void between the regularly positioned atoms only when it is substantially smaller than the parent atoms and produce an atomic distortion. The defect caused is known as interstitial defect.

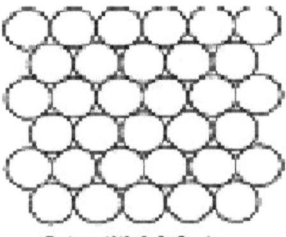

Interstitial defects

In close packed structures especially in FCC and HCP, the largest size of an atom that can fit in the interstitial void or space have a radius about 22.5% of the radii of parent atoms and thus interstitial are classified as single interstitial, di-interstitials and tri-interstitials.

Frenkel Defect

Whenever a missing atom leaves it space and occupies an interstitial the defect caused is known as Frenkel defect. Frenkel defect is therefore a combination of vacancy and interstitial defects. These defects are less in number because energy is required to force an ion into new position. This type of imperfection is more common in ionic crystals, because of their positive ions that are smaller in size get lodged easily in the interstitial positions.

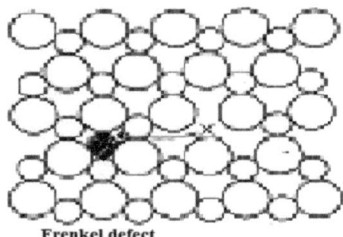
Frenkel defect

Schottky Defect

In Schotky defect, the defect occurs whenever a pair of positive and negative ions is missing from a crystal. It is similar to vacancies as shown below. This type of imperfection maintains charge neutrality. It is known that closed-packed structures have fewer interstitial and Frenkel defects than those found for vacancies and Schottky defects. This is because additional energy is required to force the atoms in their new positions.

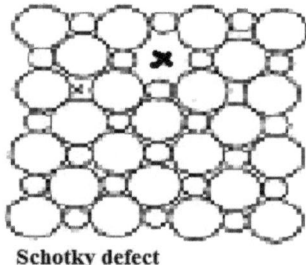
Schotky defect

Substitution Defect

This defect occurs whenever a foreign atom replaces the parent atom of the lattice and occupies the position of parent atom. The defect caused is called substitution defect and in this type of defect, the atom which replaces the parent atom may be of same size or slightly smaller or greater than that of parent atom.

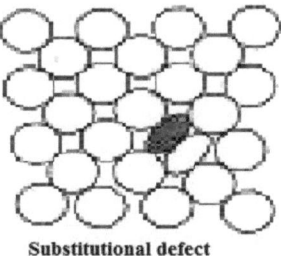

Substitutional defect

Phonon

When the temperature is raised, thermal vibrations increase in a material and this causes the symmetry of the defect and deviation to change. This defect has much effect on the magnetic and. electric properties. All kinds of point defects are important though they distort the crystal lattice and cause a certain influence on their physical properties. In commercially available pure metals, point defects increase electric resistance. In addition to point defects created by thermal fluctuations, point defects can also be created by other means that include producing an excess number of point defects at a given temperature is by quenching or quick cooling from a higher temperature. Another method of creating excess defects is by severe deformation of the crystal lattice which involves hammering or rolling. Excess point defects can be created using external bombardment by using atoms or high-energy particles. This is applied in the beam of the cyclotron or the neutrons in a nuclear reactor in which the first particle collides with the lattice atoms and displaces them, thereby causing a point defect.

Diffusion

Many processes occurring in metals and their alloys or in semiconductors especially at elevated temperatures are associated with self-diffusion or diffusion. Diffusion refers to the transport of atoms through a crystalline or glassy solid. Diffusion processes play a crucial role in many solid-state phenomena and also in the kinetics of micro structural changes during metallurgical processing and applications. This can be seen in phase transformations, nucleation, recrystallization, oxidation, creep, sintering, ionic conductivity and intermixing in thin film devices. Technological that uses diffusion include solid electrolytes for advanced battery and fuel cell applications, semiconductor chip and microcircuit fabrication.

Types of Diffusion

Self- Diffusion

It is the transition of a thermally excited atom from a site of crystal lattice to an adjacent site or interstice.

Inter Diffusion

This is observed in binary metal alloys such as the Cu-Ni system.

Volume Diffusion

This type of diffusion is caused due to atomic movement in bulk in materials.

Grain Boundary Diffusion

This type of diffusion is caused by atomic movement along the grain boundaries alone.

Surface Diffusion

This type of diffusion is caused due to atomic movement along the surface of a phase.

Diffusion Mechanisms

Various mechanisms of diffusion explain how diffusion occurs and all these explanations are based on the theory of vibrational energy of atoms in a solid. They also confine themselves onto direct-interchange or cyclic or interstitial or vacancy as diffusion based mechanisms. Diffusion is defined as the transfer of unlike atoms when a change of concentration of the components in certain zones of an alloy changes. From all the proposed mechanisms, the most probable mechanism of diffusion usually and largely accepted involve the one that explain diffusion from magnitude of energy barrier or activation energy used to be overcome by moving atoms. Activation energy is known to depend on the forces of interatomic bonds and crystal lattice defects. These defects are then known to facilitate diffusion transfer and therefore vacancy mechanism of diffusion is the most probable. For elements which have a small atomic radius like hydrogen, H, nitrogen N and carbon, C, it is believed that the interstitial mechanism explains their diffusion better.

Vacancy Mechanism

Vacancy mechanism is a very dominant process for diffusion especially in FCC, BCC and HCP metals and related solid solution alloys. In this mechanism, the activation energy comprises of the energy required to create a vacancy and that required to move the vacancy. This type of diffusion can occur when atoms move into adjacent sites that have a vacancy such that in a pure solid during diffusion process, the atoms surrounding the vacant site shift their equilibrium positions to adjust in tandem with the change in binding that accompanies the removal of a metal ion and its valence electron. The vacancies move through the lattice and produce random shifts of atoms from one lattice position to another by the atom jumping. Vacancies are continually being created and destroyed at the surface or grain boundaries or at suitable interior positions. Therefore in a pure solid, the vacancy mechanism diffusion can be illustrated by the below.

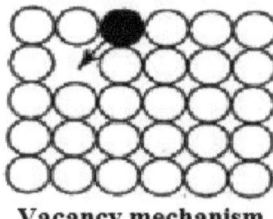
Vacancy mechanism

The rate of diffusion increases rapidly with increasing temperature as concentration changes takes place due to diffusion over a period of time. It should be noted that if a solid is composed of a single element or is a pure metal, the movement of thermally-excited atom from a site of the crystal lattice to an adjacent site or is called self-diffusion though the moving atom and the solid involved are of the same element. The self-diffusion in metals occurs mainly through vacancy mechanism.

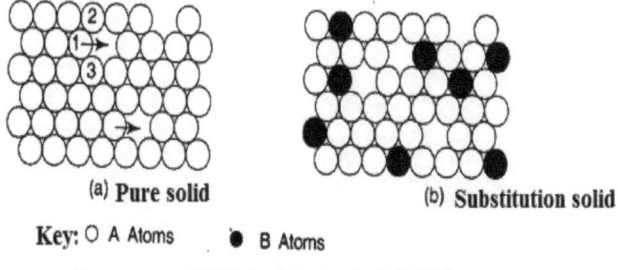

(a) **Pure solid**   (b) **Substitution solid**

Key: O A Atoms   ● B Atoms

Vacancy mechanism for atomic diffusion

Interstitial Mechanism

This is a type of diffusion mechanism where an atom changes positions using an interstitial site. In cases where a solid is composed of two or more elements diffusing elements whose atomic radii differ significantly, interstitial solution diffusion is likely to occur. Therefore, the large atoms occupy lattice sites while the smaller atoms fit into the voids which are also called as interstices. It is favored when interstitial impurities are present because of the low activation energy and can be illustrated as shown in the figure below.

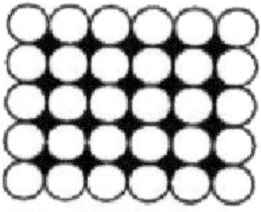

Interstitial mechanisms

These interstices are created by the larger atoms and in such cases, the activation energy is associated with interstitial diffusion based on the fact that for the diffusing atom to arrive at the vacant site, it must squeeze past the neighbouring atoms using the energy supplied by the vibrational energy of the moving atoms and that is why interstitial diffusion said to be a thermally activated process. This mechanism is used at low temperatures, oxygen, hydrogen and nitrogen can be diffused in metals easily.

Interchange Mechanism

In interchange mechanism, atoms exchange places through rotation about a mid-point and therefore its activation energy is very high. This makes this type of mechanism is highly unlikely in most systems.

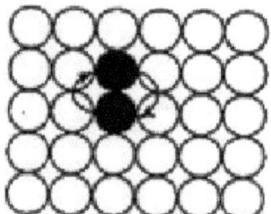

Two atoms interchange mechanism

It is also noted that two or more adjacent atoms can jump past each other and as a result exchange positions while the number of sites remains constant. This interchange may involve two-atom or four-atom forming a **Zenner ring** when in the BCC face and thus much more energy is required. As a result of these displacements of atoms surrounding the jumping pairs interchange mechanism cause severe local distortion. This is also termed as **Kirkendall's effect**. Kirkendall's effect is very important in diffusion. We may note that the practical importance of this effect is in metal cladding, sintering and deformation of metals (creep).

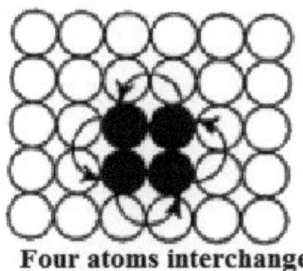

Four atoms interchange mechanism

Kirkendall was the first person to show the inequality of diffusion by using an ά brass/copper couple. Kirkendall showed that Zn atoms diffused out of brass into Cu more rapidly than Cu atoms diffused into brass. Due to a net loss of Zn atoms, voids can be observed in brass.

Fick's Laws of Diffusion

Some laws treated diffusion as a mass flow process by in which atoms or molecules change their positions relative to their neighbours in a given phase under the influence of thermal energy or a gradient variation. Gradient can be due a concentration gradient or an electric field gradient or magnetic field gradient or due a stress gradient. In Solid State Physics, we consider mass flow under concentration gradients in which thermal energy is necessary for this mass flow as the atoms have to jump from site to site during diffusion. During diffusion, thermal energy is in the form of vibrations of atoms about their mean positions in the solid and thus classical laws of diffusion like that of Fick's laws only hold for weak solutions and systems with very low concentration gradients of the diffusing substance, such that the slope of concentration gradient is given by;

$$\frac{\partial c}{\partial x} = \left(\frac{C_2 - C_1}{X_2 - X_1} \right)$$

Fick's Firfst Law

Fick's first law describes the rate at which diffusion occurs and it can be states as;

$$dn = -D\frac{dc}{dx}[a]dt$$

The quantity, **dn** of a substance diffusing at constant temperature per unit time **t** through unit surface area **a** is proportional to the concentration gradient **dc/dx** and the coefficient of diffusion (or diffusivity) **D** (m^2/s). The 'minus' (-) sign implies that diffusion occurs in the reverse direction to concentration gradient vector. This implies that it is from a higher concentration zone to a lower concentration of the diffusing element reducing the equation above to become;

$$\frac{dn}{dt} = -D\frac{dc}{dx}a$$

So that

$$J = -\frac{1}{a}\frac{dn}{dt} = -D\frac{dc}{dx}$$

where **J** is the flux or the number of atoms moving from unit area of one plane to unit area of another per unit time. The flux **J** is therefore the flow per unit cross sectional area per unit time and is proportional to the concentration gradient. The negative sign (-) in this expression means that flow occurs down the concentration gradient as shown in figure below.

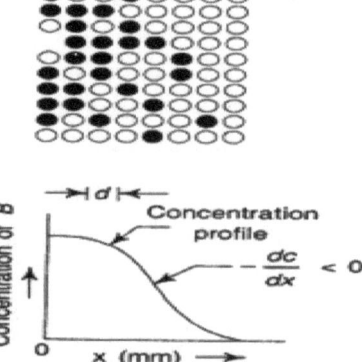

From the above curve, a large negative slope corresponds to a high diffusion rate and therefore in accordance with Fick's first law, the atoms of B will diffuse from the left side towards the right side and the net migration of B atoms to the right side means that the concentration will decrease on the left side of the solid and increase on the right as diffusion progress till a state where the concentration is uniform only at the center as shown in the figure below. Fick's first law can also be used to describe mass flow under steady state conditions. The explanation is identical in form to that given by Fourier's law for heat flow under a constant temperature gradient or that given by Ohm's law for a current flow under a constant electric field gradient. Under steady state flow, the flux is independent of time and remains the same at any cross-sectional plane along the diffusion direction.

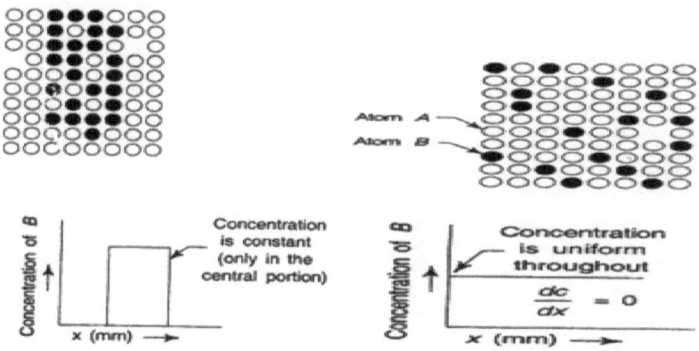

Fick's second Law

Frick's second law is an extension of Fick's first law to a non-steady flow. Frick's first law allows the calculation of the instantaneous mass flow rate or Flux past any lattice plane in a solid. It also provides no information about the time dependence of the concentration but the commonly available situations in material engineering are for non-steady in which concentrations of solute atom changes at any point with respect to time in non-steady diffusion. When the concentration gradient of various with time and the diffusion coefficient is assumed to be independent of concentration and this gives the diffusion process as described by Frick's second law when derived from the first law as;

$$\frac{dc}{dt} = \frac{dc}{dx}\left[D\left(\frac{dc}{dx}\right)\right]$$

This equation of Fick's second law is for unidirectional mass flow under non steady conditions and therefore a solution to it can be expressed as;

$$c(x,t) = \frac{A}{\sqrt{Dt}} \exp\left[-(x^2/4Dt)\right]$$

In this solution, A is constant. Consider a self-diffusion due to radioactive nickel atoms in a non-radioactive nickel block specimen. Using the solution of Fick's second law, the concentration at x = 0 falls with time as $r^{-1/2}$. As time increases, the radioactive penetrate deeper in the metal block and at time t_1 the concentration of radioactive atoms at x = 0 is $c_1 = A/(Dt_1)^{1/2}$. After some time at a distance $x_1 = 0$ is $(Dt_1)^{1/2}$ the concentration falls to 1/e of c_1. At another later time t_2 the concentration at x = 0 is $c_2 = A/(Dt_2)^{1/2}$ and this falls to 1/e and $x_2 = 2(Dt_2)^{1/2}$. This can be illustrated as shown in the figures below;

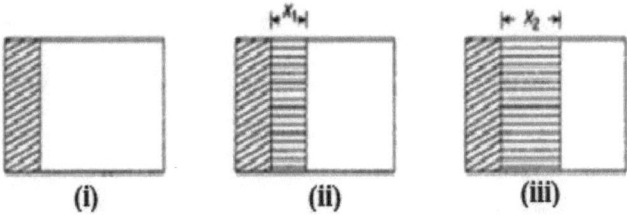

In figure, the diffusion of atoms is shown (i) for t= 0 (ii) for t_1, and (iii) t_2 with $t_2 > t_1$ and thus 1D is independent of concentration, Fick's second law is simplified to;

$$\frac{dc}{dt} = D\frac{d^2c}{dx^2}$$

It should be noted that even if D may vary with concentration, solutions to the differential above are quite simple for practical problems since it is for unidirectional diffusion from one medium to another a cross a common interface. It can thus be given in a general form as.

$$c(x,t) = A - B\,erf(x/2\sqrt{Dt})$$

where, A and B are constant to be determined from the initial and boundary conditions of a particular problem.

This equation applies for two media taken to be semi-infinite in which the interface of one end is only defined while the other two ends are at an infinite distance. Thus the initial uniform concentrations of the diffusing species in the two media are different causing an abrupt change in concentrations at the interface. Therefore a solution in *erf* of Fick's second law expressed with error function reduces to;

$$erf \frac{x}{2\sqrt{Dt}} = \frac{2}{\sqrt{\pi}} \int_{0}^{x/2\sqrt{Dt}} \exp(-\eta^2) d\eta$$

Where η, is an integration variable that gets deleted as the limits of the integral are substituted and always the lower limits of the integral is zero while the upper limit of the integral is function whose quantity is to be determined by $2\sqrt{\pi}$, which is a normalization factor. The diffusion coefficient D (m²/s) is used to determine the rate of diffusion at a concentration gradient equal to unity and depends on the composition of alloy's size of grains, and temperature. Therefore the solutions to Fick's second equations exist for a a large variety of boundary conditions which help to evaluate **D** from **c** as a function of **x** and **t**. Based on the solution given above, the time dependence of diffusion is shown in figure below.

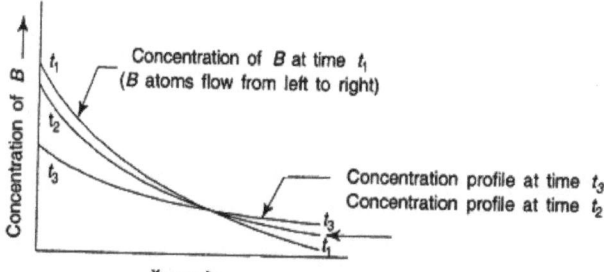

Figure: Time dependence of diffusion (Fick's second law)

A closer analysis of the curves shows that the curve corresponding to a concentration profile at a start time, t_1 is marked by time, t_1. At a later time t_2, the concentration profile has changed and this profile change is due to the diffusion of B atoms that has occurred in the time interval $t_2 - t_1$. After some time t_3, the concentration profile is marked by t_3. Thus due to diffusion process, atoms of B tend to get distributed uniformly throughout the solid salutation and this causes the concentration gradient to become less negative as time increases.

Dependence of Diffusion Coefficient on Temperature

The diffusion coefficient D is known to determine the rate of diffusion at a concentration gradient equal unity. However, it also depends on the composition of the alloy, the size of grains and its temperature. The diffusion coefficient based on temperature dependency is described by Arrhenius exponential relationship given by;

$$D = D_0 \exp(-Q/RT)$$

where D_0 is a pre-exponential frequency factor which heavily relies on the bond force between atoms of crystal lattice, Q is the activation energy of diffusion in which it is a sum of the activation energies for the formation, Q_v and motion of vacancies, Q_m such that;

$$Q = Q_v + Q_m,$$

The experimental behavior of Q for the diffusion of carbon in α-Fe is about 20.1 kcal/mole and that of D_0 is $2 \times 10^{-6} m^2/s$ when R is take as the gas constant.

Factors Affecting Diffusion Coefficient

Diffusion co-efficient is affected by concentration. The rate at which atoms jumped mainly depends on their vibrational frequency, the crystal structure. The energy required by the atom to overcome this energy barrier is called the activation of diffusion and it can be illustrated by the figure below.

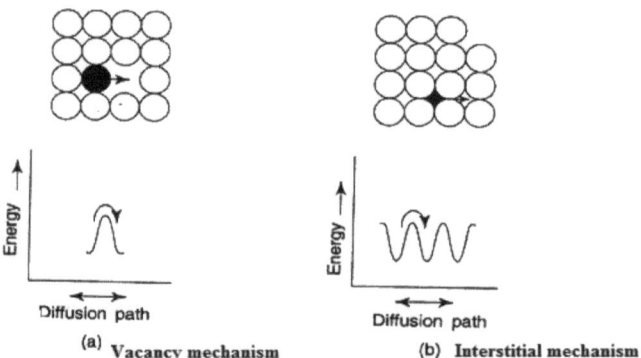

Fig. 8. Activation energy for diffusion

This activation energy is required to pull the atom away from its nearest atoms. Thus in vacancy mechanism, this energy is also required to force the atom into closer contact with neighbouring atoms as it moves along them in interstitial diffusion. When the normal inter-atomic distance is increases or decrease, this energy has to be increased. Activation energy depends on the size of the atom and it varies with the size of the atom, strength of bond and the type of the diffusion mechanism. Activation energy is high for large- sized atoms for strongly bonded materials like corundum and tungsten carbide.

Applications of Diffusion

Diffusion processes are the basis of Crystallization, recrystallization, phase transformation and saturation of the surface of alloys by other elements are the few important applications of diffusion. Other includes the following: Oxidation of metals; Doping of semiconductors; Joining of materials by diffusion bonding. Production of strong bodies by sintering i.e. powder metallurgy; Surface treatment. Diffusion is fundamental to phase changed. Some practical applications of diffusion are discussed below:

Hardening steel

Solid state diffusion is used in surface hardening of steel. Steel is used in the manufacture of gears and shafts. Steel parts made in low carbon steel are brought in contact with hydrocarbon gas like methane (CH_4) in a furnace atmosphere at about 927^0C temperature. The carbon in CH_4 diffuses into surface of steel part and the carbon concentration increases on the surface and this is what hardness the surface of steel. The concentration of carbon is higher near the surface and reduces with increasing depth as depicted in the figure below.

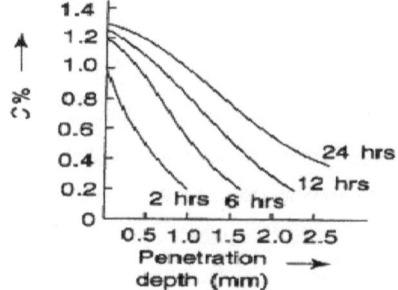

Steel carburized in 1.6% CH_4, 20% CO and 4%H.

Ionic conductivity

Anion vacancy is the dominant lattice defect responsible for the ionic conductivity in pure and doped lead chloride. Activation energy responsible for the migration of the anion vacancy ranges from 0 - 48 eV to 0 - 24 eV in lead chloride crystals, however, in single crystals of pure and doped lead chloride, the energy of formation of vacancies ranges from 1 - 66 eV and that for migration of the anion vacancies is 0-35 eV. The roles of the various point defects in pure and doped lead chlorides are yet understood.

Colour centers

Therefore a colour centre is a lattice defect, which absorbs light. A scientist known as Becquerel discovered that when transparent NaCl crystals are placed near a discharge tube they appear yellowish in colour. This discovery gave rise to the study of colour centres. Rocksalt has an infrared absorption that is caused by the vibrations of its ions and also an ultraviolet absorption that is due to the excitation of the electrons. A perfect NaCl crystal does not absorb visible light. This makes the perfectly transparent in the visible region. The colouration of crystals is therefore due to defects in the crystals.

Exposure of a coloured crystal to white light can result in bleaching of the colour and this is used give a clue of the nature of the absorption done by crystals. Experiments show that during the bleaching of the crystal the crystal becomes photoconductive. This is due to the electrons that are excited into the conduction band. Further, photo-conductivities express the quantum efficiency of a material. It shows the number of free electrons produced per incident photon resulting from the colour centres. Insulators have very large energy gaps. As a result many of them are transparent to visible light. Ionic crystals have the forbidden energy gap of about 6 eV. This energy corresponds to a wavelength of about $2000 A^0$ which is found in the ultraviolet region.

A study of the dielectric properties show that the ionic polarizability of a material resonates at a wavelength of 60 microns in the far infrared region and at that resonance levels crystals are expected to be transparent over a wide range of spectrum including the visible region. Crystals of KCl, NaCl, LiF and other alkali halides are have such dielectric properties, that is why they are used for making prisms, lenses and optical windows in optical and infrared spectrometers. Due to different various reasons, absorption bands occur in the visible, near ultraviolet and near infrared regions in some of these crystals. When the absorption band is in the visible region and the band is quite narrow. This gives a characteristic colour to the crystal and when the crystal gets coloured, it is then said to have colour centres. Therefore a colour centre is a lattice defect, which absorbs light. This theory has been used greatly to colour crystals in a number of applications as described below:

> **(i).** By the addition of suitable chemical impurities like transition element ions with excited energy levels. Hence alkali halide crystals can be coloured by ions whose salts are normally coloured.

(ii). By introducing stoichiometric excess of the cation by heating the crystal in the alkali metal vapour and then cooling it quickly. The colours produced depend upon the nature of the crystals e.g., LiF heated in Li vapour colours it pink, excess of K in KCl colours it blue and an excess of Na in NaCl makes the crystal yellow. Crystals coloured by this method on chemical analysis show an excess of alkali metal atoms, typically 10^{16} to 10^{19} per unit volume.

(iii). By exposing them to high energy radiations like X-rays or ϒ-rays or by bombarding them with energetic electrons or neutrons. Crystals can also be coloured or made darker.

Types of colour centers

They include the following types of colour centres:

F Centres

The simplest colour centre is an F centre and it is called so because its name comes from the German word Farbe which means '*colour'*. The F centres are produced by heating a crystal in the presences of an excess of an alkali vapour. It can also be grown by irradiating the crystal by X rays. Such crystals with F centers include NaCl and they have the main absorption band occurs at center at about 4650A^0 hence called the F band as shown in the figure 10 below. This absorption in the blue region is responsible for the yellow colour produced in NaCl crystal and thus it is the F band that is characteristic of the crystal. The F band in KCl or NaCl will be the same whether the crystal is heated in a vapour of sodium or of potassium as shown below.

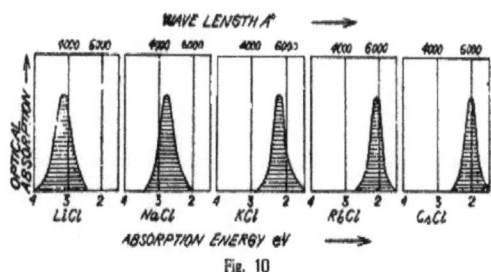

Fig. 10

Formation of F-Centres

Colour centres in crystals occur when crystals have an excess of one of its constituents due to their non-stoichiometric properties. NaCl crystal can be coloured by heating it in an atmosphere of sodium vapour and then cooling it quickly so that the excess sodium atoms absorbed from the sodium vapour causing it to split up into electrons and positive ions in the crystal as shown in the figure below.

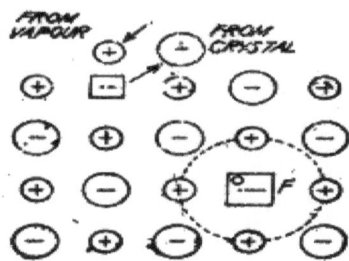

Through this process the crystal becomes slightly non-stoichiometric having excess of sodium ions than chlorine ions. This creates Cl^- vacancies and therefore the valence electron of the alkali atom is not bound to the chlorine atom. Instead it diffuses into the crystal to be bound to a vacant negative ion site at F band because a negative ion vacancy in a perfect periodic lattice has the effect of an isolated positive charge. The positive charge then traps the electron in order to maintain local charge neutrality. The excess electron captured this way at a negative ion vacancy in an alkali halide crystal is called an F centre. This model was first suggested by De-Boer and was further developed by Mott and Gurney.

Change of Density

In a normal crystal of NaCl, some Cl^- vacancies are always present in thermodynamic equilibrium. Thus if any sort of radiation is applied, it will cause some electrons to be knocked into the Cl^- vacancies. The result will be the formation of F centres. The generation of Cl^- vacancies through the introduction of excess metal atoms can be demonstrated by a decrease in the density of the crystal. This change of density is determined by X-ray diffraction measurements.

Energy Levels of F -centres

The colour centres are formed when point defects in a crystal trap electrons with the resultant electronic energy levels spaced at optical frequencies. The trapped electron has a ground state energy level that lie in the forbidden energy gap. This energy slowly progresses from a relatively widely spaced level near the valance band in an almost continuous set of levels just below the bottom of the conduction band. The absorption peak is due to transitions to excited states close to the conduction band-determined by the trapped electron as shown in the figure below. It also shows that the F absorption band is produced due to a transition from the ground state to the first excited state below the conduction band.

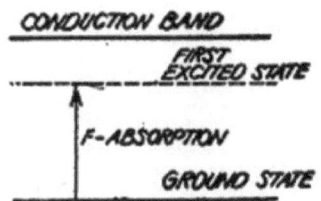

When the crystal is exposed to white light radiation, radiated energy excites the trapped electron to a higher energy level. The electron is then absorbed in the process and as a result, characteristic absorption peak near the visible region appears in the absorption spectrum. This peak has F-centres and it does not change when an excess of another metal is introduced in the crystal. This is because the foreign atoms get substituted for the metal atoms of the host crystal.

Effect of temperature on F-band

The energy levels of an F-centre depend upon the atomic surroundings of vacancy which therefore mean that the absorption peak should shift to shorter wavelengths. Thus at higher energies when the interatomic distances in the crystal are decreased this shift is observed on varying the temperature of the crystal and maximum absorption the breadth is finite even at very low temperature which also increases on increasing the temperature of the crystal. Figure below shows a graph plotted between the changes in energy of an electron in F centre and the coordination of a vacancy. The distance from centre of vacancy to nearest ions surrounding it E denotes the excited state of the electron bound to a Cl^- vacancy and G is for the ground state of that electron.

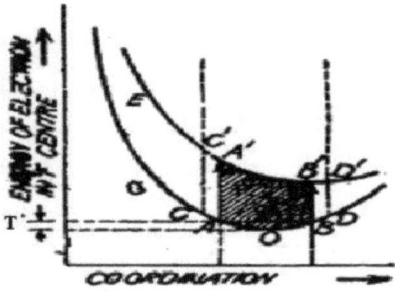

From the curve, it can be inferred that at any finite temperature, the ground state is not at 0. Thus the minimum of curve G lies above the ground state by about **kE.** This is due the coordinating ions that vibrate between the extremes of A and B as energized by thermal energy. The energy absorbed then range between the of transition A → A` or B → B`. In the event that the difference between these energies A` and B` give the width of the absorption peak. Therefore AB represents the amplitude of vibration of ions at a lower temperature. If temperature is increased the curve moves to a higher energy position so we obtain curve CD which represents the amplitude of vibration at the higher temperature. The width of the absorption peak which forms the F band eventually increases.

Magnetic Properties of F-Centres

The upper curve E as depicted above is used to determine the change in the surroundings of a vacancy when the trapped electrons are in the excited state. Mathematically, it is expressed in terms of the change in the effective dielectric constant in the neighbourhood of such a vacancy. An alkali halide crystal is diamagnetic. The ions have closed outer shells and its F-centre contains an unpaired trapped electron and hence the crystals produce additive colour with a metal due to its paramagnetic behavior. Thus the structure of F-centres can be studied by electron paramagnetic resonance experiments which tell us about the wave-functions of the trapped electron.

V-Centres

If an alkali halide crystal is heated in a halogen vapour, a stoichiometric excess of halogen ions is introduced in it. The created accompanying cation vacancies trap holes just as the anion vacancies trap electrons in F centres. A new series of colour centres are produced by excess alkali metal atoms which have holes in place of electrons. The colour centres produced in this way are called **V-centres**. The crystals having these centres show several absorption maxima which are called as V_1, V_2 bands and so on. This case uses excess bromine which enters the normal lattice positions as negative ions inducing positive holes are situated near a positive ion vacancy where they can be trapped. A hole is then trapped at a positive ion vacancy forms a V centre. The optical absorption observed is due to the transition of an electron from the filled band into the hole. The formation of V centres can be explained on the same lines as for F centres as depicted in the figure below.

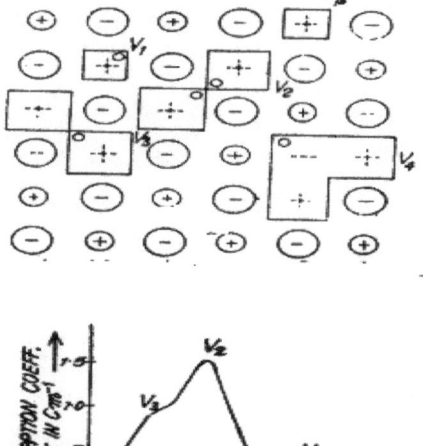

As is evident from the figures above, the V_1 centre is the counterpart of the F-centre, V_2 and V_3 are those of the R centres and V_4 is the counterpart of the M centre.

Producing Colour Centres by Particle Irradiation

Colour centres can be produced in crystals by irradiating them with very high energy radiation like X-rays or ϒ rays. X-ray quantum radiations passes through an ionic crystal and produce a fast photo-electrons having the energy nearly equal to that of the incident quantum. These high energy electrons interact with the valence electrons in the crystal. During the interaction, they lose their energy by producing free electrons and holes, or excitons which are electron hole pairs and phonons. They then diffuse into the crystal and when they come across the vacancies in the crystal, they are caught to give trapped electrons and holes. In this way both

F and V types of colour centres are produced in crystals irradiated with high energy radiations. These colours cannot be removed permanently without changing them chemically but they are easily bleached by visible light or by heating because the excited electrons and holes ultimately recombine with each other.

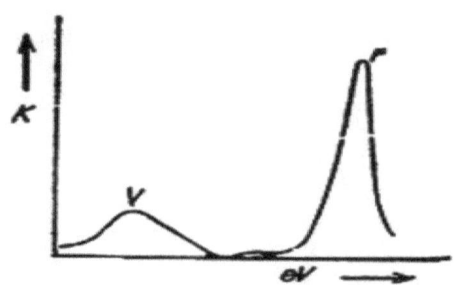

Optical transitions in semiconductors

Discrete electronic levels of individual atoms form large bands in crystals. This involve hundreds of thousands of atoms a assembled in a constant periodic structure though there are gaps between the allowed bands where no electronic states. Crystals that have a Fermi level inside one of its allowed bands are called metals while those crystals that have a Fermi level inside the gap are called semiconductors or dielectrics. The difference between semiconductors and dielectrics is quantitative base on that the materials where the band gap containing the Fermi level is narrower than about 4 eV are called semiconductors while those with wider band gaps are called dielectrics. The most essential solid features in optical transitions can be classified as shown in the table below.

Material	Essential Features		
	Fermi level	Energy gap width	Conductivity (S m^{-1})
Metals	Inside the band	any	Up to 6.3 10^7 (silver)
Semiconductors	Inside the gap	< 4 eV	Varies in large limits
Dielectric	Inside the gap	≥4 eV	Can be as low as 10^{-10}

In semiconductors, the Eigen-functions for electrons found inside the bands have a form of waves called the *Bloch waves*, a concept developed by a Swiss physicist **Felix Bloch** in 1928. This concept describes conduction of electrons in any crystalline solid. The Bloch theorem states that a wave-function of an electronic Eigen-state in an infinite periodic crystal potential $V(\mathbf{r})$ can be written in form of;

$$\Psi_{\mathbf{k},n}(\mathbf{r}) = U_{\mathbf{k},n}(\mathbf{r}) e^{i\mathbf{k}\mathbf{r}}$$

where $U_{\mathbf{k},n}$ is called the *Bloch amplitude* which has the same periodicity as the crystal potential and the **k** is so-called *pseudo-wave vector* of an electron as n form the index of the band. If this expression is substituted into the Schrödinger generalized wave equation for an electron propagating in crystal, we obtain;

$$-\frac{\hbar^2}{2m_0}\Delta\Psi_{\mathbf{k},n} + V(\mathbf{r})\Psi_{\mathbf{k},n} = E_{\mathbf{k},n}\Psi_{\mathbf{k},n},$$

In this expression, m_0 is the free electron mass. On the above Schrödinger generalized wave equation, the modified Bloch amplitude equation becomes;

$$-\frac{\hbar^2}{2m_0}\Delta U_{\mathbf{k},n} + V(\mathbf{r})U_{\mathbf{k},n} + \left(\frac{\hbar^2 k^2}{2m_0} + \frac{\hbar}{m_0}(\mathbf{k}\cdot\hat{\mathbf{p}})\right)U_{\mathbf{k},n} = E_{\mathbf{k},n}U_{\mathbf{k},n}$$

In which the momentum, p will be given by;

$$\hat{\mathbf{p}} = \frac{\hbar}{i}\nabla$$

The operators in the parentheses in the above expression as a perturbation, constitutes the $\mathbf{k}\cdot\hat{\mathbf{p}}$ method of the perturbation theory. The theory allows one to find the shape of the electronic dispersion in $\mathbf{k} = 0$ point vicinity of all bands. This then appears to be different from the free electron dispersion in vacuum which is approximated to;

$$E_{\mathbf{k},n} \approx E_{0,n} + \frac{\hbar^2 k^2}{2m_n^*}$$

This equation is called the *effective mass approximation* with in which the m_n^* is the electron effective mass in n^{th} band will be given;

$$\frac{1}{m_n^*} = \frac{1}{m_0} + \frac{2}{m_0^2}\sum_{l \neq n}\frac{|\langle U_{0,l}|\hat{\mathbf{p}}|U_{0,n}\rangle|^2}{E_{0,l} - E_{0,n}}$$

Based on frequencies and polarization, optical transitions in direct gap semiconductors are governed by the energies and dispersion of two bands closest to the Fermi level.

These bands are the conduction band or the first level above the Fermi level and the valence band or the first below the Fermi level. Semiconductors are classified into direct band gap and indirect band gap in which the indirect gap semiconductors include like Silicon and Germanium. In these, the electron and hole occupy the lowest energy states in conduction. This does not allow the valence bands to directly recombine to emit a photon as a result of wave-vector conservation. These solids emit a weak emission of light due to phonon-assisted transitions and therefore they cannot be used to fabricate light-emitting devices. Direct gap semiconductor materials include GaAs, CdTe, GaN, ZnO among others and most of them have either a zinc-blend or a wurtzite crystal lattice as shown in the figure below.

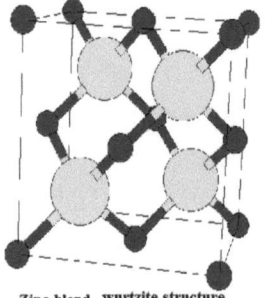
Zinc-blend wurtzite structure

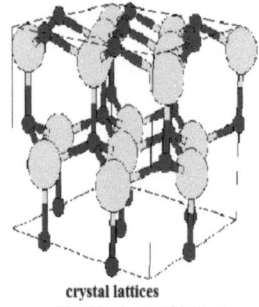
crystal lattices
Zinc-blend crystal lattices

Zinc-blend semiconductors structure have their valence band split into three sub-bands. These sub-bands are the heavy-hole, light-hole and spin-off bands. In the bulk crystals at **k** = 0, the heavy and light hole bands are degenerate. On the other hand, the wurzite semiconductors have the valence band split into three non-degenerated sub-bands. The sub-bands are referred A, B and C bands as shown in the figure below.

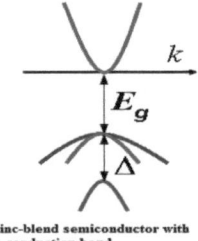
zinc-blend semiconductor with a conduction band

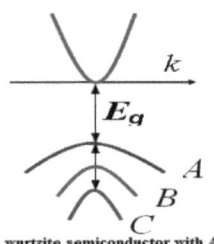
wurtzite semiconductor with A, B, and C valence subbands.

In semiconductors, the optical absorption spectrum is governed by density of electronic states in the valence and conduction bands and is given by;

$$g(E) = \frac{\partial n}{\partial E}$$

where n is the number of quantum states per unit area of the semiconductor. Inside the bands the density of states of bulk crystals behaves as $\sqrt{E}$. This results into corresponding shape of inter-band absorption spectra. At low temperatures, the absorption spectra of semiconductors exhibit very sharp peaks below the edge of their inter-band absorption and these peaks manifest the resonant light matter coupling in semiconductors. They are caused by the excitonic transitions given at frequencies of;

$$\omega < \frac{E_g}{\hbar}$$

where E_g is the band-gap energy.

Excitons in semiconductors

The narrow photoemission lines observed in the spectra of organic molecular crystals were interpreted by a Russian theorist Yakov Frenkel who introduced the concept of excitational waves in crystals in 1931 and he is honored to have invented the term exciton in 1936. He defined exciton as a coulomb-correlated electron-hole pair. According to Frenkel, he treated a crystal potential as a perturbation to the Coulomb interaction between an electron and a hole that belong to the same single crystal cell. This method is most effective in organic molecular crystals in which the energy of its ionisation to a non-correlated electron hole couple forms the binding energy in the order of 100 - 300 meV. It was in 1930s that the Swiss physicist, Gregory Hugh Wannier (1911-1983) combined efforts with an English theorist, Sir Nevil Francis Mott and developed the concept of exciton in semiconductor crystals.

The concept argue that the rate of electron and hole hopping between different crystal cells exceeds the strength of their Coulomb coupling with each other. Unlike Frenkel excitons, the Wannier-Mott excitons have a size in the order of tens of lattices constants with relatively small binding energy. In the below above, the Wannier-Mott exciton is in a solid state analogy of a hydrogen atom in which different sizes and binding energies have finite life-time excitons and therefore such excitons are described within the *effective mass approximation* described earlier which allow us to neglect the periodic crystal potential and instead describe electrons and holes as free particles having a parabolic dispersion. The effective masses of carriers are usually lighter than the free electron mass in vacuum m_0 and hence they are characterized using the effective masses that dependent on the crystal material used.

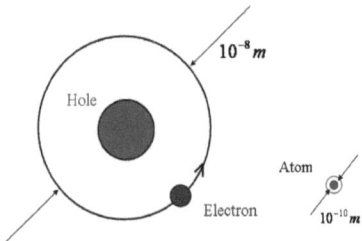

In GaAs extrinsic semiconductor, the effective electron mass is given by $m_e = 0.067 m_0$ while the heavy-hole mass is given by $m_{hh} = 0.45 m_0$. Therefore, considering an electron-hole pair that is bound by Coulomb interaction in a crystal having a dielectric constant ε, the wave-function $f(r)$, relative to electron-hole motion derived from Schrödinger wave equation describing an electron state in a hydrogen atom can be given by;

$$-\frac{\hbar^2}{2\mu}\Delta f(r) - \frac{e^2}{\varepsilon r} f(r) = E f(r)$$

where the Laplacian operator is given by;

$$\Delta \equiv \frac{\partial^2}{\partial x^2} + \frac{\partial^2}{\partial y^2} + \frac{\partial^2}{\partial z^2}$$

The reduced mass, μ will be given by;

$$\mu = \frac{m_e m_h}{m_e + m_h},$$

and the distance between electron and hole is obtained as;

$$r = \sqrt{x^2 + y^2 + z^2}$$

Deriving the expression from the distance between electron and hole, the corresponding solution to the Eigen-functions of the hydrogen atom with two substitutions give;

$$m_0 \to \mu, \ e^2 \to e^2/\varepsilon$$

The wave-function of the 1^s state of exciton can then is expressed as;

$$f_{1S}(r) = \frac{1}{\sqrt{\pi a_B^3}} e^{-r/a_B}$$

where the Bohr radius a_B is obtained from the equation given as;

$$a_B = \frac{\hbar^2 \varepsilon}{\mu e^2}$$

The binding energy of the ground exciton state at this Bohr radius a_B will be expressed as;

$$E_B = \frac{\mu e^4}{2\hbar^2 \varepsilon^2} = \frac{\hbar^2}{2\mu a_B^2}$$

One can estimate the exciton binding energy to be about three orders of magnitude less than the Rydberg constant when the difference between the reduced mass μ and the free electron mass are substituted at constant dielectric in the denominator of the equation above. When this equation is used, the binding energies and Bohr radii for Wanier-Mott excitons in different semiconductor materials, Band gap energy (E_g), electron effective mass, binding energy and Bohr radius of excitons in different semiconductor crystals are shown in the table below.

Semiconductor crystal	E_g, eV	m_e/m_0	E_B, meV	a_B, Å
PbTe*	0.17	0.024/0.26	0.01	17000
InSb	0.237	0.014	0.5	860
Cd$_{0.3}$Hg$_{0.7}$Te	0.257	0.022	0.7	640**
Ge	0.89	0.038	1.4	360
GaAs	1.519	0.066	4.1	150
InP	1.423	0.078	5.0	140
CdTe	1.606	0.089	10.6	80
ZnSe	2.82	0.13	20.4	60
GaN***	3.51	0.13	22.7	40
Cu$_2$O	2.172	0.96	97.2	38****
SnO$_2$	3.596	0.33	32.3	86****

Key : ** *In the presence of magnetic field of 5T,*

*** *A exciton in hexagonal GaAs;* **** *The ground state corresponds to an optically forbidden transition, data given for n = 2 state.]*

The exciton excited states form a number of hydrogen-like series of excitonic transitions in the photoluminescence spectra of CuO_2 in 1951 was the first experimental evidence for Wannier-Mott excitons as made by a Russian spectroscopist, Evgeniy Gross obtained an hydrogen-like yellow series in emission.

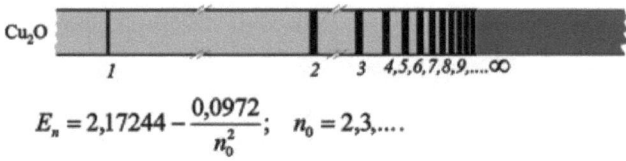

$$E_n = 2{,}17244 - \frac{0{,}0972}{n_0^2}; \quad n_0 = 2,3,\ldots.$$

Excitons in confined systems

The technological progress in the growth of semiconductor hetero-structures allow Wannier-Mott excitons in confined systems. This in essence includes quantum wells, quantum wires and quantum dots. The invention of hetero-structures created artificial potential wells and barriers for electrons and holes. The conduction and valence band potentials form a shape determined by the positions of the corresponding band edges as well as the structural geometry. Large sizes of Wannier-Mott excitons are sensitive to nanometer-scale variations of the band edges positions. The energy spectrum and wave-functions of quantum confined excitons are thus different from those of bulk excitons as shown in the figure below having excitons in the quantum wells, wires and dots. This is based on effective mass approximation and neglect the complexity of the valence band structure and consequent anisotropy of the holes effective mass which strongly affect the excitonic spectrum in real semiconductor systems as shown in the curves below.

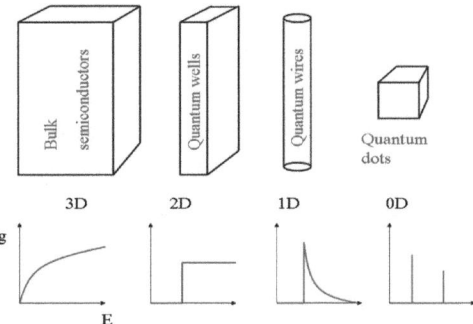

As shown in the reduction of the dimensionality of a semiconductor system from 3D to 0D in the above figure from a bulk semiconductor to a quantum dot, the electronic density of states is given as;

$$g(E) = \frac{dN}{dE},$$

where dN is the number of electron quantum states within the energy interval dE. As observed from the curves, it changes drastically with systems of different dimensionalities; a variation of the density of states is applied in light emitting semiconductor devices.

Excitons in Quantum wells

As derived from the Schrödinger wave equation for an exciton in a quantum well (QW), the Lalacian form of Schrödinger wave equation can be expressed as;

$$\left\{ \hat{K}_e + \hat{K}_h + V_e(z_e) + V_h(z_h) - \frac{e^2}{\varepsilon |\vec{r}_e - \vec{r}_h|} \right\} \Psi_{exc} = E \Psi_{exc}$$

where,

$$\hat{K}_{e,h} = -\frac{\hbar^2}{2m_{e,h}} \Delta_{e,h}$$

And the Laplacian, $\Delta_{e,h}$ depends on electron and hole coordinates respectively.

$$V_{e,h}(z_{e,h})$$

The confined potential for the electron and the hole in the z-axis is the growth axis of the structures. Thus solving the electron and the hole in the z-axis varying over a class of trial functions having a form given as;

$$\Psi_{exc} = F(R)f(\rho)U_e(z_e)U_h(z_h)$$

where exciton center of mass coordinate is given by;

$$\vec{R} = \frac{m_e \vec{r}_e + m_h \vec{r}_h}{m_e + m_h}$$

And is the in-plane radius-vector of electron and hole relative motion;

$$\vec{\rho} = \vec{\rho}_e - \vec{\rho}_h \quad \text{where,} \; (\vec{r} = (\vec{\rho}, z)).$$

Looking at the four components of the above function, they ideally describe the exciton center of mass motion, the relative electron-hole motion in the plane of the quantum well, the electron and hole motions normal to the plane direction. Factorization of the exciton wave function in the QW width is less or comparable to the exciton Bohr diameter in the bulk semiconductor. This causes the electron and hole to be quantized independently of each other. Larger QWs assume excitons confined as whole particles to keep the internal structure of a 3D hydrogen atom. The four terms that compose the exciton wave-function when normalized are given by;

$$\int dz_e |U_e(z_e)|^2 = \int dz_h |U_h(z_h)|^2$$
$$= \int_0^\infty 2\pi\rho d\rho |f(\rho)|^2$$
$$= \int_0^\infty 2\pi R dR |F(R)|^2$$
$$= 1$$

Substituting the electron and the hole in the z-axis and integration over, $\vec{R}$, we get;

$$\left\{ \begin{array}{l} \hat{K}_e^z + \hat{K}_h^z + \hat{K}_\rho + V_e(z_e) \\ + V_h(z_h) - \dfrac{e^2}{\varepsilon\sqrt{\rho^2 + (z_e - z_h)^2}} \\ - E - K_{exc} \end{array} \right\} f(\rho) U_e(z_e) U_h(z_h) = 0$$

Where the components are given as;

$$K_{exc} = \frac{P_{exc}^2}{2(m_e + m_h)},$$

$$\hat{K}_e^z = -\frac{\partial}{\partial z_e} \frac{\hbar^2}{2m_e} \frac{\partial}{\partial z_e},$$

$$\hat{K}_h^z = -\frac{\partial}{\partial z_h}\frac{\hbar^2}{2m_h}\frac{\partial}{\partial z_h},$$

$$\hat{K}_\rho = -\frac{1}{\rho}\frac{\partial}{\partial \rho}\left(\frac{\hbar^2}{2\mu}\rho\frac{\partial}{\partial \rho}\right),$$

And the reduced mass is given by;

$$\mu = \frac{m_e m_h}{m_e + m_h}$$

In the equations above, P_{exc} is the excitonic momentum such that $P_{exc} = 0$ for the ground state. The $\check{K}$ components can be transformed into a system of three coupled differential equations. Each of these transformed equations, $f(\rho)$ is obtained by multiplying both parts of the equation by $U^*_e(z_e)U^*_h(z_h)$ and integration over z_e and z_h which then yields;

$$\left\{\hat{K}_\rho - \frac{e^2}{\varepsilon}\iint dz_e dz_h \frac{|U_e(z_e)|^2 |U_h(z_h)|^2}{\sqrt{\rho^2 + (z_e - z_h)^2}}\right\}f(\rho)$$
$$= -E_B^{QW} f(\rho)$$

where the component E_B^{QW} is the exciton binding energy.

Therefore the electron and hole confinement energies expressed as E_e, E_h and their corresponding wave-functions, $U_{e,h}(z_{e,h})$, can be obtained by multiplying the integrated equation over z_e and z_h by $f*(\rho)U^*_{h,e}(z_{h,e})$ and integration over $z_{e,h}$ and ρ to get;

$$\left\{\hat{K}_{e,h} + V_{e,h} - \frac{e^2}{\varepsilon}\iint 2\pi\rho d\rho dz_{h,e}\frac{|f(\rho)|^2|U_{h,e}(z_{h,e})|^2}{\sqrt{\rho^2 + (z_e - z_h)^2}}\right\}U_{e,h}(z_{e,h})$$
$$= E_{e,h}U_{e,h}(z_{e,h})$$

In cases where the ideal 2D is taken into account, we get that;

$$|U_{e,h}(z_{e,h})|^2 = \delta(z_{e,h})$$

and this condition equation transforms the integration over z_e and z_h into;

$$\left\{-\frac{\hbar^2}{2\mu}\frac{1}{\rho}\frac{\partial}{\partial \rho}\left(\rho\frac{\partial}{\partial \rho}\right) - \frac{e^2}{\varepsilon\rho}\right\}f(\rho) = E_B^{2D} f(\rho)$$

which is an exactly solvable 2D hydrogen atom problem. For the ground state

$$f_{1s}(\rho) = \sqrt{\frac{2}{\pi}} \frac{1}{a_B^{2D}} \exp(-\rho/a_B^{2D})$$

The Bohr radius of the three-dimensional exciton is given by;

$$a_B^{2D} = \frac{a_B}{2}$$

In normal realistic QWs excited states, the binding energy of the two-dimensional exciton exceeds by a factor of 4 the bulk exciton binding energy and can be expressed as;

$$E_B^{2D} = 4E_B$$

This equation now allows one to find the functions of $U_{e,h}(z_{e,h})$ independently from each other and $f(\rho)$ by solving function $f(\rho)$ as;

$$f(\rho) = \sqrt{\frac{2}{\pi}} \frac{1}{a} \exp(-\rho/a)$$

where a is a variation parameter

Combining all the expression above and integration over z_e and z_h, the binding energy can then be expressed as;

$$E_B^{QW}(a) = -\frac{\hbar^2}{2\mu a^2} + \int_{-\infty}^{\infty} dz_e \int_{-\infty}^{\infty} dz_h \int_0^{\infty} 2\pi\rho d\rho \frac{|f(\rho)|^2 |U_e(z_e)|^2 |U_h(z_h)|^2}{\sqrt{\rho^2 + (z_e - z_h)^2}}$$

When this energy $E_B^{QW}(a)$ is maximized, the exciton binding energy obtained in the QW ranges from E_B to E_B^{2D} and also depends on the QW width and barrier heights for both the electrons and the holes. It is also observed that binding energy increases if the exciton confinement strengthens also increases as shown in the curve below.

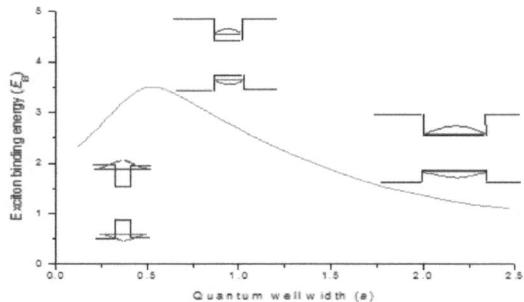

Quantum wires and dots

The variation of the ground exciton state energy and their wave-function in quantum wires or dots is calculated using the method of separation of variables and also the decoupling of equations as for the QWs. For a quantum wire, Schrödinger wave equation for the wave-function of electron-hole relative motion, $f(z)$ in the z-axis can be given as;

$$\left\{ \hat{K}_z - \frac{e^2}{\varepsilon} \iint d\vec{\rho}_e d\vec{\rho}_h \frac{|U_e(\vec{\rho}_e)|^2 |U_h(\vec{\rho}_h)|^2}{\sqrt{z^2 + (\vec{\rho}_e - \vec{\rho}_h)^2}} \right\} f(z) = -E_B^{QWW} f(z)$$

In this expression, $\hat{K}_z$ is the kinetic energy operator for motion along the axis of the wire, $U_{e,h}(\vec{\rho}_{e,h})$ is the electron/hole wave-function in the plane normal to the wire axis, E_B^{QWW} is an exciton binding energy in the wire. As a quantum particle in 1D, the Coulomb potential has no ground state with a finite energy. This means that the exciton binding energy in a quantum wire is dependent on the function, $U_{e,h}(\vec{\rho}_{e,h})$. It is theoretically between E_B and infinity. The Gaussian function have realistic quantum wires that are not cylindrical symmetry and computing of $U_{e,h}(\vec{\rho}_{e,h})$ is unrealistic because they have a finite extension in z-direction which is comparable with the exciton Bohr-diameter. If an exciton is strong and fully confined in a QD, its wave function can be represented as a product of the electron and the hole wave-functions given by;

$$\Psi = U_e(\vec{r}_e) U_h(\vec{r}_h),$$

From the equation, the single-particle wave-functions $U_{e,h}(\vec{r}_{e,h})$ are coupled Schrödinger wave equations expressed as;

$$\left\{\hat{K}_{e,h} + V_{e,h} - \frac{e^2}{\varepsilon}\int d\vec{r}_{h,e}\frac{\left|U_{h,e}(\vec{r}_{h,e})\right|^2}{\left|\vec{r}_e - \vec{r}_h\right|}\right\}U_{e,h}(\vec{r}_{e,h})$$
$$= E_{e,h}U_{e,h}(\vec{r}_{e,h})$$

where $\hat{K}_{e(h)}$ is the operator for electron or hole kinetic energy, $V_{e(h)}$ is the Quantum Dot (QD) potential for an electron or hole.

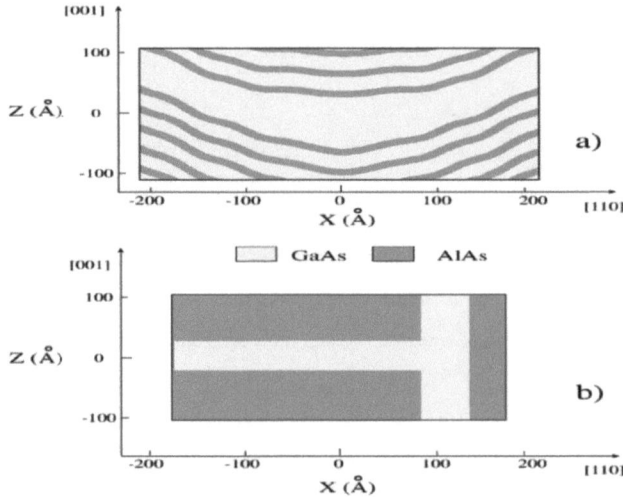

Derived from the above equation, the exciton binding energy can be reduced and be defined as;

$$E_B^{QD} = E_e^0 + E_h^0 - E_e - E_h$$

In this case E_e^0 and E_h^0 are the energies for non-interacting electron and hole pair. The Eigen-energy of the Hamiltonian are then below the Coulomb terms, in such small QDs Coulomb interaction a perturbation to the quantum confinement potential for electrons and holes through the exciton binding energy is estimated through the perturbation theory to be equivalent to;

$$E_B \approx \frac{e^2}{\varepsilon}\iint d\vec{r}_e d\vec{r}_h \frac{\left|U_e(\vec{r}_e)U_h(\vec{r}_h)\right|^2}{\left|\vec{r}_e - \vec{r}_h\right|}$$

In the quantum dot state, the exciton binding energy is dependent on the spatial extension of the electron and hole wave-functions. The bulk exciton binding energy varies to infinity and in realistic wires and dots, the binding energy rarely exceeds $4E_B$. Small QDs are developed by Stransky-Krastanov method that uses molecular beam epitaxy. In large quantum dots that exceed exciton Bohr diameter are confined as whole particles and their binding energy is equal to the bulk exciton binding energy those found in photonic dots.

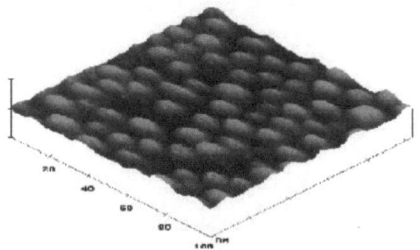

self-assembled QDs of GaN

Point defects consist of missing ions called vacancies, excess ions called interstitials or the wrong kind of ions which are substitution impurities. A new type of an ion defect in a perfect crystal that differs from others is in an excited electronic state. Such a "defect" is called a Frenkel exciton. Since any ion is capable of being excited and that ionic coupling between outer electronic shells is strong, the excitation energy can be transferred from ion to ion. A Frenkel exciton can move through the crystal with or without the ions themselves change positions. This creates mobile vacancies, interstitials, or substitution impurities. This behaviour takes the same relation to specific excited ions as that of Bloch tight – binding levels in individual atomic levels in the theory of band structures. An exciton is regarded as one of the more complex manifestations of electronic band structure that as a crystal defect.

Consider the electronic ground state of an insulator in the independent electron approximation. The lowest excited state of the insulator will be obtained by removing one electron from the highest level in the highest occupied band (the valence band) and placing it into the lowest lying level of the lowest unoccupied band (conduction band). Such a rearrangement of the distribution of electrons does not alter the self- consistent periodic potential in which they move. Since the Bloch electrons are not localized, the change in local charge density produced by changing the level of a single electron will be negligibly small. The electronic energy levels do not have to be recomputed for the excited configuration and the first excited state will lie an energy $\varepsilon_c - \varepsilon_v$ above the energy of the ground state, where ε_c is the conduction band minimum and ε_v the valence band maximum since $|\psi_{nk}(r)|^2$ is periodic.

Suppose we form a one-electron level by superposing enough level near the conduction band minimum to form a well- localized wave packet. To produce the wave packet, the energy ε_c of the wave packet will be somewhat greater than ε_c. If we depopulate wave packet in addition that the valence band level, the formed levels in the neighborhood of the valence band maximum energy ε_v is less than ε_v. This is chosen so that the center of the wave packet is spatially very near the center of the conduction band wave packet. If electron – electron interactions are ignored, the energy required to move an electron from valence to conduction band wave packet would be given as;

$$\varepsilon_c - \varepsilon_v > \varepsilon_c - \varepsilon_v,$$

Additional of negative electrostatic energy reduce the total excitation energy to an amount that is less than $\varepsilon_c - \varepsilon_v$. Evidence for this is the onset of optical absorption at energies below the inter band continuum threshold the following elementary theoretical argument, indicating that one always does better by exploiting the electron hole attraction. Consider a case in which the localized electron and hole levels extend over many lattice constants. If we regard the electron and hole as particles to be isotropic of mass m_c and m_v, they interact through an attractive Coulomb interaction screened by the dielectric constant $\in$ of the crystal. Assuming it hydrogen atom problem in which the hydrogen atom has reduced mass of μ given by;

$$(1/\mu = 1/M_{proton} + 1/m_{electron} \approx 1/m_{electron})$$

Then replacing the reduced effective mass m* given by;

$$(1/m^* = 1/m_c + 1/m_v)$$

and by applying the electronic charge replaced by;

$$e^2/\in,$$

there will be bound states, the lowest of which extends over a Bohr radius given by;

$$a_{ex} = \frac{\hbar^2}{m^*(e^2/\in)} = \in \frac{m}{m^*} a_0$$

The energy of the bound state will thus be lower than the energy ($\varepsilon_c - \varepsilon_v$) as illustrated earlier for the non-interacting electron and hole hence expressed as;

$$E_{ex} = \frac{(e^2/\in)}{2a_0^*} = \frac{m^*}{m} \frac{1}{\in^2} \frac{e^2}{2a_0}$$
$$= \frac{m^*}{m} \frac{1}{\in^2} (13.6)eV$$

This model is only valid when a_{ex} is large on the scale of $a_{ex} \gg a_0$. Since insulators with small energy gaps tend to have small effective masses and large dielectric constants, semiconductors have no difficult to achieve this energy states. The hydrogen spectra demonstrate optical absorption that occurs below inter- and the threshold. The exciton described here by this model is known as the Mott- Wannier exciton. As the atomic levels out of which the band levels are formed become more tightly bound, the value of $\in$ decreases m* increases, a_0* decrease, the exciton becomes more localized and thus the Mott- Wannier picture eventually break down. It is noted that the Mott- Wannier exciton and the Frenkel exciton are opposite extremes of the same phenomenon. Frenkel exciton is based on a single excited ionic level, the elelctron and hole are sharply localized on the atomic scale just like the exciton spectra of gases.

Luminescence

A substance can absorb and fraction of the absorbed energy can be re-emitted in the form of electromagnetic radiation especially in the visible or near-visible region of the spectrum. This is called luminescence. Luminescence does not include blackbody radiation emission which obeys the Kirchhoff laws and Wien laws. In solids, Luminescent solids are referred to as phosphors. Luminescence process involves at least two steps: the excitation of the electronic system of the solid and the subsequent emission of photons. Excitation may be achieved through bombardment with photons which can be referred as photoluminescence or it can be achieved through the electron which is referred to as cathode-luminescence or with other particles. Finally, luminescence can also be induced through a chemical reaction which is called chemiluminescence. It can also achieved by an application of an electric field hence referred as electro-luminescence.

On the other hand fluorescence put in the emission of light during excitation. Fluorescence can be defined therefore as the emission of light for which the decay time is temperature independent and phosphorescence as the temperature-dependent part. Once this excitation emission of light has ceased, then it is termed as phosphorescence or afterglow. The lifetime of an atom in an excited state require about 10^{-8} second for it to return to the ground state and it is accompanied by dipole radiation. This means that for a forbidden transitions that involve quadrupole or higher-order radiations, this lifetime attains 10^{-4} second or longer. A decay time of about 10^{-8} second is acceptable as the demarcation line between fluorescence and phosphorescence.

The ability of a material to exhibit luminescence is associated with the presence of activators and thus these activators are impurity atoms at relatively small concentrations in the host material. However, the presence of a certain type of impurity can inhibit the luminescence hence referred to as killers. Therefore small amounts of impurities play an important role in determining the luminescent properties of solids. Some important groups of luminescent crystalline solids include the following.

(i). Compounds which luminesce in the "pure" state. Examples are the manganous halides, samarium and gadolinium sulfate, molybdates and platinocyanides.

(ii). Alkali halides activated with thallium or other heavy metals.

(iii). ZnS and CdS activated with Cu, Ag, Au, Mn, or with an excess of one of their constituents (self-activation).

(iv). Silicate phosphors that include zinc orthosilicate (willernite, Zn_2SiO_4) activated with divalent maganese, which is used as oscilloscope screens.

(v). Oxide phosphors such as self-activated ZnO and Al_2O_3 activated with transition metals.

(vi). Organic crystals such as anthracene activated with naphtacene these materials are often used as scintillation counters.

CHAPTER TWO
THEORIES FOR MATERIALS PROCESSING

Material Technologies

Materials Science and Engineering is a field of engineering that encompasses a wide spectrum of materials types. These materials span form metals, ceramics, polymers or plastics, semiconductors and composites which are material combinations. The world now is both dependent upon certain specific limited materials. Every product we see or use is made of certain materials. These ranges from cars, airplanes, computers, refrigerators, microwave ovens, TVs, dishes, silverware, athletic equipment of all types, and biomedical devices such as replacement joints and limbs. All of them are made of materials tailored for their specific application. Their specific application is fine-tuned result of a careful selection of materials and controlled manufacturing processes to convert the materials into the final engineered product. A great understanding of material properties is thus desirable.

New materials technologies developed through engineering and science continue to make startling changes in our lives in this 21st century. These engineers deal with the science and technology of producing materials that have properties and shapes suitable for practical use. Engineering activities range from primary materials production through design and development of new materials for reliable and economical final product manufacturing and such activities are found in aerospace, transportation, electronics, energy conversion and biomedical systems industries. New and improved materials are an "underpinning technology" and one that stimulate innovation and product improvement. High quality products result from improved processing and more emphasis will be placed on reclaiming and recycling. For these many reasons, most surveys name the materials field as one of the careers with excellent future opportunities.

CD-ROMs, dessert plates, basketballs, car engines, telephones, and audiocassettes are made of materials. Although the field deals with materials, it encompasses an incredible diversity of topics especially those in solid state physics and problems constituting the four elements of the field; processing, structure, properties, and performance depend on a clear understanding of material properties hence history of material science is measured by innovations made in materials. Developments in metals like iron and bronze enabled advances in civilization thousands of years ago due to a clear understanding of quantum beahviour of materials.

The synergy continues today in the fiber optics and has created the World Wide Web or in the development of biomaterials that has developed a mimic of living tissues. It is noticeable that as you explore the quantum field of materials, it is useful to be familiar with some generic categories of materials. Metals are materials that are a combination of "metallic elements" and these elements when combined have electrons that are non-localized. As a consequence, these materials have generic properties. Metals are good conductors of heat and electricity and quite strong but deformable bond and structure that tend to have a lustrous look when polished.

Ceramics are compounds between metallic and nonmetallic elements. They include compounds as oxides, nitrides, and carbides. They are insulating in nature and resistant to high temperatures as well as harsh environments. Plastics on the other hand are also known as polymers. They are organic compounds based upon carbon and hydrogen with very large molecular structures, low density and are not stable at high temperatures. Semiconductors are materials that have electrical properties intermediate between metallic conductors and ceramic insulators. Their electrical properties strongly depend upon the presence of small amounts of impurities. We also have materials we call composites that consist of more than one material type. An example is the fiberglass which is a combination of glass and a polymer. Concrete and plywood are other composites. New composites include ceramic fibers in either metal or polymer matrixes.

Material processing

The term processing refers to the way in which a final material is achieved. Advances in technology have made it possible to create a material atomic layer by atomic layer through the quantum level understanding of materials. Thus, there are four categories of useful processing techniques: *solidification* processing, *powder* processing, *deposition* processing and *deformation* processing. Metals are formed by creating an alloy in the molten state. This process is also utilized in glasses and some polymers in which once the proper temperature and composition is achieved, the melt is cast. Castings can be divided into two types, depending on the subsequent processing steps. The first type is shape casting, which takes advantage of the fluidity of liquid metal to form complex shapes directly. Therefore any changes in microstructure or properties must either be achieved first during solidification or through subsequent heat treatments.

Powder processing involves consolidation, or packing, of particulate to form a `green body' in which densification follows by sintering giving to two basic methods of consolidating powders: either dry powder can be compacted in a die, a process known as dry-pressing, or the particles can be suspended in a liquid and then filtered against the walls of a porous mold in a process known as slip-casting or filter pressing. Bulk ceramics are processed in powder form since their high melting points and low formability prohibit other types of processing. Metals and polymers can also be processed from powders. Deposition processing modifies a surface chemically by depositing a chemical vapor or ions onto a surface. It is used in semiconductor processing and for decorative or protective coatings.

Vapor source methods require a vacuum to transport the gaseous source of atoms to the surface for deposition. Common vapor sources are thermal evaporation which is similar to boiling water to create steam, sputtering which uses energetic ions to bombard a source and create the gas state and laser light in which the light ablates or removes atoms from surface to create the gaseous state. Other sources use carrier media such as electrochemical mixtures which involves ions in a solution being transported by an electrical field to the surface for depositions or spray coating in which ions or small particles are transported virtually by gases, liquids, and/or electrical field. All these processes are basically well understood at atomic level and this requires a clear understanding of the quantum theory of materials.

One of the most common processes is the deformation of a solid to create a desired shape. A large force is used to accomplish the deformation and many techniques heat the material in order to reduce the force necessary to deform it. Sometimes a mold is used to define the shape. Forging which is an old method which utilizes heated on metal to cause the metal to be deformed the metal by hammer blows. Rolling to reduce the thickness of a plate is another common process. Some glasses when heated can be formed with tools or molds. Structure refers to the arrangement of a material's components from an atomic to a macro scale. Understanding the structure of a substance is therefore key to understanding the state or condition of a material. These in turn provide information which is correlated with processing of the material in tandem with its properties. Understanding these relationships is an intrinsic part of materials science engineering, as it allows engineers to manipulate the properties of a material.

Materials engineers must frequently reconcile the desired properties of a material with its structural state to ensure compatibility with its selected processing and this requires the quantum level understanding of material properties. Solid state physics is typically concerned with material properties in the following classes: Mechanical Properties which include the tensile strength, fracture toughness, fatigue strength, creep strength and hardness. The Electrical Properties in which conductivity or resistivity, ionic conductivity, semiconductor and conductivity (mobility of holes and electrons) are of great concern. The Magnetic Properties where the quantum theory is required to explain magnetic susceptibility, Curie Temperature, Neel Temperature and saturation magnetization. The optical and Dielectric Properties in which polarization, capacitance, permittivity, refractive index and absorption are properties utilized in many electronic devices as well as in thermal Properties where coefficient of thermal expansion, heat capacity and thermal conductivity are very important properties.

The evaluation of performance is an integral part of the solid state device or field. The Analysis of any failed product is used to obtain feedback on processing and its control as well as to assist in the initial selection of the material and in the stages of processing. This utilizes the quantum approach which is associated with some quantum property test and/or a structural characterization. In the market now we have products ranging from cellular phones to artificial hip joints to lightweight bicycles, materials engineers apply the quantum theory to develop such products that improve lives. The quantum theory has brought advances in many areas that include the auto-machines, aerospace, construction industry, manufacturing, electronics industry, computer telecommunications industries by developing improved materials from metals, plastics, ceramics, semiconductors and composites and this has increase the strength of steel, toughen ceramics, lower the cost of composites and make faster computer circuits. Materials are involved in almost every engineering product, and materials engineers are needed to select the best material, improve its properties, lower processing cost and increase its durability.

Applications of Material in Devices

High Speed Semiconductor Devices are commonly used in a microwave or millimeter wave analyzer, in RF frequencies analyzers and optical frequencies analyzers among others. High Speed Semiconductor Devices are also used in communication devices so as to make point to point broadcasting of signal processing, interconnection and networking of communication gadgets. These gadgets include the RADAR, the satellite communication, mobile communication, metropolitan and Wide Area Network (WAN) communication, transmitting, decoding, encoding, receiving and retrieving devices. These high speed devices are also found in oscillators, amplifiers, mixers, modulations and demodulation gadgets which employ the use of MESFETs, BJTs, HEMTs, Klystrons, Magnetrons, IMPATT Diodes, Gunn diodes, Schottky diodes, PIN diodes and Varactor diodes. This shows that certain device properties make these devices applicable for high speed applications. Consider the circuit diagram in figure below;

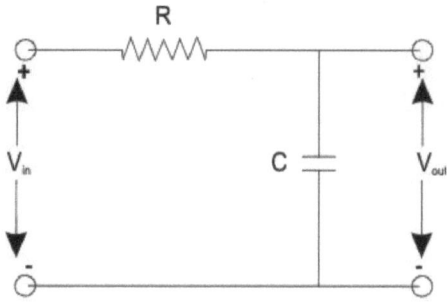

RC circuit

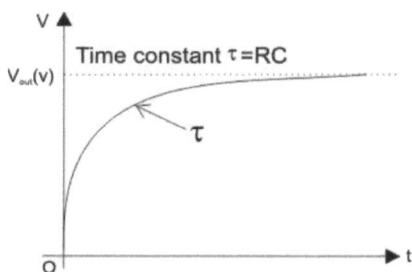

Sketch of an RC time constant curve

Assume that when there is no charge on the capacitor, C for t < 0. In this condition, V_{in} = unit step at t = 0 and therefore. V_{out} will be given as;

$$V_{out} = V_{in}(1-e^{-t/RC}) \text{ for } t > 0$$

And the time constant will be given as;

$$\tau = RC$$

On this condition then, when RC becomes very small, the steady state is reached more quickly and therefore further analysis on the frequency domain give V_{in} as;

$$v_{in} = \cos(\omega_0 t)$$

$$v_{out} = |H(j\omega_0)|\cos(\omega_0 t + \angle H(j\omega_0))$$

$$H(j\omega) = \frac{1}{1+j\omega\tau} \qquad \tau = RC$$

Further analysis can graphically display a high frequency response curve shown.

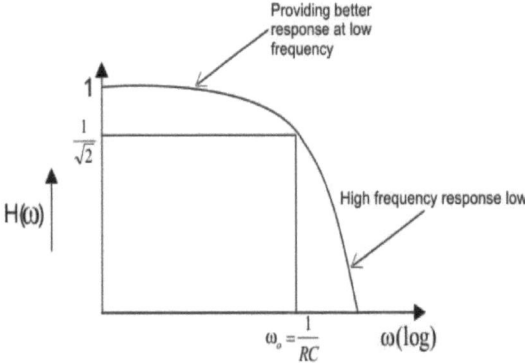

High frequency response curve

Cut off frequency (ω_c)

Based on the circuit above, it is also noted that as RC decreases, cut off frequency, (ω_c) increases and circuit becomes capable to carry and handing high frequency and high speed signals given by;

$$\omega_c = 1/RC$$

From expression 1.4, it can be noted that the high speed or frequency becomes low as RC increase and vice versa as RC reduce in these devices. Therefore there is a different delay mechanism required to explain it as transit time delay increases. Thus if we assume that in any normal circuit, as soon as the applied voltage or Electric Field is applied, current flows immediately in the external circuit. The carriers present which are influenced by the electric field applied has to start from one terminal of the device and reach the other terminal of the same device to show any observable current in the external circuit. Due to finite velocities of these carriers, a delay occurs and thus is called the transit time delay.

Resistivity and cut off frequency

Consider a block semiconductor as shown in figure 1.4 in which some current, **I** is applied by a certain voltage, **V**. Let the cross – section area, **A** and the length of the block be **L**. Think about a piece of semiconductor carrying current **I**, then;

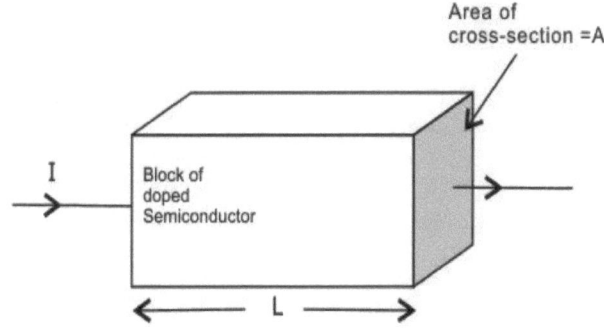

Block Semiconductor

Taking A = area of cross-section, L = length and ρ = resistivity, the expression for capacitance becomes;

$$C = \frac{\epsilon_r \epsilon_o A}{L}, \quad R = \frac{\rho L}{A}$$

since I = JA, where J is the current density, current density can also be give as;

$$J = q(v_n n + v_p p) = qE(\mu_n n + \mu_p p)$$

where, v = drift velocity, µ = mobility for (n) electrons and (p) holes, respectively and q the electronic charge. Hence,

$$\rho = \frac{1}{q(\mu_n n + \mu_p p)}$$

Therefore, cut off frequency associated with this device and carrying, μ= mobility for (n) electrons and (p) holes is given as:

$$\omega_c = \frac{1}{RC} = \frac{1}{\rho \epsilon_o \epsilon_r} = \frac{q(\mu_n n + \mu_p p)}{\epsilon_o \epsilon_r}$$

It is noted that from this expression, to increase ω_c we must also increase the quantity;

$$\frac{q(\mu_n n + \mu_p p)}{\epsilon_o \epsilon_r}$$

q and ϵ_0 are therefore fundamental constants and for Si, the fundamental constant will be

$$\epsilon_r = 11.8$$

This likewise gives the fundamental frequency f_c as;

$$f_c = \frac{1.53 \times 10^{11}}{\rho(\Omega - cm)}$$

If calculate based on silicon, we obtain;

$$\rho = 100 \Omega - cm, \, f_c = 1.53 \, GHz$$

From this calculation, we must increase mobility and so as to increase drift velocity and this can be done by increase doping. However, increase in doping reduces mobility & velocity. Therefore optimization needed and it needs a study the basics of solid state physics to understand about mobility, carrier concentration and other parameters.

Quantum Theory

Started in 1901 when Max Planck explained blackbody radiation. Solid objects emit radiation when heated. Planck attempted to derive a formula using microscopic considerations of Statistical Mechanics developed by Boltzmann. He succeeded and re-obtained such a formula in December 1900. He assumed that the radiation could be subdivided into discrete chunks (*quanta*) of energy, "ε". This was a common trick of continuum mechanics: assume discreteness and then pass to the continuous limit. Probably, this was also Planck's intention. But he found that his formula could be obtained *only if ε was taken to be a specific function* of the radiation's frequency v:

$$\varepsilon = hv$$

where he found a value for the constant h to be close to its value;

$$h = 6.626 \times 10^{-27} \text{ erg.s}$$

and that we accept today (1 erg = 1g s^{-2} cm² as a true and applicable conversion. *This constant h is the famous Planck constant.*

In 1906, Planck wrote "Books on the Theory of Thermal Radiation", where, among other things, he gave a detailed derivation of his formula. It is often said that since the time Planck derived that formula in 1900 using the assumption of discrete nature of radiation, he was trying to get rid of that assumption. Overcoming the discreteness assumption was *not* a central theme of Planck's research. While in his original microscopic derivation in December 1900 he assumed that the molecules of the walls emitted quanta of energies $n\varepsilon = nh\nu$, when he first published that derivation in 1901, he wrote that the molecules could emit radiation with energies *lying between* $nh\nu$ and $(n+1)h\nu$. Both derivations gave an almost identical result, as we will show in the last section of this Book. Thus, Planck did *not* consider the radiation being *intrinsically* discrete, although he had to introduce discreteness into his derivations.

The realization that Planck's derivation *did mean* that the radiation *actually consists* of discrete quanta, appeared gradually as a result of contributions to this issue by a number of notable scientists, including: Hendrik A. Lorentz (*of Holland*), Paul Ehrenfest (*of Austria*), Max von Laue (*of Germany*), and Albert Einstein (at that time, Einstein was young and little-known). In particular, Lorentz gave a new derivation of Planck's formula in 1910, and Einstein gave yet another, totally different derivation of it in 1916 when he wrote a ground-laying paper on induced and spontaneous radiation. On the other side of the barricades in 1900 - 1905 were two prominent opponents of Planck's formula: Lord Rayleigh and an English physicist James Jeans in 1900, Rayleigh proposed a formula;

$$K_\lambda(T) = b\lambda^{-5}T \cdot e^{(a/\lambda T)}$$

where the pre-exponential factor had some phenomenological explanation, while the $e^{(a/\lambda T)}$ was brought in "by brute force" to provide agreement with the experiment. In 1905, Jeans used Maxwell's electrodynamic equations to rigorously derive that;

$$K_\lambda(T) = b\lambda^{-4}T$$

where he also derived a value for constant *b*. He argued that for small λ, for which his formula was in blatant contradiction with both the experiments and common sense as we will show in a later section, Maxwell's equations were not applicable for some unknown reason. It may be interesting to note that later Jeans not only converted to the Quantum theory, but also became one of its first proponents in England.

After *1911*, the black-body radiation theory was overshadowed by a newly emerged topic of specific heat calculation and measurement for solid-state substances. Planck's formula, however, found experimental confirmation there as well. A critical development that eventually propelled Planck's discovery into its prominent place occurred in *1913*. In that year, a young and later, the great Danish physicist Niels Bohr related Planck's hypothesis of discreteness of radiation with two then-unexplainable phenomena inside the atom: the atom's stability and radiation spectra emitted by atoms.

A couple years before that, in *1911*, Ernst Rutherford, based on the results of his experiments carried out at the University of Manchester, proposed the *planetary model* of an atom. An earlier model, proposed by J.J. Thompson in 1904, considered an atom as a pudding, with electrons being included there as raisins. There was a problem with Rutherford's planetary model, however. An electron rotating about the nucleus has centripetal acceleration. According to Maxwell's electromagnetic theory, any accelerating charged particle must emit radiation. Therefore, a rotating electron would constantly emit radiation and hence lose energy, so that eventually it would fall into the nucleus. Rutherford was well aware of this problem, but insisted that in spite of it, an atom still had to look like the Solar System.

Niels Bohr, who was a young researcher in Rutherford's laboratory, came up with a generously simple solution: An electron cannot emit continuously, but only by quanta. Therefore, when it orbits the nucleus, the electron does not emit at all because it cannot emit part of a quantum, and hence the atom is stable. The only possibility for an electron to emit a quantum is when it goes for whatever reason from one stationary orbit to another. Calculations that Bohr did using this principle yielded the first-ever theoretical explanation of experimentally observed atomic radiation spectra. Thus, Bohr's ingenious idea of the connection between Planck's quantum hypothesis and atomic physics paved the way to the creation of Quantum Mechanics.

Density of states of the radiation

The derivation of Planck's formula was done in the actual development of this theory in the 19th century which "*change the variable*" from the wavelength, λ, to the angular frequency, ω, of the radiation are used and characterized by wavelength λ, the speed c, the period as;

$$\tau = \lambda/c,$$

the frequency $v = 1/\tau = c/\lambda$, or, equivalently, by the angular frequency,

$$\omega = 2\pi v = 2\pi c/\lambda.$$

In what follows we will refer to both the frequency v and the angular frequency ω as simply the frequency, since this will not lead to a confusion of the two. Let us recall that our final goal is to derive Eq. above for the radiation spectral density $K_\lambda(T)$. Since we now are using the frequency instead of the wavelength in our description, we need to relate $K_\omega(T)$ with $K_\lambda(T)$. The energy of radiation emitted within a wavelength interval $[,\lambda, \lambda + \Delta\lambda]$ can be written in two ways:

$$K_\lambda(T) = K_\omega(T)\, dw$$

The l.h.s. of this equation is merely the definition of the spectral density $K_,(T)$. The r.h.s. expresses the fact that λ and ω are related by a one-to-one function. Then, Equations above provide a relation between $K_\lambda(T)$ and $K_\omega(T)$

$$K_\lambda(T) = K_\omega(T) \left[\frac{\omega d}{d\lambda}\right]$$
$$= K_\omega(T) \times [2\pi c]/\lambda^2]$$

To begin the derivation of $K_\omega(T)$, consider a large box with some dimensions L_x; L_y; L_z, as shown below. The density of the radiated energy then equals: $K_\omega(T)\, dw = \{$no of frequencies in interval of $[\omega, \omega_+ d\omega]$ in box$\} \times \{$average energy of one radiation mode of frequency $\omega]\}$.

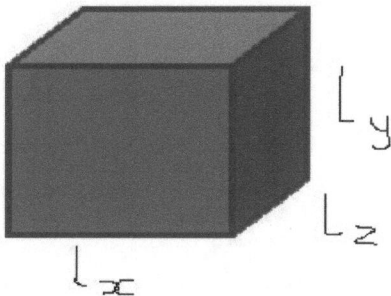

The qualifier shown above "average", is used because the radiation is in equilibrium with its source *on average,* over some macroscopic interval of time spans. In this Section we will estimate the *first term* on the r.h.s. It is possible to compute this term exactly using the Maxwell equations for the radiation, but this will not be required for our purposes. First we note that μ number of frequencies in the interval $[\omega, \omega + d\omega]$ in the box;

{no of frequencies in interval of $[\omega, \omega + d\omega]$ in the box = $[\frac{dZ}{d\omega}] d\omega$

where $Z(\omega)$ is the number of frequencies *up to* ω that can exist in this box. From Maxwell's equations describing the electromagnetic radiation in vacuum (and our box is assumed to contain vacuum, i.e., no matter), it can be shown that allowed frequencies of the radiation propagating in any one direction are spaced *evenly*. Then;

$$Z(\omega) = \text{const} \times (\omega/\omega_{\min, x}) \times (\omega/\omega_{\min, y}) \times (\omega/\omega_{\min, z})$$

= (number of waves in *x*-direction)

×(number of waves in *y*-direction)

×(number of waves in *z*-direction)

× *constant*

where $\omega_{\min, x}$ is the minimum frequency of radiation that can propagate in the box in the *x*-direction; and similarly for $\omega_{\min, y}$ and $\omega_{\min, z}$. This minimum frequency exists because there is the maximum wavelength,

$$\lambda_{,\max;x} = 2L_x$$

that can exist between the walls located L_x units apart. The illustrating figure on the left assumes that the wave is zero at the walls, but a similar result can also be obtained for other boundary conditions.

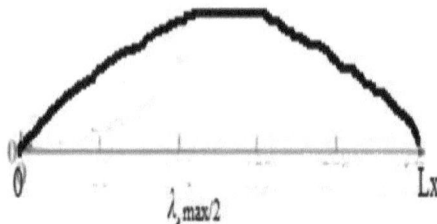

Now, from above,

$$\omega_{\min;j} = 2\pi c/2L_j = \pi c/L_j \quad \text{where} \quad j = \{x; y; z\}$$

From the Figure above one can conclude that the next two largest wavelengths are $2L_x/2$ and $2L_x/3$, where, respectively, two and three semi-periods of the wave fit between the walls. The corresponding frequencies, in analogy are $2\omega_{\min;x}$ and $3\omega_{\min;x}$. This illustrates the statement, made that the frequencies of the radiation in a box are spaced evenly. Next, one has:

$$Z(\omega) = \text{const} \times \omega^3 /[(\pi c)^3/(L_x L_y L_z)]$$
$$= \text{const} \times \omega^3 /[(\pi c)^3 \times L_x L_y L_z]$$

An exact solution of Maxwell's equations for the radiation's electromagnetic field with two transverse polarizations yields the value of the constant to be:

$$\text{const} = \frac{\pi}{3}$$

Then, accordingly, No. of frequencies in range;

$$\{[\omega, \omega + d\omega]\} = \omega^3 d\omega / \pi^3 c^3 \times L_x L_y L_z$$
$$= \frac{\text{No. of frequencies in the range }\{[\omega, \omega + d\omega]\}}{volume\ of\ the\ box}$$
$$= \omega^3 d\omega / \pi^3 c^3$$

We have derived the first term on the r.h.s. of the above equation. To proceed with the derivation of the second term, we need some elementary background in the statistical mechanics of gases.

Black body

A blackbody is defined as a perfect emitter and absorber of radiation. At a specified temperature and wavelength, no surface can emit more energy than a blackbody. A blackbody is a *diffuse emitter* which means it emits radiation uniformly in all direction. Also a blackbody absorbs all incident radiation regardless of wavelength and direction. The radiation energy emitted by a blackbody per unit time and per unit surface area can be determined from the *Stefan-Boltzmann* Law:

$$E_s = \sigma T^4 \ (W/m^2)$$

where

$$\sigma = 5.67 \times 10^{-8}\ w/m^{-2}\ K^{-4}$$

where T is the absolute temperature of the surface in K and E_b is called the blackbody emissive power.

A large cavity with a small opening closely resembles a blackbody. *Spectral blackbody emissive power* is the amount of radiation energy emitted by a blackbody at an absolute temperature T per unit time, per unit surface area, and per unit wavelength as;

$$E_{b\lambda}(T) = C_1\{1/\lambda^5 [\exp C_2/\lambda T - 1]\}$$

$$C_1 = 2\pi hc_0^2$$
$$= 3.742 \times 10^8 \ W\mu m^4 \ m^{-4}$$
$$C_2 = hc_0/k$$
$$= 1.439 \times 10^4 \ \mu mK$$

$K = 1.3805 \times 10^{-23}$ J/K is, Boltzmann's constant.

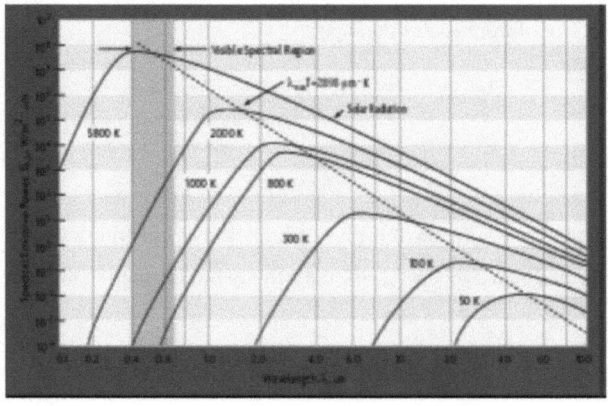

Variation of blackbody emissive power with wavelength

This is called *Plank's distribution law* and is valid for a surface in a vacuum or gas. For other mediums, it needs to be modified by replacing C_1 by C_1/n^2, where n is the index of refraction of the medium. The wavelength at which the peak emissive power occurs for a given temperature can be obtained from Wien's displacement law:

$$(\lambda T)_{max \ power} = 2897.8 \ \mu m K$$

It can be shown that integration of the spectral blackbody emissive power $E_{b\lambda}$ over the entire wavelength spectrum gives the total blackbody emissive power E_b:

$$E_b(T) = \int_0^\infty E_{b\lambda}(T) \, d\lambda - \sigma T^4 \ (W/m^2)$$

The Stefan-Boltzmann law gives the total radiation emitted by a blackbody at all wavelengths from 0 to infinity. But, we are often interested in the amount of radiation emitted over some wavelength band. To avoid numerical integration of the Planck's equation, a non-dimensional quantity f_λ is defined which is called the *blackbody radiation function* as;

$$f_\lambda(T) = \frac{1}{\sigma T4} \int_0^\infty Eb\lambda(T) \, d\lambda$$

The function f_λ represents the fraction of radiation emitted from a blackbody at temperature T in the wavelength band from 0 to λ. Therefore, one can write:

$$f_{\lambda 1 - \lambda 2}(T) = f_{\lambda 2}(T) - f_{\lambda 2}(T)$$

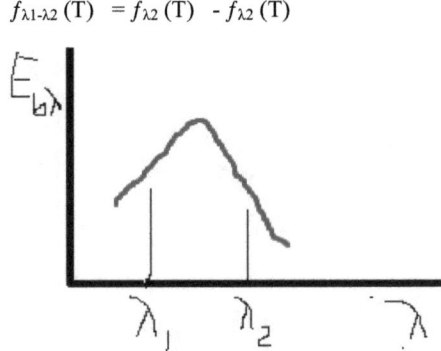

Fraction of radiation emitted in the wavelength λ_1 to λ_2

Example

The temperature of the filament of a light bulb is 2500 K. Assuming the filament to be a blackbody, determine the fraction of the radiant energy emitted by the filament that falls in the visible range. Also determine the wavelength at which the emission of radiation from the filament peaks.

Solution

The visible range of the electromagnetic spectrum extends from 0.4 to 0.76 micrometer. Using the table below, we get that :

$$\lambda_2 T = 0.76 \ \mu m \ (2500K)$$
$$= 1900 \rightarrow f_{\lambda 2} = 0.053035$$
$$\lambda_1 T = 0.4 \ \mu m \ (2500K)$$
$$= (1000) \rightarrow f_{\lambda 1} = 0.000321$$

which means only about 5% of the radiation emitted by the filament of the light bulb falls in the visible range. The remaining 95% appears in the infrared region or the "invisible light".

Radiation Properties

A blackbody can serve as a convenient reference in describing the emission and absorption characteristics of real surfaces.

Emissivity

The emissivity of a surface is defined as the ratio of the radiation emitted by the surface to the radiation emitted by a blackbody at the same temperature. Thus,

$$0 \leq \varepsilon \leq 1$$

Emissivity is a measure of how closely a surface approximate a blackbody, $\varepsilon_{blackbody} = 1$. The emissivity of a surface is not a constant; it is a function of temperature of the surface and wavelength and the direction of the emitted radiation, $\varepsilon = \varepsilon(T, \lambda, \theta)$ where θ is the angle between the direction and the normal of the surface. The *total emissivity* of a surface is the average emissivity of a surface overall direction and wavelengths:

$$\delta(T) = \frac{E(T)}{Eb(T)}$$

$$= \frac{E(T)}{\sigma T^4}$$

$$\rightarrow E(T) = \delta(T) \sigma T^4$$

Spectral emissivity is defined in a similar manner:

$$\delta_\lambda(T) = \frac{E(T)}{Eb(T)} b\lambda$$

where $E_\lambda(T)$ is the spectral emissive power of the real surface. As shown, the radiation emission from a real surface differs from the Planck's distribution.

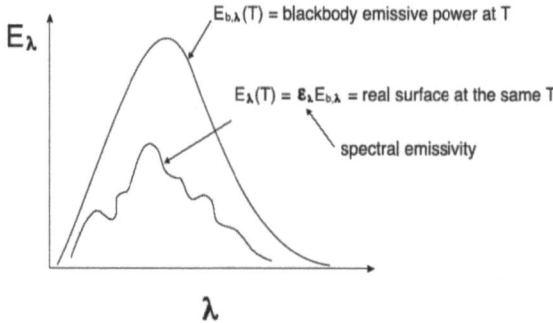

Comparison of the emissive power of a real surface and a blackbody

To make the radiation calculations easier, we define the following approximations:

Diffuse surface: is a surface which its properties are independent of direction.

Gray surface: is a surface which its properties are independent from wavelength.

Therefore, the emissivity of a gray, diffuse surface is the total hemispherical (or simply the total) emissivity of that surface. A gray surface should emit as much as radiation as the real surface it represents at the same temperature:

$$\delta(T) = \frac{1}{\sigma T^4} \int_0^\infty Eb\lambda(T) \, d\lambda$$

Emissivity is a strong function of temperature. Other radiation properties include those that analyze the radiation energy incident on a surface per unit area per unit time is called *irradiation*, G. and are:

Absorptivity, α

This is the fraction of irradiation absorbed by the surface.

Reflectivity, ρ

This is the fraction of irradiation reflected by the surface.

Transmissivity, τ

This the fraction of irradiation transmitted through the surface.

Radiosity, J

This is defined as the total radiation energy streaming from a surface, per unit area per unit time.

It is the summation of the reflected and the emitted radiation. absorptivity:

$$\text{Absorptivity, } \alpha = \frac{absorbed\ radiation}{incident\ radiation}$$

$$= \frac{G\ (absorbed)}{G} \quad\quad 0 \leq \alpha \leq 1$$

$$\text{Reflectivity, } \rho = \frac{reflected\ radiation}{incident\ radiation}$$

$$= \frac{G\ (reflected)}{G} \quad\quad 0 \leq \rho \leq 1$$

$$\text{Transmissivity, } \tau = \frac{transmitted\ radiation}{incident\ radiation}$$

$$= \frac{G\ (transmitted)}{G} \quad\quad 0 \leq \tau \leq 0$$

Applying the first law of thermodynamics, the sum of the absorbed, reflected, and the transmitted radiation radiations must be equal to the incident radiation:

$$G_{abs} + G_{ref} + G_{tr} = G$$

Divide by G:

$$\alpha + \rho + \tau = 1$$

For opaque surfaces $\tau = 0$ and thus:

$$\alpha + \rho = 1.$$

The above definitions are for total hemi-spherical properties (over-all direction and all frequencies. Note that the *absorptivity* α is almost *independent of surface temperature* and it strongly depends on the temperature of the *source* at which the incident radiation is originating. For example α of the concrete roof is about 0.6 for solar radiation and 0.9 for radiation originating from the surroundings and show absorption, reflection and transmission of irradiation by a semitransparent material.

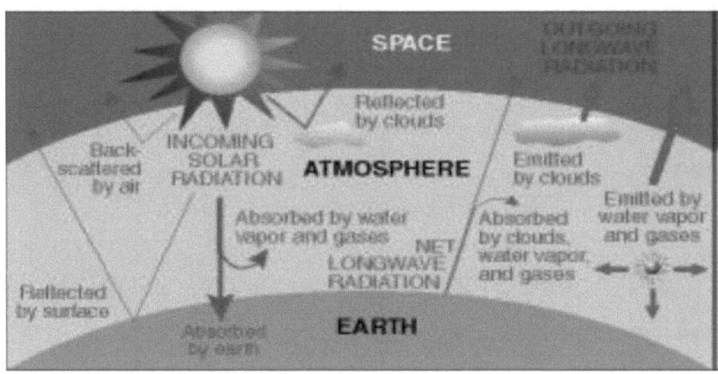

Behaviour of earth radiation

Black body radiation curves showing peak wavelengths at various temperatures. Through the above analysis, Planck resolved that vibrating atoms in material could only radiate or absorb energy in discrete packets and for a given atomic oscillator vibration at a frequency υ (Hz), Planck postulated that the energy of the oscillator was restricted to quantized values given by the expression;

$$E_n = nh\nu = n\hbar\omega, \quad \text{Where} \quad n = 0,1,2......$$

where ω, is the circular frequency and **h**, is the Planck's constant with a value of

$$h = 6.63 \times 10^{-34} \text{ js}, \hbar = h/2\pi$$

Based on Planck's approach, one quantum of energy is expressed as

$$= h\nu = \hbar\omega = hc/\lambda,$$

where λ = Wavelength of the radiation in metres and 'c' is the velocity of light in free space or vacuum.

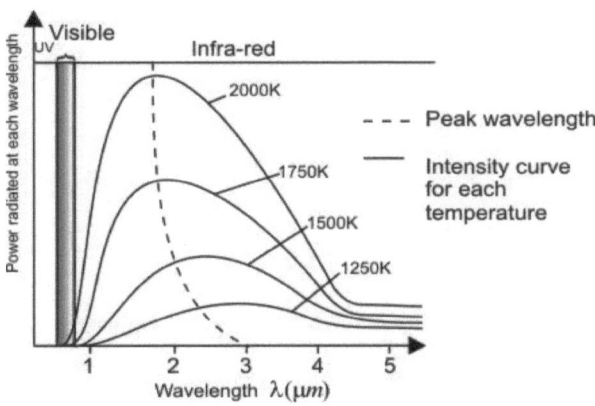

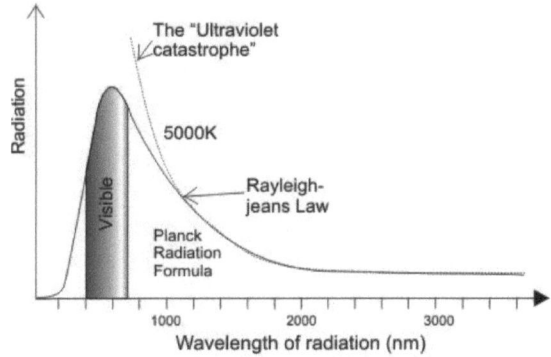

Bohr's Atomic Theory

Heat and electric charge remained phenomena that could not be explained for a long time but Bohr's theory was able to describe it comprehensively. The theory explains why when atoms are energized through heat or electric discharge emits discrete spectral lines, with integer separation distances in terms of wavelengths as shown. Through these observations, Bohr concluded that electrons in an atom move in well-defined orbits with each orbit having a fixed energy level or amount and angular momentum.

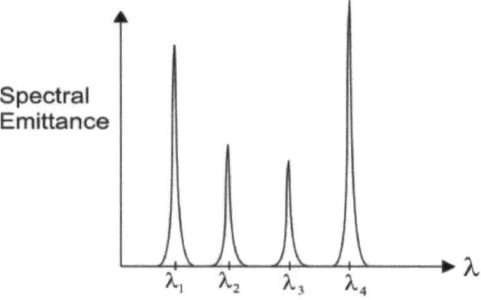

Spectra lines due to electric discharge

Taking hydrogen atoms which have atomic number of 1 and are known to have only one electron, their angular momentum of this hydrogen atom is given by;

$$L_n = m_0 v r_n = n\hbar \qquad n = 1,2,3.....$$

where, m_0 is electron rest mass, v is the linear electron velocity and r_n is the radius of the orbit for a given **n**. Electric charges are known to exert force and at equilibrium, centrifugal force of the charges is equal to electric force i.e,;

Centrifugal force = to electric force

$$\frac{m_0 v^2}{r_n} = \frac{q^2}{4\pi \epsilon_0 r_n^2}$$

Hence, this implies that;

$$\rightarrow m_0 v^2 r_n = \frac{q^2}{4\pi \epsilon_0}$$

Further analysis intended to reduce the above expression results into;

$$\frac{(n\hbar)^2}{m_0 r_n} = \frac{q^2}{4\pi \epsilon_0}$$

This reduces further giving the value of the radius, **r**, at integer **n** as;

$$\rightarrow r_n = \frac{4\pi \epsilon_0 (n\hbar)^2}{m_o q^2}$$

For the nth orbit or radius of orbiting electrons, the kinetic energy is expressed as;

$$KE = \frac{1}{2} m_o v^2 = \frac{1}{2}\left(\frac{q^2}{4\pi \epsilon_o r_n}\right)$$

While potential energy of the electron at the same orbit will be expressed as;

$$PE = -\frac{q^2}{4\pi \epsilon_o r_n}$$

The total energy for this system remains constant and is the sum of both the kinetic and potential energies. It is then expressed as;

$$KE + PE = E_n = -\frac{1}{2}\left(\frac{q^2}{4\pi \epsilon_o r_n}\right)$$

$$KE + PE = E_n = -\frac{1}{2}\left(\frac{q^2}{4\pi \epsilon_o r_n}\right)$$

$$= -\frac{m_o q^4}{2(4\pi \epsilon_o n\hbar)^2} = -\frac{13.6}{n^2} eV$$

It is therefore noted that the light energy emitted by a hydrogen atom when heated or excited discrete in nature and equal to the difference between the two orbits in question and is given as;

$$E_{n_1} - E_{n_2} \qquad n_1 > n_2$$

This can be illustrated as shown in Fig .1.8 below where orbits are represented by 1,2,... 3 etc. At this level the quantum approach is required to determine the spectrum of He and Li using the wave particle duality nature that is based on the De Broglie postulates. Form Albert Einstein's energy expression, the total energy of an electron is given by;

$$E = h\nu = mc^2$$

where **m** is the mass of a photon, h, is Planck's constant and c, is the speed of light in free space. Thus from this expression, the mass of the photon can be expressed as;

$$\frac{h\nu}{c^2}$$

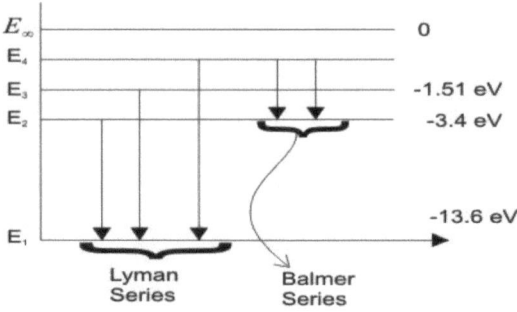

Energy differences between orbitals

And therefore the photon momentum can equally be given by the expression;

$$p = mc = \frac{h\nu}{c} = \frac{h}{\lambda}$$

Quantum theory analysts argument that an electron with a mass, m_0 will have a momentum also in terms of wavelengths and can thus can be represented by wave, where for this particle momentum p can be associated with a wave of wavelength, λ,

$$p = \frac{h}{\lambda}$$

where p, is the particle momentum. Both Schrödinger and Heisenberg used quantum mechanics approach to describe the behavior of small particles like electrons, protons and photons.

Postulates of Quantum Mechanics

There exist a wave function, $\psi = \psi(x,y,z,t)$ of a particle where x, y and z are the space variables and t is the time. From this ψ one can find the dynamic behaviour of a particle in system, ψ is complex. The ψ for a given system is determined by Schrödinger equation:

$$-\frac{\hbar^2}{2m}\nabla^2 \Psi + U(x,y,z)\Psi = -\frac{\hbar}{i}\frac{\partial \Psi}{\partial t}$$

where, m is the mass of the particle and U(x,y,z,t) is the potential energy operator for the system. At these boundary conditions, the values of ψ and $\Delta \psi$ must be finite, continuous and single-valued for all x, y, z and t. Thus the expression;

$$\Psi^*\Psi dV = |\Psi|^2 dV =$$

is the probability that the particle is in a spatial volume element dV though very small and in this event the integral of this expression is equal to unity (1) as;

$$\int |\Psi|^2 \, dV = 1$$

This integration is done over all space in which the electron orbits. A mathematical operator, A_{op} can be associated with each dynamic system variable that includes position, momentum and velocity. The expectation value, when this operator operates on ψ is given as;

$$<A_{op}> = \int \Psi^* A_{op} \Psi \, dV$$

As examples, the space variables x, y, z are given as;

$$<x> = \int \Psi^* x \Psi \, dV, \quad <y> = \int \Psi^* y \Psi \, dV, \text{ and } <z> = \int \Psi^* z \Psi \, dV$$

The momentum, p_x, p_y, p_z are given as;

$$<p_x> = \int \Psi^* \frac{\hbar}{j} \frac{d\Psi}{dx} dV, \quad <p_y> = \int \Psi^* \frac{\hbar}{j} \frac{d\Psi}{dx} dV, \text{ and } <p_z> = \int \Psi^* \frac{\hbar}{j} \frac{d\Psi}{dx} dV$$

where,

$$j = \sqrt{-1}$$

Similarly, the energy E is derived with an operator, $(\frac{\hbar}{j} \frac{\partial}{\partial t})$. Therefore the expected value of energy from postulate 5 is;

$$<E> = \int_v \Psi^* \left(-\frac{\hbar}{j} \frac{\partial}{\partial t}\right) \Psi \, dv$$

$$<E> = E_o$$

If the energy, $E_{0(constant)}$ or the value of energy can be expressed by;

$$\int_v \Psi^* \left(-\frac{\hbar}{j} \frac{\partial}{\partial t}\right) \Psi \, dv = E_o . 1 = E_o \int_v \Psi^* \Psi \, dv$$

$$\therefore \quad -\frac{\hbar}{j} \frac{\partial}{\partial t} \Psi = E_o \Psi \text{ or } \frac{\partial \Psi}{\partial t} = \frac{-j}{\hbar} E_o \Psi$$

This gives the value, $\psi(x,y,z,t)$ thus forms;

$$\Psi(x,y,z,t) = \psi(x,y,z) e^{-jE_o t/\hbar}$$

Schrödinger Equation

This is just like time harmonic case for electromagnetic field where the time dependence is given by;

$$e^{j\omega t}$$

Therefore, now Schrödinger equation can be written as;

$$\nabla^2 \psi + \left(-\frac{2m}{\hbar^2}\right) U(x,y,z)\psi = \left(-\frac{2m}{\hbar^2}\right) E_o \psi$$

Or it can be further reduced to the condensed expression given below;

$$\nabla^2 \psi + \frac{2m}{\hbar^2}(E_o - U)\psi = 0$$

Where, $\psi = \psi(x,y,z)$.

This final expression above is then known as time independent Schrödinger equation and is much easier to solve and it is used to solve simple problems in quantum theory that involve free particle, particle in a potential well, the infinite well or the finite well.

Solution of Schrödinger Equation

In the quantum nature of particles, a free Particle is taken as an electron that id free and alone in the universe. This particle has a mass, m, and exists has a fixed total energy E. this particle at this alone state has no force on the particle nor on its environment and thus it has a constant potential energy everywhere. Thus; $U(x, y, z)$ is equal to a constant, say zero and then taking the universe as a one dimensional only and taking the, x, variation as;

$$\psi = \psi(x)$$

Thus the time independent Schrödinger equation is therefore given by the expression;

$$\frac{\partial^2 \psi}{\partial x^2} + \frac{2m}{\hbar^2} E \psi = 0$$

This time Schrodinger equation can also be written as;

$$\frac{\partial^2 \psi}{\partial x^2} + k^2 \psi = 0$$

and this is in one dimensional differential equation where the value of the Eigen constant k, is expressed or given by;

$$k = \sqrt{\frac{2mE}{\hbar^2}}$$

Or the quantum total energy is obtained from the expression given by;

$$E = \frac{\hbar^2 k^2}{2m}$$

Since the total energy, E has a parabolic dependence on Eigen constant, k, then the general solution for the simple equation based on Schrodinger one dimensional equation can be written as;

$$\psi(x) = A_+ e^{jkx} + A_- e^{-jkx}$$

Where, A_+ and A_- are unknown constants. Therefore, total wave function in space and time dependence form will be given by the expression;

$$\Psi(x,t) = \left[A_+ e^{j(kx - Et/\hbar)} + A_- e^{-j(kx + Et/\hbar)} \right]$$

Compare this with a time-harmonic electromagnetic wave in free-space where

$$f(x,t) = \left(A e^{j(kx - \omega t)} + B e^{-j(kx + \omega t)} \right)$$

Where, k is the constant of propagation given by;

$$k = \frac{2\pi}{\lambda}$$

This analysis is based on a wave function of free particle that consists of a traveling wave and if particle is assumed to be moving in the positive $+x$ then, the wave function will be given by the expression when $A_- = 0$, the expression reduces to;

$$\psi(x,t) = A_+ e^{j(kx - Et/\hbar)}$$

where, $\frac{E}{\hbar} = \omega$ and becomes a simplified wave finction travelling in the positive direction. Thus , nnormalizing the wave equation, we get;

$$\rightarrow \psi^* \psi = A_+^* A_+$$

as equal to a constant for all values of x in the positive direction of wave motion. Thus the probability of finding the particle in any *dx* is equal or the same. If we also take the universe to be infinite as tending its probability to zero, = 0, and that the universe is finitely large, then no such difficulty is experienced. The momentum <p_x> will then be given by;

$$<p_x> \text{ operator} = (\frac{\hbar}{j}\frac{\partial}{\partial t}).$$

Therefore, expected value of momentum, <p_x> will be obtained from the expression;

$$<p_x> = \int_{-\infty}^{\infty} \psi^* \frac{\hbar}{j}\frac{\partial}{\partial x}\psi dx = \hbar k \int_{-\infty}^{\infty} \psi^*\psi dx = \hbar k = \frac{h}{\lambda}$$

This expression is the same as that of the classical theory and therefore the expression reduces to,

$$<p_x> = \frac{h}{\lambda} \rightarrow$$

and become the De Broglie relationship for a finite well represented by the curve in the figure below.

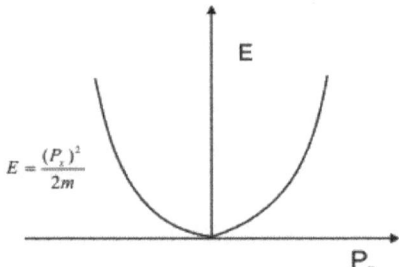

Particle in an infinite potential well

Particle of mass m and fixed total energy E confined to a relatively small segment of one dimensional space between $x = 0$ and $x = a$. In terms of the potential energy we can view that the particle is trapped in an infinitely deep one dimensional potential well, with, $U = 0$ is constant for range values of $0 < x < a$.

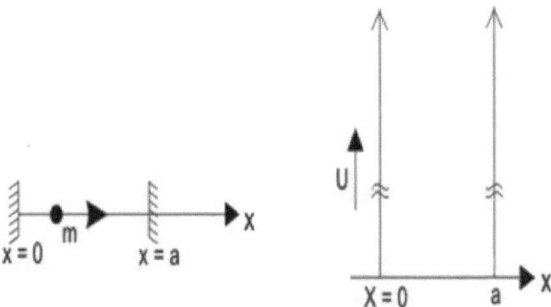

At the condition that U = 0, the time independent Schrödinger equation can be expressed as;

$$\frac{d^2\psi}{dx^2} + k^2\psi = 0 \qquad 0 < x < a$$

$$k = \sqrt{\left(\frac{2mE}{\hbar^2}\right)} \quad \text{or} \quad E = \frac{\hbar^2 k^2}{2m}$$

and, the values of Eigen constant, k and total energy, E, then applying the boundary conditions below;

$$\psi(0) = 0 \quad \text{as} \quad \psi = 0 \text{ for } x < 0$$
$$\psi(a) = 0 \quad \text{as} \quad \psi = 0 \text{ for } x > a$$

Solving for a standing wave function shown,

$$\psi(x) = A \sin kx + B \cos kx$$

can also be represented as a complex expression given as;

$$A_+ e^{+jkx} + A_- e^{-jkx}$$

So that at these boundary conditions, the value of **B** reduces to zero as;

$$\psi(0) = 0 \Rightarrow B = 0.$$
$$\psi(a) = 0 \Rightarrow A \sin ka = 0.$$

Since the value of $A \neq 0$, as ψ, vanishes for all x, so it implies that;

$$\rightarrow ka = n\pi, n = \pm 1, \pm 2, \pm 3 \ldots$$

Forming discrete values resulting into the value of k as;

$$\rightarrow k = n\pi/a = k_n$$

Therefore, the wave function forms and reduces to;

$$\psi(x) = \psi_n(x) = A_n \sin\left(\frac{n\pi x}{a}\right)$$

The Eigen functions k, at orbital n reduces to

$$k_n = \frac{n\pi}{a}$$

and for Energy, E, at orbital, n as;

$$E = E_n = \frac{\hbar^2 n^2 \pi^2}{2ma^2}$$

Looking at the expression above, it is clear that the particles can have only a few discrete energy states or energy levels that can be graphical be represented as shown in the figure below.

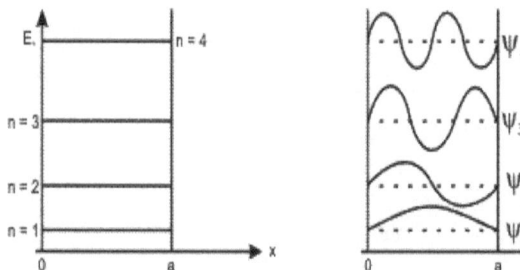

The wave function represented by the wave equation;

$$\psi_n(x) = A_n \sin\left(\frac{n\pi x}{a}\right)$$

Signify that it is a standing wave in which the particle or electron is bouncing back and forth between the walls of the potential well shown in the figure above. Therefore the average value of momentum at any x is 0 as $<p_x> = 0$. This finally confirms that to calculate momentum, we have to isolate forward wave going in $+x$ or backward wave going in $-x$ and calculate them independently.

For $+x$ wave function average momentum will be given by,

$$p_{x+} = \frac{n\pi\hbar}{a}$$

and -x going wave function, the average momentum will be given by;

$$p_{x-} = \frac{-n\pi\hbar}{a}$$

Therefore the discrete shown in figure below and quantized (E - p_x) points lie on the continuous parabola of a free particle that can be expressed using the equation;

$$E = \frac{(p_x)^2}{2m}$$

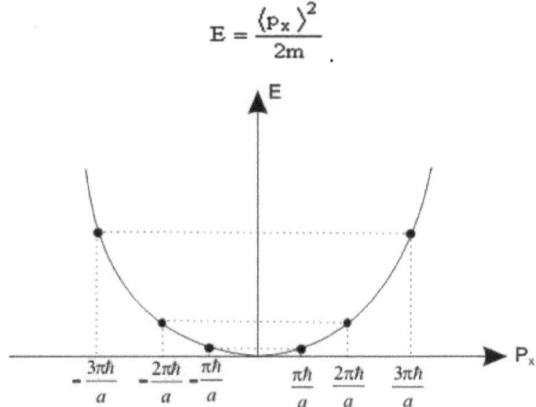

The integer *n* is called the quantum number and the value of A_n from the expression;

$$\int_v \psi_n^* \psi_n \, dv = 1$$

Gives the value of A_n as;

$$A_n = \sqrt{\frac{2}{a}}$$

Finite Potential well

The figure below shows a sketch or an illustration of a potential well. The figure shows a well of diameter, a, in the x direction and a potential U. The particle or electron is allowed to propagate in the positive x direction as shown below.

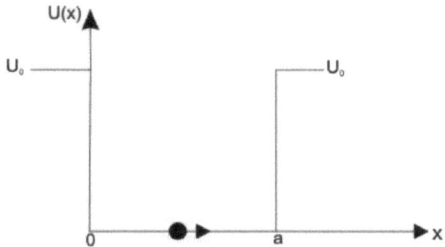

$$U(x) = 0 \text{ for } 0 < x < a$$

At all times, $U(x) = U(0)$ otherwise, if we let a particle with energy;

$$E < U_0$$

Be confined with the potential well, the for $0 < x < a$, we get that;

$$\frac{d^2 \psi_0}{dx^2} + k^2 \psi_0 = 0, \ k = \sqrt{\frac{2mE}{\hbar^2}}$$

And for $x < 0$, we get the expression reducing to;

$$\frac{d^2 \psi_-}{dx^2} - \alpha^2 \psi_- = 0, \ \alpha = \sqrt{\frac{2m}{\hbar^2}(U_0 - E)}$$

In the case where, $x > 0$, we have;

$$\frac{d^2 \psi_+}{dx^2} - \alpha^2 \psi_+ = 0$$

From the above equations, the three sets of solution are can be expressed;

$$\left. \begin{array}{ll} \psi_-(x) = A_- e^{\alpha x} + B_- e^{-\alpha x}, & x < 0 \\ \psi_0(x) = A_0 \sin kx + B_0 \cos kx, & 0 < x < a \\ \psi_+(x) = A_+ e^{\alpha x} + B_+ e^{-\alpha x}, & x > a \end{array} \right\}$$

We can now set up the new boundary conditions and far from the potential well, the wave functions must go to zero as the probability of finding the particle away from potential region = 0. Therefore,

$$\psi_-(-\infty) = 0, \ \psi_+(+\infty) = 0$$

And with this condition, ψ must be continuous at well boundary to give;

$$\psi_-(0) = \psi_o(0), \psi_+(a) = \psi_o(a)$$

and thus

$$\frac{d\psi}{dx}$$

must be continuous at well boundaries $x = 0$ & $x = a$

$$\frac{d\psi_-}{dx}\bigg|_0 = \frac{d\psi}{dx}\bigg|_0 \quad \& \quad \frac{d\psi_+}{dx}\bigg|_a = \frac{d\psi_o}{dx}\bigg|_a$$

From the other boundary conditions given by;

$$B_- = 0, A_+ = 0$$

The resulting four simultaneous equations can be given as follows;

$$A_- = B_o$$

$$A_o \sin ka + B_o \cos ka = B_+ e^{-\alpha a}$$

$$aA_- = kA_o$$

$$kA_o \cos ka - kB_o \sin ka = -\alpha B_+ e^{-\alpha a}$$

From the four expression, elimination of A-, B0 and B+, we obtain the expression;

$$A_o \left[(k^2 - \alpha^2) \sin ka - 2\alpha k \cos ka \right] = 0$$

If the values of

$$A_o = 0, \psi = 0$$

Are identical, then the expression reduces to zero as,

$$(k^2 - \alpha^2) \sin ka - 2\alpha k \cos ka = 0$$

which can also be given as

$$\tan ka = \frac{2\alpha k}{k^2 - \alpha^2}$$

This is a transcendental equation which can also be used to define a dimensionless quantity of frequency f_0 as;

$$f_o = \sqrt{\frac{2mU_o}{h^2}}$$

Where f_0 is constant of the system and

$$\xi = \frac{E}{U_o} \qquad (0 < \xi < 1)$$

So that;

$$\alpha = f_o\sqrt{1-\xi}, \; k = f_o\sqrt{\xi}$$

As obtained from the definition of value of **k** and α. Thus as shown in figure, the Eigen value equation becomes;

$$\tan(f_o a \sqrt{\xi}) = \frac{2\sqrt{\xi(1-\xi)}}{2\xi - 1} = S(\xi)$$

$$f_o a \sqrt{\xi} = \theta$$

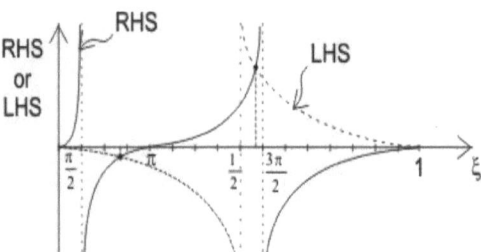

Then if also we have;

$$f_o a < \pi, \qquad \text{where } \tan\theta$$

Intersect with

$$S(\xi)$$

at one point then likewise;

$$\pi < f_o a < 2\pi, \; \tan\theta \; S(\xi)$$

intersect also at 2 points.

$$2\pi < f_o a < 3\pi, \tan\theta$$

Intersect at 3 points with

$$S(\xi)$$

So for any potential well, it can be generalized that;

$$m\pi < f_o a < (m+a)\pi$$

Where there are *m* points. Thus in the limit;

$$U_o \to \infty,$$

where the E is finite.

$$S(\xi) = 0.$$

And thus;

$$\tan ka = 0$$
$$\to k = k_n = \frac{n\pi}{a}, \qquad n = \pm 1, \pm 2, \ldots$$

Finite potential

In finite potential wells, the Eigen value equation is given by;

$$\tan\left(f_o a \sqrt{\xi}\right) = \frac{2\sqrt{\xi(1-\xi)}}{2\xi - 1}$$

for a given system constant, we also have;

$$f_o a = a\sqrt{\frac{2mU_o}{\hbar^2}}$$

From the two expressions, we have a discrete number of solutions where LHS or RHS of the first equation intersect as illustrated in the figure below.

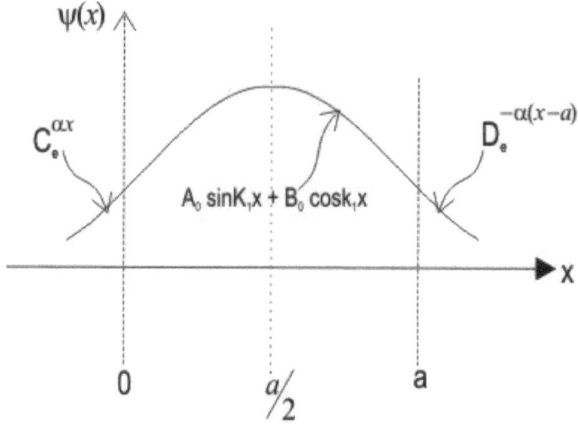

Wave Function of Finite Potential Wells

For each intersection a value of

$$\xi_1 = \frac{E_1}{U_o} \rightarrow \alpha_1, k_1, \psi \neq 0$$

particular outside the well of potential, it can be seen from this that there is a finite probability of existence outside in the classically forbidden region classically a particle with an energy, E < U_0 cannot exist outside the well, if the potential well is slightly modified. The wave function will be nonzero in region C also. Thus the particle will have a finite probability of existing or coming in region C past the potential barrier B. In C the particle can appear as a free particle. This quantum mechanical phenomenon of passing through a barrier is known as tunneling.

The wave function is different inside and outside. Therefore there exists a finite probability of reflection at the well walls, called quantum mechanical reflection at the well walls. In analogy to optics it may be looked at as two partially reflecting mirrors, where an infinite potential well could be visualized as two 100 % reflecting mirrors.

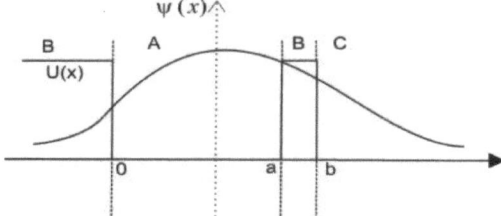

It is used in Tunnel diodes and operation of many other solid-state devices. One of the usage of this phenomenon for high frequency device is called Resonant Tunneling Diode (RTD), which would be used as an oscillator and even as an amplifier.

Linear Harmonic Oscillator

A harmonic oscillator is a particle which is bound to an equilibrium position by a force which is proportional to the displacement from that position. Thus we have,

$$\text{force} = -\gamma x = -\frac{dU}{dx}$$

where, γ is the spring constant. The potential is expressed as,

$$U(x) = \frac{1}{2}\gamma x^2$$

The linear harmonic oscillator can then be visualized on a mass connected to a spring of spring constant, γ on shown in figure below.

Since the time independent Schrödinger equation is given by,

$$\frac{\hbar^2}{2m}\frac{d^2\psi}{dx^2} + \frac{1}{2}\gamma x^2 \psi = E\psi$$

or,

$$\frac{d^2\psi}{dx^2} + \frac{2m}{\hbar^2}\left(E - \frac{1}{2}\gamma x^2\right)\psi = 0$$

Then to solve this time independent Schrodinger's equation, we consider a dimensionless quantity, we get

$$y = \left(\frac{m\gamma}{\hbar^2}\right)^{1/4} x$$

and

$$\lambda = \frac{2}{\hbar}\left(\frac{m}{\gamma}\right)^{1/2} E$$

Combining the above two expression, we finally get the expression;

$$\frac{d^2\psi}{dy^2} + (\lambda - y^2)\psi = 0$$

For large values of y we can neglect, λ, we reduce the above equation to,

$$\frac{d^2\psi}{dy^2} - y^2\psi = 0$$

Thus this equation can be satisfied approximately by the solution given by,

$$\psi(y) = e^{-y^2/2}$$

In which the two equations can be substituted to give the expression below;

$$\frac{d^2\psi}{dy^2} + (1 - y^2)\psi = 0$$

This indicates that the two expressions above can approximate to have exact solution as,

$$\psi(y) = e^{-y^2/2} \xi(y)$$

So as to given;

$$\frac{d^2\xi}{dy^2} - 2y\frac{d\xi}{dy} + (\lambda - 1)\xi = 0$$

To linearize the above equation, we can use the power series method and the trail solution will be,

$$\xi(y) = \sum_{n=0}^{\infty} a_n y^n$$

$$\frac{d^2\xi}{dy^2} = \sum_{n=0}^{\infty} (n+1)(n+2) a_{n+2} y^n$$

$$-2y\frac{d\xi}{dy} = \sum_{n=0}^{\infty} -2n a_n y^n$$

$$(\lambda-1)\xi = \sum_{n=0}^{\infty} (\lambda-1) a_n y^n$$

The above four expressions combined form a summarized expression as;

$$\sum_{n=0}^{\infty} \left[(n+1)(n+2) a_{n+2} - (\lambda - 1 - 2n) a_n \right] y^n = 0$$

This equation must hold for all values of ζ and therefore the coefficient of each power of ζ must vanish separately. This gives in the recursion relation between, $a_n + 2$, and a_n,

$$a_{n+2} = \frac{2n - \lambda + 1}{(n+1)(n+2)} a_n$$

It seems that knowing the values of a,

$$a_0, a_1, a_1, a_3, a_4, a_5 \ldots$$

can help to calculated and obtain them as,

$$\rightarrow a_2 = -\frac{(\lambda-1)}{2!} a_0 \qquad a_3 = -\frac{(\lambda-3)}{3!} a_1$$

$$\rightarrow a_4 = +\frac{(\lambda-1)(\lambda-5)}{4!} a_0 \qquad a_5 = +\frac{(\lambda-3)(\lambda-7)}{5!} a_1$$

Thus, from these equations, we can write equation as,

$$\xi(y) = a_0 \left[1 - \frac{(\lambda-1)}{2!} y^2 + \frac{(\lambda-1)(\lambda-5)}{4!} y^4 + \ldots \right] + a_1 \left[y - \frac{(\lambda-3)}{3!} + \frac{(\lambda-3)}{5!} y^5 + \ldots \right]$$

If in the above equation,

$$\lambda - 1 - 2n$$

should be zero for some value of the index n, then

$$a_{n+2} = 0$$

But since a_{n+4} is a multiple of a_{n+2} so on,

Then all the succeeding coefficients which are related to a_n by the recursion relation would vanish and one or the other bracketed series in equation above would terminate to become a polynomial of degree *n*. This occurs, when,

$$\lambda - 1 - 2n = 0$$

$$\lambda = 2n + 1$$

Energy Quantization

We have obtained the condition when the wave function is acceptable for n = 0, 1, 2..... ... as;

$$\lambda = 2n + 1$$

This makes λ to vary as;

$$\lambda = \frac{2}{\hbar}\left(\frac{m}{\gamma}\right)^{1/2} E$$

And for any n value, the total energy will be given by;

$$E_n = \frac{\hbar}{2}\omega(2n+1) \qquad \left[\text{As } \omega = \left(\frac{\gamma}{m}\right)^{1/2}\right]$$

Thus;

$$E_n = \left(n + \frac{1}{2}\right)\hbar\omega$$

This variation of the energy levels can be graphically be represented by the curve shown in the figure below.

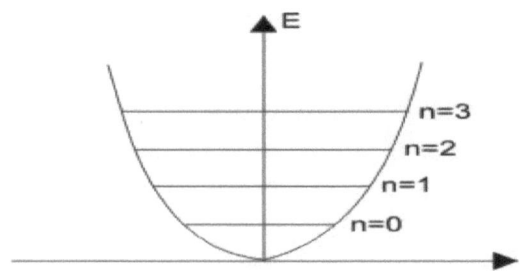

CHAPTER THREE
ENERGY BAND THEORY

Energy Bands Theory in Solids

From quantum mechanics it is made clear that the electrons of an isolated atom can have only discrete energy levels. Atoms constitute a crystal. When a number of atoms are brought together to form a crystal their discrete energy levels split to form a band. Most semiconductors occur in solid form. These solids are made up of atoms which in turn contains electrons. As the inter-atomic spacing decreases these bands merge together to form a single band. When the distance between these atoms approaches equilibrium distance the band splits again into two bands separated by a region called the forbidden gap, E_g. The upper band is called the conduction band (CB) and the lower band is called the valence band (VB). In conductors (i.e. metals) or the conduction band is either partially filled or overlaps the valence band so that there is no band-gap.

This courses the uppermost electrons in the partially filled band or electrons at the top of the valence band to move to the next higher available energy level when they gain kinetic energy. Through this current conduction readily occur in conductors. In semiconductors the bonds between neighbouring atoms are only moderately strong and therefore thermal vibrations will break some bonds and a free electron along with a free hole is produced. That is why the band gap of a semiconductor is not as large as that of an insulator. Therefore some electrons will be able to move from the valence band to the conduction band leaving holes in the valence band.

If an electric field is applied then both the electrons in the conduction band and the holes in the valence band will gain kinetic energy and conduct electricity. The valence electrons form strong bonds between neighbouring atoms in insulators. These strong bonds are difficult to break. This means that there are no free electrons to participate in current conduction. These courses the band gap in an insulator to be large. When you consider the band gap in an insulator, thermal energy or any external applied electric field cannot raise the uppermost electron in the valence band to the conduction band. Using the energy parameters we get that in the donor case;

$$E_C - E_F = KT \ln \{N_C/N_D\}$$

and in the acceptor case;

$$E_F - E_V = KT \ln \{N_V/N_A\}$$

where the symbols have their usual meanings. Electron and hole densities in extrinsic semiconductors can therefore be expressed as:

$$n = n_i \exp\{(E_F - E_i)/KT\}$$
$$p = n_i \exp\{(E_i - E_F)/KT\}$$

It is also noted that when both the donor and acceptor impurities are present simultaneously, the Fermi level adjusts itself to preserve charge neutrality according to the mass action law. Between conductors and insulator materials lie semiconductors that have revolutionized electronic technology in the world and contributed a lot to clean energy. In solid state electronics materials can be classified as insulators, semiconductors and conductors based on their electrical conductivity. The electrical conductivity of semiconductors at room temperature is in the range of $10^2 - 10^{-9}$ Ω-cm^{-1} (i.e. electrical resistivity, ρ from $10^{-2} - 10^9$ Ω-cm). Conductivity of semiconductors is generally sensitive to temperature, illumination, magnetic field and minute amount of impurity atomss. The energy gap, E_g of semiconductor materials is about 1. In the periodic table of elements we have elements that exist as semiconductors. These elemental semiconductors are composed of single species of atoms such as silicon (Si) and germanium (Ge).

Bond and Structures

Accurate knowledge of solid structures is important in solid state physics because these structures influence the physical properties of solids. Solids are classified into two types in terms of their structures. They are either, crystalline solids that include single and polycrystalline solids, or they are non-crystalline solids. A great number of useful semiconductors have diamond or zinc-blende lattice structures which belong to the tetrahedral phases. In the above mentioned structure, each atom is surrounded by four equidistant nearest neighbour atoms that lie at the corners of a tetrahedron.

The bonds between any two nearest neighbour atoms are formed by two electrons with opposite spins. Diamond and zinc-blende lattices are considered as two interpenetrating face-centred cubic lattices. Most of the III-V compounds are crystalline in the zinc blende/ diamond structure while others are crystalline in the wurtzite or rock-salt structures. The wurtzite structure is tetrahedral with four equidistant nearest neighbour atoms similar to a zinc blende structure. In the rock-salt structure each atom has six nearest neighbours. Some compounds such as zinc sulphide and cadmium sulphide can crystallize in both zinc blende and wurtzite structures.

Energy band structure Theory

In the figure (a) the maximum in the valence band occurs at p = 0 while the minimum in the conduction band occurs along the [100] direction at $p = p_C$. The electron at the conduction band (minimum) with zero kinetic energy can have crystal momentum different from zero (i.e. $p = p_C$). Materials are classified as direct or indirect semiconductors depending on the band structure. When an electron makes a transition from the valence band to the conduction band it requires then, not only an energy change greater than E_g but also some change in the crystal momentum, P.

This makes such semiconductors to be indirect semiconductors because a change in crystal momentum is required in their transitions. In figure (b) the maximum in the valence band and the minimum in the conduction band occurs at the same crystal momentum, p = 0 and in this case, an electron making a transition from the valence band to the conduction band can do so without a change in their momentum, P value. Such semiconductors are classified as direct semiconductors because their transition from the valence band to the conduction band does not require a change in crystal momentum for the electron.

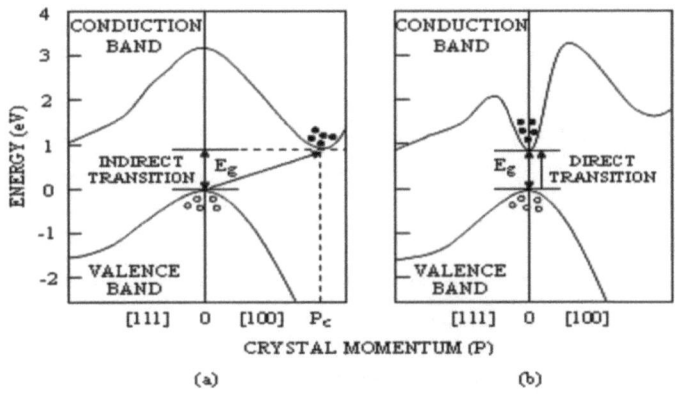

Direct and indirect transition

Semiconductors

Conduction in these materials is due to the intrinsic processes without the influence of impurities. These are pure crystals of semiconductor materials. It occurs due to the electrons that are excited from the top of the valence band to the bottom of the conduction band by thermal energy. The promotion of some of the electrons across the band gap leaves some vacant electron sites in the valence band. These are called holes. The number of electrons excited across the gap can then be calculated from the Fermi-Dirac probability distribution as:

$$F_{(E)} = 1/[1+\exp(E-E_F)/KT]$$

The Fermi level (E_F) for an intrinsic semiconductor lies midway in the forbidden gap. The probability of finding an electron here is 50 % even though energy levels at this point are forbidden. This means then that $(E - E_F)$ in the denominator of equation above is equal to $E_g/2$, where E_g is the magnitude of energy gap.

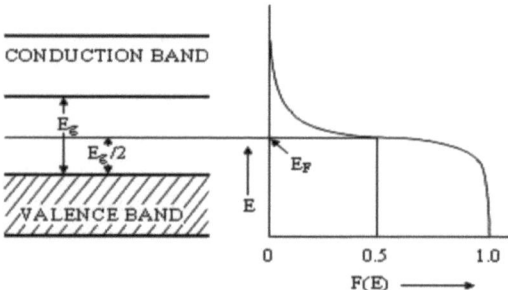

The Fermi level in an intrinsic semiconductor

The unity (1) factor in the denominator can be ignored in comparison to the exponential term because of the fact that $(E - E_F)$ value is larger than the thermal energy (KT) at room temperature. Therefore the probability $F_{(E)}$ of an electron occupying energy level E becomes;

$$F_{(E)} = \exp[-E_g/2KT]$$

Using the bottom of conduction band as E_C and the top of valance band as E_V, the electron density in the conduction band can then be given by;

$$n = N_C \exp[(E_C-E_F)/KT]$$

in a similarly way, hole density in the valence band can be given by;

$$p = N_V \exp[-(E_F - E_V)/KT]$$

where, N_C, N_V, are the effective density of states in the conduction and valence bands respectively.

As each excited electron leaves behind one hole an intrinsic semiconductor contains an equal number of holes in the valence band and electrons in the conduction band (i.e. $n = p = n_i$) where n_i is intrinsic carrier density. Since the mass action law is given by,

$$np = n_i^2$$

This means that the product of the two types of carriers will remain constant at a given temperature. This law is valid for both intrinsic and extrinsic semiconductor under thermal equilibrium condition. Using equation earlier discussed, the intrinsic carrier density can be written as;

$$n_i = (N_c N_v)^{1/2} \exp[-E_g/2KT]$$

$$\text{where } E_g \equiv (E_C - E_V).$$

When an external field is applied the electrons that are excited into conduction band by thermal means accelerate using the vacant states available in the conduction band. At the same time the holes in the valence band move but in a direction opposite that of the electrons.

Extrinsic Semiconductors

The process of deliberate addition of controlled quantities of impurities to a pure semiconductor is called doping. If a dopant (impurity) is introduced into an intrinsic semiconductor then it changes and becomes an extrinsic semiconductor. This is because impurity energy levels are introduced. If the doping process results in *n*-type semiconductors due to the addition of negative charges (electrons) then the impurity atoms are called donors and the energy levels of these electrons are called the donor energy levels (E_D). If the doping process results into a *p*-type semiconductor then the impurity atoms are called acceptors and the energy levels of these holes are called the acceptor energy levels (E_A).

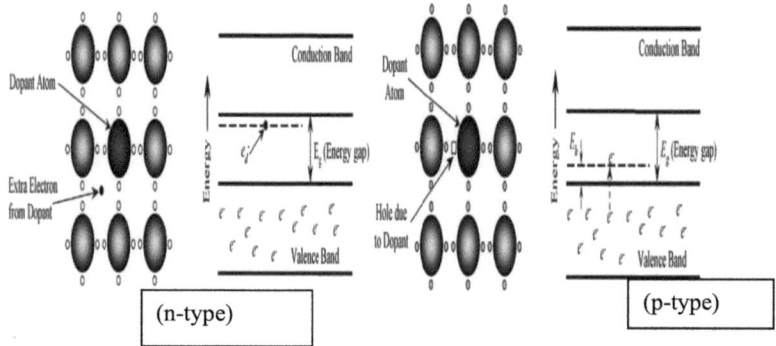

The addition of impurities (doping) greatly increases the conductivity of most semiconductors as illustrated in the figure below;

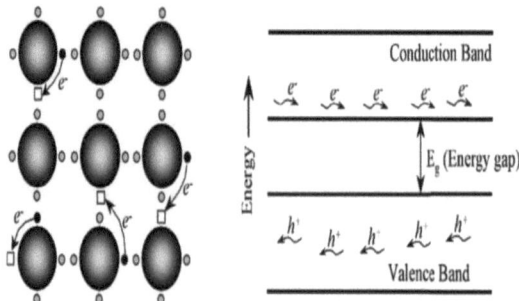

Movement of electrons and holes during conduction

In general when an extrinsic semiconductor is analysed in terms of all the energy parameters used to describe semiconductors [(E_c, E_D, E_F, E_A, E_i, and E_V, it is observed that their interactions can be demonstrated as shown in the figure below;

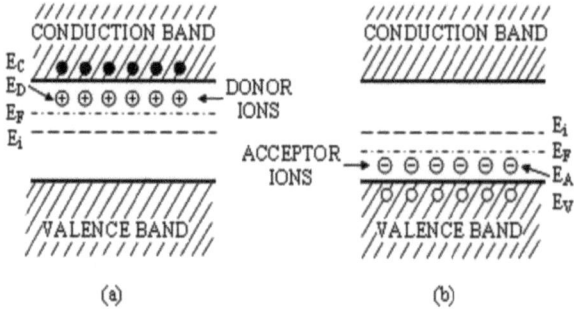

Energy interactions in (a) n-type, (b) p-type doped materials

To excite electrons from the donor level into conduction band (CB) or holes from acceptor level into valence band (VB), energy known as ionization energy is required. The ionization energy (E_i) of donor electrons is approximately the same as the ionization energy of acceptor holes in the same crystal. When it is compared to the energy gap, the ionization energy of an impurity atom is very small. This means then that at room temperature a large fraction of the donor level electrons as well as acceptor level holes are excited into the conduction and valence band respectively. This fraction is much larger than the fraction of electrons excited from the valence band or that of holes created by these electrons due to the intrinsic process. The product of the number of electrons in the conduction band and the number of holes in the valence band must be constant for any semiconductor according to the law of mass action.

For an *n*-type semiconductor this condition drastically reduces the number of holes and therefore electrons in the conduction band become the majority charge carriers. In a similar way, the number of electrons in the *p*-type semiconductor is reduced causing the holes in the valence band to become the majority charge carriers. From the figure it can be seen that the Fermi level (E_F) with higher donor concentration (E_D) will move closer to the bottom of the conduction band and in the same manner the one with a higher acceptor concentration will move closer to the top of valence band. Therefore when the semiconductor is under complete ionization condition (i.e. when $n = N_D$ and $p = N_A$) we obtain the Fermi level in an extrinsic semiconductor as follows:

In donor case,
$$E_C - E_F = KT \ln [N_C/N_D]$$

In acceptor case,
$$E_F - E_V = KT \ln [N_V/N_A]$$

Also electron and hole densities in extrinsic semiconductors can then be expressed as:
$$n = n_i \exp[(E_F - E_i)/KT]$$

and
$$p = n_i \exp[(E_i - E_F)/KT]$$

If both the donor and acceptor impurities are present simultaneously, then, the Fermi level must adjust itself in such away so as to preserve charge neutrality. All solids are classified as conductors, semiconductors or insulators according to their conduction electrons in their solid structures. Band theory is a theory that gives an explanation for differences in electrical properties. As defined by quantum theory, individual atoms have certain permitted energy levels for their electrons. When large groups of atoms are incorporated into a solid mass, then those energy levels reorganize themselves in such a way so that bands have possible energy levels. This is thus commonly referred to as the Tight Binding Approximation Theory.

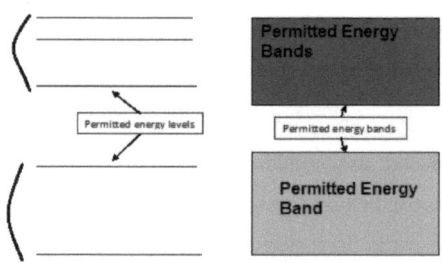

Discrete energy levels within an individual atom (left) and bands of permitted energy levels within a solid (right)

In reality, there exist a large number of electrons in any solid mass. Although the bands in actually sense consist of a very large number of closely packed discrete energy levels, these bands behave as essentially continuous bands. As a result, there are several permitted energy level bands, though analysis only considers the two uppermost bands known as the *valence band* and the *conduction band* as shown in the figure below. The figure shows the conduction and valence bands particularly in an insulator and it the bands contain the only permitted energy levels. The valence band is full while the conduction band is virtually empty and thus there is no net movement of electrons can occur within the material.

This makes the gap separating the bands quite large. In this case, electrons with lower energy levels occupy the valence band and the innermost electrons in any atom are less influenced by their neighbouring atoms since they occupy discrete energy levels hence considered to be bound. At higher levels in the valence band electrons can, these electrons conduction electrons move from atom to atom and only up to the top of the valence band. They are permitted only to swap places with other valence electrons in neighbouring atoms and do not make it to the conduction band for conduction purposes.

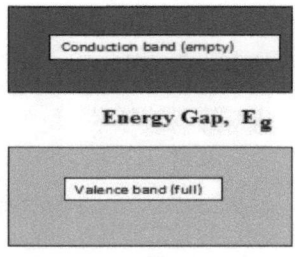

Conduction and valence bands in an insulator

In any atom, electrons fill the valence band from the lowest level to the highest band such that at the top of the valence band of any material is the highest level to be filled by electrons within an atom of that material at absolute zero kelvin (0K) temperature. In insulators and semiconductors alike, their valence bands are completely filled with electrons while the conduction band is empty. These electrons fill energy levels in order by obey the Pauli Exclusion Principle and therefore they cannot occupy identical energy levels. Therefore the only way an electron moves from one atom to another in an insulator or semiconductor is by occupying a slightly different energy level in a neighbouring atom. Since all those energy levels are already full, the electrons effectively swap places and leap up to the conduction band only when some space or vacant energy levels exist for there to be any net movement of electrons within the material.

A material can only conduct electricity if it has electrons in its conduction band or spaces in its valence band and these spaces or electronics must move into a partially filled band. This motion creates a wavelike behaviour for electrons within atoms and materials exhibit a certain range of 'forbidden' energy levels as shown in the figure below. This means that it is not possible for an electron to exist with an energy level that would place it in this range. This leaves insulators and semiconductors with a gap between the two bands as depicted below. It is noted that the energy gaps in insulators and semiconductors shown below have no gap but are simply a continuous and partially filled conduction band.

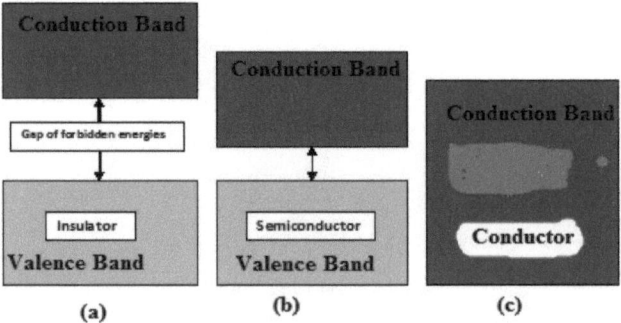

In insulators the forbidden zone is very substantial and it separates the valence band and the conduction band more clearly than in metals since it is in the order of a few electron volts. Thermal excitations and conventional electric circuit voltages within an insulator material provide energies that are smaller than 1 eV on an atomic scale. It means that you can expose an insulator to electric fields of the order of 10^{10} V m^{-1} in order to give the valence electrons enough energy to jump across the gap to the conduction band and only this happens when there is dielectric breakdown in the material.

Conductors have only one band and the top of this band is only partially filled permitting electrical conductivity which therefore means that there are plenty of nearby energy levels available for electrons to move into. They can flow easily from one atom to another when a potential difference is applied across the conducting material. Like insulators, semiconductors have a completely full valence band and so electrons are not able to facilitate conduction at low temperatures. For semiconductors, the forbidden energy level zone between the two bands is sufficiently small to make it much easier for significant numbers of electrons to move across this gap and go from the valence band to the conduction band. This can happen if there is some thermal excitation. Semiconductors also exhibit increased conductivity with increasing temperatures. In many semiconductors, a temperature increase of 10 K will permit a doubling of the numbers of electrons in the conduction band. In order to increase the conductivity in semiconductors, small amounts of a doping material is used and this significant increases conductivity as a result of the narrowing of the gap between the conduction and valence bands.

Intrinsic semiconductors

Pure or un-doped silicon and germanium are intrinsic semiconductors and both are Group IV elements in the Periodic Table. They exhibit a tetrahedral crystalline structure that is similar to diamond in which each atom in silicon and germanium has four electrons in its outermost electron shell. Also each of these electrons is used in a covalent bond with one of the four neighbour atoms. When each individual atom of silicon is considered, it is found that each of its four valence electrons is used to maintain covalent bonds with the atom's neighbours.

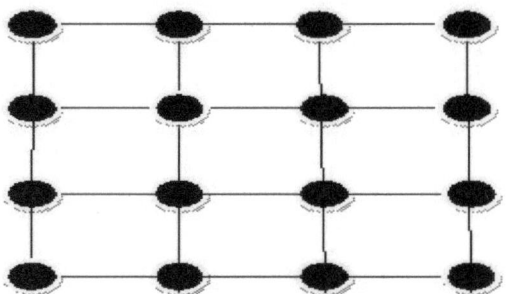

Two-dimensional illustration of a crystal of pure un-doped Si.

Semiconductors like pure silicon or germanium are intrinsic semiconductors and always contain equal numbers of conduction electrons and holes. If an electron can move from its place then it must leave behind a hole. All valence electrons in pure silicon and germanium are used for bonding making them good insulators but only relatively small energies are required to move a valence electron across the energy gap to the conduction band. For silicon the energy is about 1.1 eV and 0.7 eV for germanium and this moves a significant number of electrons in the conduction band even at room temperature. As shown below, only significant numbers of electrons to cross the energy gap in semiconductors. When this happens most thermal excitation involves energies much less than even 0.7 eV and as explained by the theory of quantum mechanics, a small significant probability of an electron is able to jump across the energy level gap at relatively low temperatures.

Once an electron jumps up to the conduction band in the crystal lattice, it leaves behind a 'hole' in the covalent bond which in turn enables another neighbouring valence band electron to move into it. As such, a hole behaves rather like a positive charge carrier though it is actually a vacancy for an electron. A hole can travel through the crystal lattice of the semiconductor. A helpful analogy might be to consider a queue of cars on a road. If a space appears at the front of the queue, cars may move forward in turn. Each time a car moves forward, it leaves a space behind it, into which the next car may now move. An observer from above might consider that the cars are moving forwards or that the space is moving backwards.

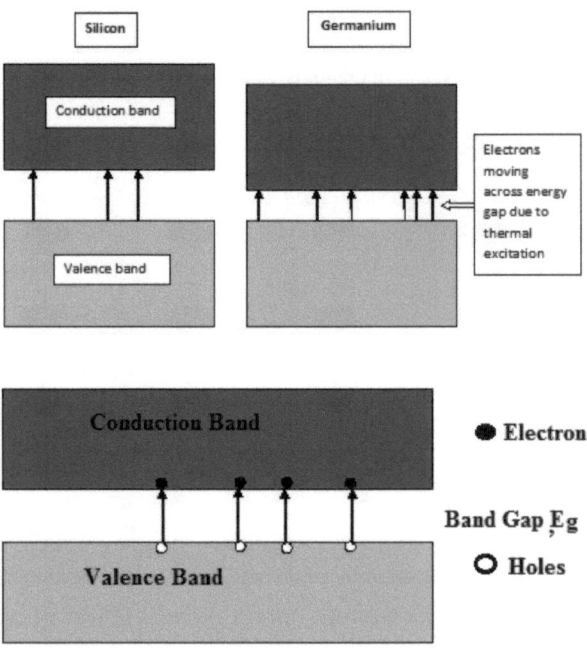

Intrinsic semiconductor

Extrinsic Semiconductors

In a practical sense, it is useful to control the properties of Group IV semiconductors. This is done by deliberately introducing very small proportions of a Group III or Group V element commonly referred to as dopants and this process is known as doping forming an extrinsic semiconductor. Extrinsic semiconductors therefore have majority charge carriers that may be either electrons or holes. Consider a semiconductor that is doped with a Group III element. Each atom of the doping agent will have only three electrons in its outer shell. This is insufficient to form the four covalent bonds with its Group IV neighbours and therefore results in a hole. Countless holes are now built into the semiconductor's crystal lattice. It may be referred to as a p-type semiconductor as the majority charge carriers are positively charged holes. As a result of the doping process, it will require much less energy to allow charge to flow through the semiconductor and so its conductivity is greatly enhanced and leaves holes built into the valence band.

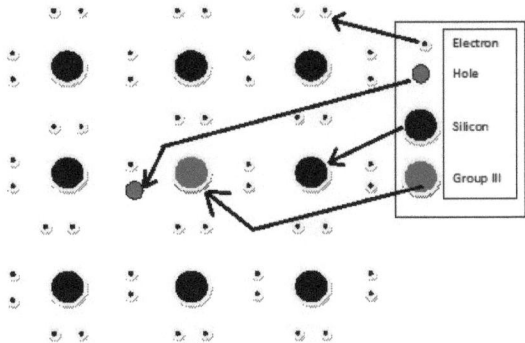

Doiping silicon lattice with a Group III atom

The doping agent adds an extra energy level just above the valence band called an acceptor band though some small degree of intrinsic behaviour is retained as electrons leave behind holes. A similar process can be carried out if a Group V element is used in doping that gives an extra electron in surplus to the covalent bonding for each atom by the doping agent. These electrons are negatively charged resulting into *n*-type semiconductor. In n-type semiconductors, the majority charge carriers are electrons enhancing conductivity greatly through electron conduction that introduces an extra energy level just below the conduction band called the donor band. Conduction in **p**-type and **n**-type semiconductors occurs easily because there are unfilled spaces within either the valence or the conduction band. A specialist application for this type of semiconductors is in detection of magnetic fields using the Hall Effects.

The p–n junction

When a single semiconducting crystal is doped in a way such that one of its ends is p-type and the other is an n-type, then a p-n junction is formed. This definition is based on the interface between the **p**-type and **n**-type sections, hence a *p–n* junction. It should be noted that in this boundary region, electrons from the n-type material diffuse across the boundary to combine with holes from the p-type material and results into a lack of majority charge carriers in the immediate vicinity of the junction. A region formed is thus known as the *depletion zone.*

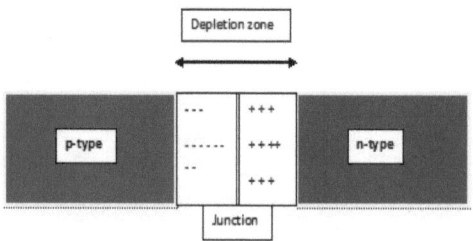

Behaviour of holes and electron Depletion zone in p-n junction

The p–n junction affects the conductivity of the semiconductor and when electrons from the n-type material diffuse into the p-type material, they form negative ions as they combine with holes. Positive ions are also left behind in the n-type material. Eventually, this process results in there being no further diffusion of electrons or holes as a result of Coulomb attraction and repulsion. Band model theory is used to explain what happens at a p–n junction A p–n junction only allow current to flow if it is forward biased and this is done by connecting the negative terminal of a power supply to the n-type material. Electrons are pushed across the depletion zone if the supply has a sufficient potential difference to overcome the Coulomb repulsion especially of the order of 0.7 V.

Once the depletion zone is crossed, conduction is easily facilitated by the majority charge carriers in each of the semiconducting materials. The p–n junction is reverse biased by connecting the n-type material to the positive terminal of a power supply and the result is that the depletion zone becomes greater and greater barrier for conduction. The junctions only allow a tiny leakage current to flow because of the intrinsic semiconductor's electrons and holes. If the reverse voltage continues to increase, the semiconductor will break down and may result in damage to the junction.

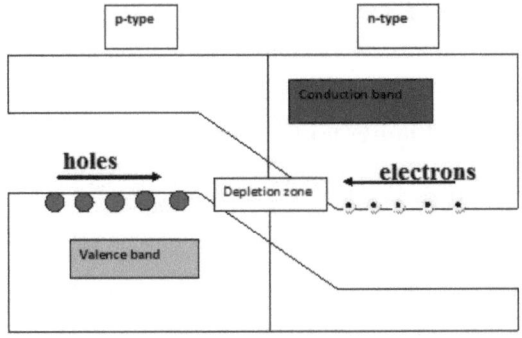

Flow of electrons and holes in the deletion zone

Photovoltaic cells

A photovoltaic cell or a reference solar cell consists of a very thin layer of p-type semiconductor in contact with a layer of n-type material. Conduction electrons are freed through the action of photons of light when they provide sufficient energies to the electrons to enable them jump up across the energy gap to the conduction band leaving behind a hole. In the case of silicon, the band gap energy for silicon is of the order of 1.1 eV and so only photons with at least 1.1 eV energy release conduction electrons. The electric field produced by the depletion layer at this junction forces the electron and hole apart. This creates a potential difference which causes a current to flow if the cell is connected to a circuit. In such a cell, the p-type layer must be very thin, perhaps 1μm (micro meter) thick to prevent conduction electrons from being captured and immobilized by holes.

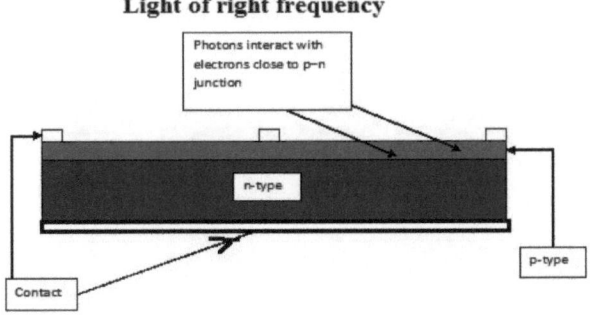

Cross-section of a simplified photovoltaic cell

Solar cells also have a layer of antireflective coating to improve efficiency. Typical efficiencies of solar cells stand close to 15% and the greatest efficiencies rarely exceed 25%. By reducing the band gap energy in the semiconducting material, photons with longer wavelengths and lower frequencies may be harnessed to free electrons and holes. It turns out that a band gap energy of about 1.4 eV is close to ideal, maximising the current and voltage, and therefore the power, of the cell.

Light-emitting diodes

If a diode is forward biased, electrons move from the n-type semiconductor across the junction and combine with holes in the p-type within the higher energy conduction band. As they cross the junction, they move briefly into the empty conduction band of the p-type material. Since the lower energy p-type valence band is only partially filled, however, the electrons rapidly fall into an energy level within that valence band. In effect, electrons fall into holes, and as this happens energy is released in the form of emitted photons. For ordinary diodes, these photons have a relatively low frequency and long wavelength, which means that they fall outside the visible spectrum.

In constructing LEDs, semiconducting materials used to engineer them result in photons having sufficiently high frequencies that fall within the visible light range. The emitted frequency of the light from LEDs is controlled by the size of the energy gap between the conduction and valence bands. A bigger gap will result in a larger energy change ($E = hf$), with a higher frequency being emitted. A small energy gap results in red light and a much larger energy gap is thus required for green or blue light emission.

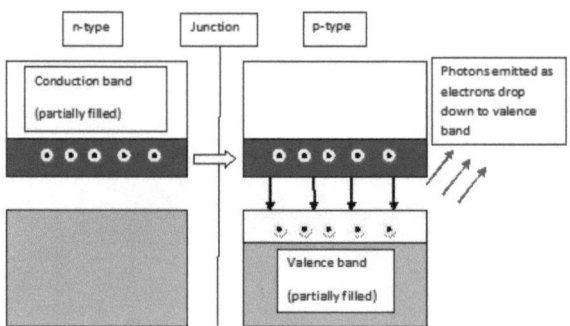

LED electrons cross the junction from n to p in the conduction band

As shown in the figure below, LED electrons cross the junction from n- to p- occur in the conduction band. When on the p- side of the junction, they fall back across the energy gap to the valence band and this releases photons of light. Since electrons usually drop from the bottom of the conduction band into the top of the valence band, light from LEDs tends to be monochromatic with a narrow emission spectrum. To alter the energy gap in an LED, different doping agents are used that include indium, gallium and nitrogen to produce blue light, gallium and phosphorous for green light, and gallium, phosphorous and arsenic for red light. By using combinations of red, green and blue it is possible to produce any colour of light and this has led to the advent of LED televisions.

By varying the proportions of the doping agents, a single intermediate colour can be produced. In the context of this book, LEDs have extremely high switching speeds. This therefore means that they are useful for applications where light sources must be pulsed, stroked or simply switched on and off rapidly with reliability. They also do well at low temperatures, are shock resistant and contain fewer hazardous materials than energy-saving light bulbs. LEDs have many advantages over other light sources. LEDs have efficiency of around 80% which is far superior to energy-saving compact lights. This is major implications for reducing carbon dioxide emissions and fuel bills. LEDs also have much greater longevity than conventional light sources. LED and OLED TVs are expensive though they have lower power usage at high reliabilities. Infrared LEDs are used in remote controls because of their reliability, high switching speeds and low power consumption. Ultraviolet LEDs are used for detecting counterfeit notes and also for sterilization.

Semiconductors Models for Heat Capacity

The Debye Model

The amount of energy required to raise the temperature of one kilogram of the substance by one kelvin is referred to as heat capacity. The SI unit for specific heat capacity is the joule *per* kilogram kelvin, $J \cdot kg^{-1} \cdot K^{-1}$ and, it is often referred that heat capacity at constant volume. The heat capacity at constant volume is defined as;

$$C_v = T \sum (\partial S / \partial T)_V = (\partial U / \partial T)_V$$

where S is the entropy, U is the energy, and T is temperature. In room temperature range the value of the heat capacity of nearly all monoatomic solids is close to 3Nk and at lower temperatures the heat capacity drops rapidly and approaches zero as T^3 in insulators and T in metals. The Debye model was developed by Peter Debye in 1912 in which he estimated the phonon contribution to the heat capacity in solids by treating the vibration of the lattice as phonons in a box.

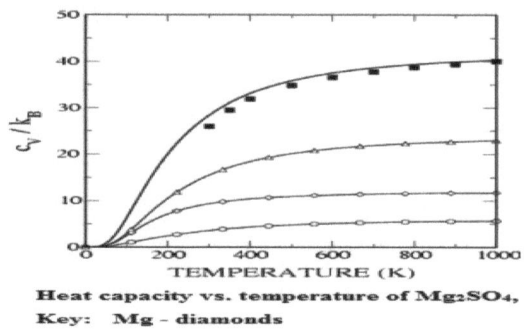

Heat capacity vs. temperature of Mg_2SO_4,
Key: Mg - diamonds
       Si - circles
       O - triangles

The Einstein Model

The average energy of an oscillator having a frequency, ω, is given as $\langle n \rangle \hbar \omega$ and for a number of N oscillators in one dimension having the same frequency will have a thermal energy is given as;

$$U = N \langle n \rangle \hbar \omega = \frac{N \hbar \omega}{e^{\hbar \omega / \tau} - 1}$$

Such that, $\tau = kT$, k Boltzmann constant and $\langle n \rangle$ is the thermal avarage of the number of phonons in an elastic wave of given frequency. Then the heat capacity of oscillator will finally be expressed as;

$$C_V = \left(\frac{\partial U}{\partial T} \right)_V = Nk \left(\frac{\hbar \omega^2}{\tau} \right) \frac{e^{\hbar \omega / \tau}}{(e^{\hbar \omega / \tau} - 1)^2}$$

So that at high temperatures we have;

$$\hbar \omega \ll kT \text{ and } e^{\frac{\hbar \omega}{kT}} \approx 1 + \hbar \omega / kT$$

The thermal energy U will hence be given as;

$$U = \frac{3N\hbar}{\hbar\omega/kT} = 3NkT = 3RT$$

Also it can be seen that, C = dU/dT = 3R, in agreement with experiment.

Likewise at low temperatures we have;

$$\hbar\omega \gg kT \text{ and } e^{\frac{\hbar\omega}{kT}} \gg 1$$

And this makes the thermal energy to be given as;

$$U = 3N\hbar\omega e^{-\frac{\hbar\omega}{kT}}$$

So heat capacity, C can be calculated as;

$$C = 3Nk\left(\frac{\hbar\omega}{kT}\right)^2 e^{-\frac{\hbar\omega}{kT}}$$

$$= 3R\left(\frac{\theta_E}{T}\right)^2 e^{-\frac{\theta_E}{T}}$$

Einstein model predicts that heat capacity, C decreases with decreasing T and if low frequencies are present, $\hbar\omega$ will be much smaller than kT even at low temperatures.

The Debye Model

The Debye model uses wide spectrum of frequencies to describe very complicated patterns of lattice vibrations. It assumes that oscillators generate simple sine waves throughout the crystal that displace atoms in such a way from their equilibrium positions by an amount equal to the amplitude of the sine wave. Therefore if a whole set of oscillators generate sine waves of certain known frequencies and amplitudes, the superposition of such waves will simulate complicated patterns of actual atomic vibrations. This requires that the density of states function is required and it is needed that the range of frequencies over which the integration is to be performed. The distribution of oscillators is quasi-continuous and given as;

$$E(\omega) = \sum_k \hbar\omega_k \bigg/ (e^{\frac{\hbar\omega_k}{kT}} - 1)$$

$$U = \int N(\omega)E(\omega)d\omega$$

Density of States

Considering a cube of material of side L having a standing wave, the boundary condition for vibrational waves to be reflected from free surfaces forms the antinode of the vibration amplitude at each surface. It corresponds to an integral number of half-wavelengths of the standing wave along the length of the cube and the allowed values of the standing wave vectors are given by;

$$k_i^s = n_i\left(\frac{\pi}{L}\right), \quad i = x, y, z; \quad n_i = 1, 2, 3, \ldots$$

The figure above shows a schematic 1D illustration of a standing wave set up between the free surfaces of a cube of an elastic continuum with antinodes at the free surfaces in which each allowed standing-wave solution of the wave equation consistent with the boundary conditions is represented by a point in the reciprocal space containing the k-vectors.

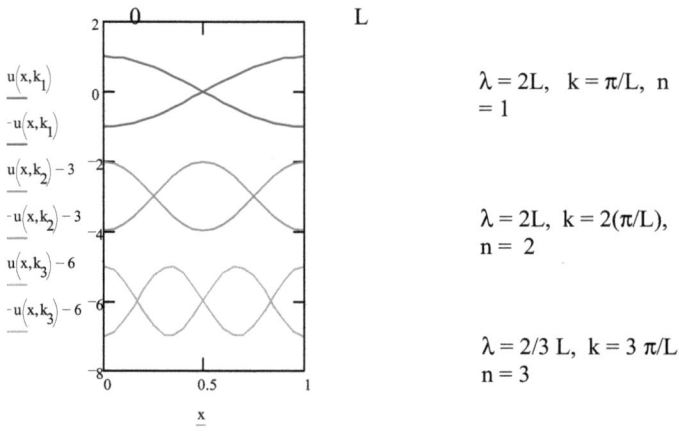

λ = 2L, k = π/L, n = 1

λ = 2L, k = 2(π/L), n = 2

λ = 2/3 L, k = 3 π/L, n = 3

The spacing between allowed k-values is $\Delta k^s = \pi/L$, and so the volume of k-space corresponding to the one k-value (standing-wave state) is as;

$$V_k^s = \left(\frac{\pi}{L}\right)^3.$$

The number of k-values contained in a unit volume of k-space is given as;

126

$$\rho_k^s = \frac{V}{\pi^3}$$

where the space sample volume, V given as; $V=L^3$. The density of k states is uniform in **k** space and depends only on the sample size where **k** can be taken as a continuous variable rather than as a discrete quantity so as;

$$g_i(k)dk = \frac{1}{8}4\pi k^2 dk \frac{V}{\pi^3} = \frac{Vk^2 dk}{2\pi^2}.$$

It makes the density of states for mode **i** in terms of frequency, $g_i(\omega)$, to be obtained using the linear dispersion relation valid for long-wavelength acoustic modes from;

$$g_i(\omega)d\omega = \frac{V\omega^2}{2\pi^2 v_i^3}d\omega,$$

In this expression, v_i is the appropriate sound velocity for mode **i**. Three acoustic modes obtainable are the longitudinal (LA) mode and also two other degenerate transverse (TA) modes. When they propagate in continuous elastic media, their total vibrational density of states is given by;

$$g(\omega)d\omega = \frac{3V\omega^2}{2\pi^2 v_o^3}d\omega$$

where v_o is an appropriate average of the LA- and TA-mode velocities and this reduces their quadratic frequency dependence which is commonly referred to as Debye density of vibrational states to be expressed as;

$$3N = \frac{3V}{2\pi^2 v_o^3}\int_0^{\omega_D}\omega^2 d\omega$$

And this gives the Debye frequency, ω_D to be given by;

$$\omega_D = \left(\frac{6\pi^2 N}{V}\right)^{1/3} v_o$$

This then gives the density N, to be found from;

$$N(\omega) = \frac{V}{8\pi^3}\int \frac{dS_\omega}{\nabla_k \omega} = \frac{V}{8\pi^3 v_g}\int dS_\omega$$

$$\int dS_\omega = 4\pi k^2$$

$$\omega = v_g k$$

$$N(\omega) = \frac{V}{8\pi^3 v_g^3} 4\pi k^2 = \frac{V}{2\pi^2} \frac{\omega^2}{v_g^3}$$

Therefore the total density becomes the integration to the maximum frequency as;

$$\int_0^{\omega_{max}} N(\omega)d\omega = N = \frac{V}{6\pi^2} \frac{\omega_{max}^3}{v_g^3}$$

$$\Rightarrow \omega_{max}^3 = \frac{6N\pi^2}{V} v_g^3$$

This will give the reduced mass U, as;

$$U = \int N(\omega) E(\omega) d\omega$$

$$= \int \frac{V}{2\pi^2} \frac{\omega^2}{v_g^3} \frac{\hbar\omega}{e^{\frac{\hbar\omega}{kT}} - 1} d\omega$$

$$= \frac{V}{2\pi^2} \frac{\hbar}{v_g^3} \int_0^{\omega_{max}} \frac{\omega^3}{e^{\frac{\hbar\omega}{kT}} - 1} d\omega$$

Where Z and change in frequencies is given as;

$$Z = \frac{\hbar\omega}{kT}; \quad d\omega = \frac{kT}{\hbar} dZ$$

Using this change in frequency and substituting them to obtain U we get;

$$U = \frac{V}{2\pi^2} \frac{kT}{v_g^3} \left(\frac{kT}{\hbar}\right)^3 \int_0^{Z_{max}} \frac{Z^3}{e^Z - 1} dZ$$

And this gives maximum frequency as;

$$\omega_{max}^3 = \frac{6\pi^2 N v_g^3}{V}$$

And reduced mass U as'

$$U = \frac{3N}{\omega_{max}^3} \frac{(kT)^4}{\hbar^3} \int_0^{Z_{max}} \frac{Z^3}{e^Z - 1} dZ$$

This equation can be used to obtain the what is called as Debye temperature as;

$$\theta_D = \frac{\hbar \omega_{max}}{k}$$

And when this Debye temperature is substituted in the expression for reduced mass, we obtain;

$$U = 3N\,kT\left(\frac{T}{\theta_D}\right)^3 \int_0^{Z_{max}} \frac{Z^3}{e^Z - 1}dZ$$

The constant value of U obtained is for one branch. To obtain for the three branches, the expression is multiplied by three (3) i.e. 2 transverse acoustical and 1 longitudinal acoustic to obtain;

$$\boxed{U = 9N\,kT\left(\frac{T}{\theta_D}\right)^3 \int_0^{Z_{max}} \frac{Z^3}{e^Z - 1}dZ}$$

The Debye Specific Heat Capacity will then be given as;

$$U = \frac{V}{2\pi^2}\frac{\hbar}{v_g^3}\int_0^{\omega_{max}} \frac{\omega^3}{e^{\frac{\hbar\omega}{kT}} - 1}d\omega$$

It can also be expressed as;

$$\frac{dU}{dT} = \frac{V}{2\pi^2}\frac{\hbar}{v_g^3}\frac{-\hbar\omega}{kT^2}\int_0^{\omega_{max}} \frac{-e^{\frac{\hbar\omega}{kT}}\omega^3}{\left(e^{\frac{\hbar\omega}{kT}} - 1\right)^2}d\omega$$

$$= \frac{V}{2\pi^2}\frac{1}{v_g^3}\frac{\hbar^2}{kT^2}\int_0^{\omega_{max}} \frac{e^{\frac{\hbar\omega}{kT}}\omega^4}{\left(e^{\frac{\hbar\omega}{kT}} - 1\right)^2}d\omega$$

$$= \frac{3N}{\omega_{max}^3}\frac{\hbar^2}{kT^2}\left(\frac{kT}{\hbar}\right)^4\frac{kT}{\hbar}\int_0^{Z_{max}} \frac{e^Z\,Z^4}{\left(e^Z - 1\right)^2}dZ$$

For a single acoustic branch, the Debye specific heat capacity will be given as;

$$C_{V_D} = 3Nk\left(\frac{T}{\theta_D}\right)^3 \int_0^{Z_{max}} \frac{Z^4 e^Z}{\left(e^Z - 1\right)^2}dZ$$

The total specific heat for the two transverse acoustic and one longitudinal acoustic branch (three branches) is therefore given as;

$$C_{V_D} = 9Nk\left(\frac{T}{\theta_D}\right)^3 \int_0^{Z_{max}} \frac{Z^4 e^Z}{\left(e^Z - 1\right)^2}dZ$$

At low temperatures as Z_{max} approaches infinity (∞), we have;

$$T \ll \theta_D$$

$$\int_0^\infty \frac{Z^3}{e^Z - 1} dZ = \frac{\pi^4}{15}$$

And this gives U as;

$$\Rightarrow U = \frac{3NkT\pi^4}{5}\left(\frac{T}{\theta_D}\right)^3$$

$$= \frac{3Nk\pi^4}{5\theta_D^3} T^4$$

Debye specific heat capacity will then be given as;

$$C_{V_D} = \frac{dU}{dT} = \frac{12}{5}\pi^4 Nk\left(\frac{T}{\theta_D}\right)^3$$

And this becomes the Debye T^3 approximation at low temperature as in the experimental results. At high temperature where Z approaches infinity, we have;

$$e^Z \to 1 + Z \quad \text{or} \quad 1$$

So as to obtain specific heat capacity at Debye temperature as;

$$C_{V_D} = 9Nk\left(\frac{T}{\theta_D}\right)^3 \int_0^{Z_{max}} \frac{Z^4}{Z^2} dZ$$

$$= 9Nk\left(\frac{T}{\theta_D}\right)^3 \int_0^{Z_{max}} Z^2 dZ$$

$$= 9Nk\left(\frac{T}{\theta_D}\right)^3 \frac{Z_{max}^3}{3}$$

$$= 3Nk$$

For sufficiently low temperatures, the Debye approximation is fairly accurate as only long as the wavelength acoustic modes are excited and treated as in an elastic continuum with macroscopic elastic constants.

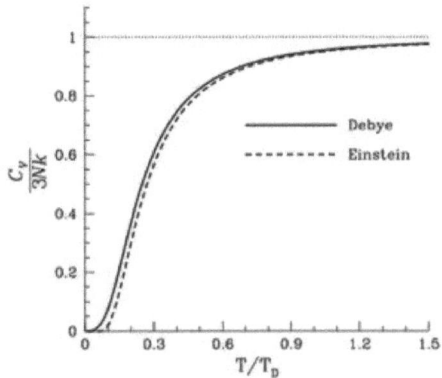

Curve Comparision between Debye and Einstein model

Semiconductor transport carriers

Carrier transport in semiconductors thin films is temperature sensitive. At room temperature electrons in a semiconductor are considered to be moving in all directions. The thermal velocity then at this temperature for any individual electron is caused by random scattering from collisions with the lattice atoms, or with impurity atoms, or with other scattering centres or all of them simultaneously. If an electric field ($\bar{E}$) is applied to the semiconductor thin film, then each electron will experience a force from this field and it will be accelerated along the field (say on x-axis) and a hole in the opposite direction (- x-axis) of it. The electrons under this field will have a velocity called the drift velocity, V_d. The relationship between the drift velocity (V_d) of electrons and the electric field ($\bar{E}$) applied to a semiconductor is then obtained as:

$$v_d = \mu_n \bar{E}$$

where μ_n is the electron mobility and it is given by:

$$\mu_n = q\tau_c / m_n$$

where τ_c is the mean free time (the average time between collisions) and m_n is the effective mass of electrons. Research show that electron mobility is greater than hole mobility because of their effective masses i.e. an electron is smaller than a hole. This equation also applies to drift velocity for holes as:

$$v_p = \mu_p \bar{E}$$

where μ_p is the hole mobility. When a transport carrier is under the influence of an applied electric field ($\bar{E}$) it produces a current called the drift current (J). Therefore electron current density (J_n) caused by electron mobility is written as:

$$J_n = -qpv_n = qn\mu_n \bar{E}$$

and, the holes current density (J_p) due to hole mobility is likewise given as:

$$J_p = qpv_p = qp\mu_p \bar{E}$$

where, q is the electronic charge, n and p are the concentration of electrons and holes per unit volume respectively.

The total current density (J) in the semiconductor sample is given by the sum of electron and hole mobility current densities:

$$J = J_n + J_p = (qn\mu_n + qp\mu_p)\bar{E}$$

This total current summed up can also be written as:

$$J = \sigma \bar{E}$$

where, σ is given by;

$$\sigma = qn\mu_n + qp\mu_p$$

where, σ is the conductivity of the semiconductor which depends on the concentration of charge carriers n and p. The numbers of charge carriers are dependent on temperature in an exponential way and therefore conductivity increases exponentially and hence the corresponding resistivity which is a reciprocal of σ is given by:

$$\rho = 1/\sigma = [1/\{qn\mu_n + qp\mu_p\}]$$

This equation can be used to describe the conductivity of extrinsic semiconductor especially the doped ones. The only difference is that the number of electrons in the conduction band and the number of holes in the valence band are not equal in the case of an extrinsic semiconductor. One of the two (p or n) dominates depending on the type of the extrinsic process but the mass action law is maintained ($np=n_i^2$).

If both the donor and acceptor impurities are present simultaneously, then the impurity that is present in a greater concentration determines the type of conductivity in the semiconductor. Therefore, it can be written for n-type semiconductor as;

$$\rho = 1/qn\mu_n$$

and for p-type semiconductor as:

$$\rho = 1/q p \mu_p$$

Sheet resistance, R_s is an important parameter in thin films and it is used to characterize both wafers and thin doped layers. It is easier to measure the sheet resistance rather than the resistivity of thin film materials. The sheet resistance of a uniformly doped layer) with a resistivity, ρ, and thickness, t, is given by their ratio;

$$R_s = \rho/t$$

while the unit of the sheet resistance is Ohms, it is usually referred to as Ohms per square (Ωcm^{-1}). This is because when the resistance of a rectangular piece of thin film material with length, L, and width, W, is to be obtained, it is taken to be equal to the product of the sheet resistance and the number of squares:

$$R = R_s L/W$$

where the number of squares equals the length L, divided by the width, W.

Transport carriers move from a region of high concentration to a region of low concentration when a spatial variation of carrier concentration exists in any semiconductor material. Diffusion currents emanates from this random thermal motion of carriers due to a concentration gradient. This motion causes a certain current to flow without the influence of an electric field and therefore it diffuses freely. This current is called diffusion current. This current for electron carriers (n) can be given by:

$$J_n = [qD_n] \, dn/dx$$

where, q is the electronic charge, D_n is the diffusivity and the term dx/dn is the spatial derivative of the electron density The diffusivity, D_n according to Einstein relation is given as:

$$D_n = [KT/q]\mu_n$$

From the above equation the diffusion current is proportional to the spatial derivative of the electron density. When both electric field and concentration gradient are present, both drift and diffusion currents flow. The total electrons current density will then be the sum of these two current components:

$$J_n = qn\mu_n E + [qD_n]dn/dx$$

and for total holes current density:

$$J_p = qp\mu_p E - [qD_p]dp/dx$$

The total conduction current density (J_{cond}) is given by the sum of the two equations as:

$$J_{cond} = J_n + J_p$$

Optical excitation is the injection of carriers by shining light on a semiconductor. Carrier injection is a process of introducing excess carriers to a semiconductor thin film by using various methods among them optical excitation and it is a process that affects the carrier transport. When this happens and if the photon energy (hv) of the illumination is greater than the energy band-gap (E_g) of the semiconductor, then photons are absorbed by the semiconductor and electron-hole pairs are generated. Photodiodes are designed and function based on optical excitation and that is why their electron and hole concentrations are increased above their equilibrium values.

Injection level is determined by the magnitude of the excess carrier concentration relative to the majority carrier concentration. The mean free time and the mobility of the carriers in a semiconductor are affected by lattice and impurity scattering mechanisms. Lattice scattering is caused by thermal vibrations of lattice atoms which disturb lattice periodic potential and allow energy to be transferred between the carriers and the lattices. Since the lattice vibration increases with increasing temperature, lattice scattering becomes dominant at high temperatures and hence mobility decreases with increasing temperature. On the other hand impurity scattering results when a charge carrier travels past an ionized dopant.

3D semiconductors Crystal structures

Considering the 3D semiconductors nature of Si, Ge, GaAs, InP and CdTe are all in crystalline solids in regular arrangement of atoms called lattice. Some elemental semiconductors such as Si and Ge have only one kind of material at their lattice sites, whereas GaAs, InP and those related are compound semiconductors that are composed of two interpenetrating 3D lattices. There may be defects in the lattice and each atom in a lattice vibrates. Such defects and vibrations are second order phenomenon. If we have a charged ion at $x = 0$ and an electron outside the attractive force between the ion and the electron will be given by;

$$F = \frac{z_q^2}{4\pi \epsilon_o r^2}$$

If the ion involved is given by the equation

$$\text{ion} = -\frac{z_q^2}{4\pi \epsilon_o r} \text{ or } -\frac{z_q^2}{4\pi \epsilon_o x}$$

then the potential energy (PE) of the ion is given as;

PE = 0, at $x = \pm\infty$

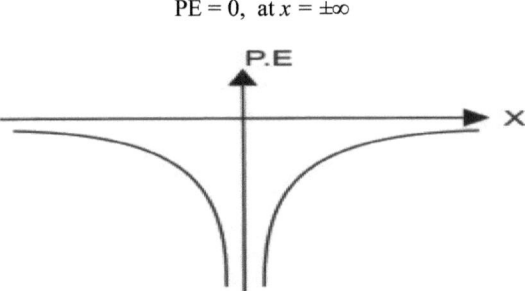

In case where there two ions and if we consider that the two ions separated by a distance, 'a', then the potential profile will be as shown below.

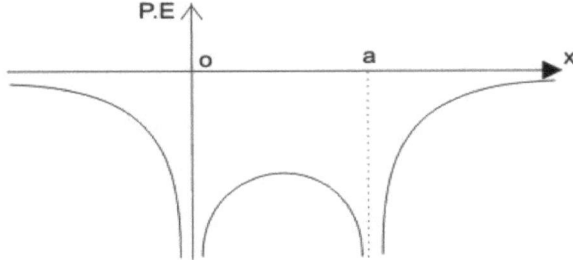

The simplest lattice structure is a 1 D lattice. To get a feel for the electron energies in a crystal semiconductor let us consider an electron in a 1D lattice of atoms. A regular array of atoms placed periodically.

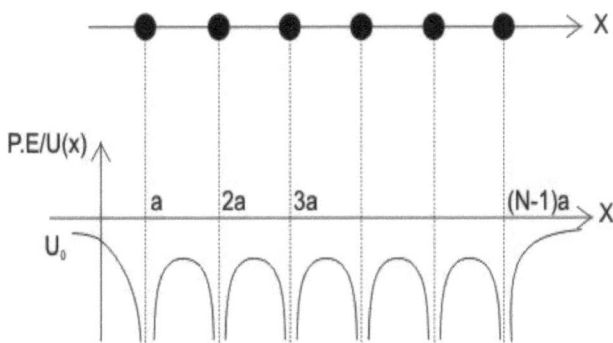

It is thus noted that at each lattice site (x = 0, a, 2a, ...) there exists an ion with some net charge Z_q.

CHAPTER FOUR
BRILLOUN ZONES

Energy in Brilloun Zone representation

We learned about energy bands E or ζ values for which $S(\zeta) = \cos k(a+b)$ has a solution and between these bands where it is not possible to find a value of E or ζ are called forbidden gaps. Therefore, the discrete number of such bands is all separated by band gaps. Thus for the case where;

$$\cos k(a+b) = \pm 1$$

$$k = 0 \text{ or } \pm \frac{\pi}{a+b}, \pm \frac{2\pi}{a+b} = \cos m\pi$$

$$m = 0, \pm 1, \pm 2, \ldots \text{ for these band edges}$$

Using these values of *m* and plotting the allowed values of energy as a function of *k*, we obtain the *E - k* diagram for the one dimensional lattice. When expanded *E-k* diagram of Periodic potential perturbation is plotted it is noticed that there is a dissimilarity with the free space E-K diagram (given by the dotted line) to given a free Particle Solution. What remains is how in particular can the periodic potential solution with an adjustable k approach the free particle solution with a fixed k in the limit where $E >> U_0$.

It has to be noted here that the wave function for an electron in a crystal is the product of two e^{-ikx} and Q(x) where Q(x) is the wave function in the unit cell. Q(x) is also a function of k. Increasing or decreasing k by *(2π/all length)* modifies both e^{-ikx} and Q(x) in such a way that the product of Q(x) e^{-ikx} approach the free particle. The k-value associated with given energy band is called a Brillouin Zone. From this approach, we have the 1 st Brillouin Zone given by the range of;

$$\rightarrow -\frac{\pi}{a+b} \text{ to } +\frac{\pi}{a+b}$$

The second 2nd Brillouin Zone range is given by;

$$\rightarrow -\frac{2\pi}{a+b} \text{ to } -\frac{\pi}{a+b} \ \& \ \frac{\pi}{a+b} \text{ to } \frac{2\pi}{a+b}$$

In drawing the *k* between the ranges given $\pm 2\pi/(a+b)$ in basically *(2π/(all length)* range, the Eigen value equation is noticed to be increasing or decreasing *cos k(a+b)* by *2π/(a+b)* has no effect on the allowed electron energy of E(k) is periodic with a period of *2π/(a+b)* as;

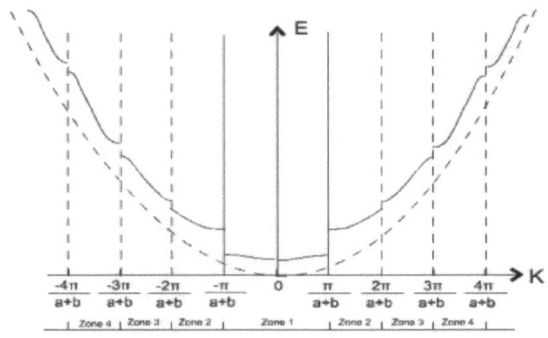

Therefore all the electron energies can be represented within;

$$-\frac{\pi}{a+b} \leq k \leq +\frac{\pi}{a+b}$$

by changing the k values by $2\pi n/(a+b)$, where n is an integer. This representation of the electron is called the reduced Brillouin zone representation as shown where the bands of energies are identified. As the number of electrons in the system increases the bands starts to be filled up from the lowest available energies. Normally most of the bands are completely full of electrons as allowed by Pauli's Exclusion Principle. At low temperature, there could be a band completely empty. The one below it is usually completely full, called the valence band. None of these electrons can now conduct electricity. If now a condition arises that some of the electrons from the completely filled band can be excited into the completely empty band, then current can be conducted by the electrons in the empty band called the conduction band.

Also according to the Bloch theorem there are two and only two k values associated with each allowed energy, one for the electron moving in the +ve direction and the other for the electron moving in the –ve direction. Also note that;

$$\frac{dE}{dk} = 0$$

so that at;

$$k = \pm\frac{\pi}{a+b}$$

And this gives the value of k = 0 and this is a property of all E-k plots.

Bloch Theorem

The Block theory relates value of wave function within any unit cell of a periodic potential to an equivalent point in any other unit cell and helps to concentrate on the behavior of any unit cell in the whole array. For a 1D system, the Bloch theorem says that if a particle or electron takes a wave function;

$$\psi(x) = u_k(x)e^{jkx}$$

So that when U(x) is the periodic such that U(x + a) = U(x) then;

$$\psi(x + a) = e^{jka}\psi(x)$$

It becomes equivalently equated as,

$$\psi(x) = e^{jkx}\phi(x)$$

where, $\Phi(x)$ = wave function for an unit cell and thus

$$\Phi(x+a) = \Phi(x).$$

Therefore when the boundary conditions are imposed at end points of the periodic potential, the wave number **k** in periodic potential set has several properties which can be explained as:

It can be shown that and only two distinct values of k exist for each and every allowed values of E i.e.

$$\left(k^2 = 2mE/\hbar^2\right).$$

For a given E, values of k differing by a multiple of $2\pi/a$ give rise to one and same wave function solution. As the Δk value is periodic or multiple-valued with a range $2\pi/a$. Usually we take Δk ranges as

$$\rightarrow -\pi/a \leq k \leq \pi/a.$$

If the periodic potential is assumed to be in extent, running from, $x = \infty$ to $x = -\infty$, then there are restrictions on **k**. The, k value can take a continuum of values. So **k** must be real otherwise e^{jkx} or thus $\psi(x)$ will blow up at either $x = \infty$ or $x = -\infty$.

In dealing with crystals of finite extent, information about the boundary conditions on crystal surface may be lacking. To avoid this we assume a periodic boundary condition assuming the lattice is a ring of N atoms.

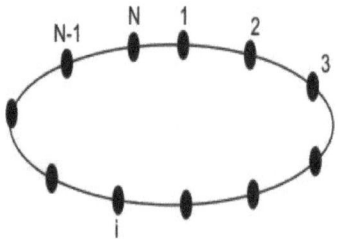

$$\psi(x) = \psi(x + N_a) = e^{jkN\,a}\,\psi(x)$$

Therefore,

$$e^{jkN\,a} = 1$$

or

$$PE = -\frac{z_q^2}{4\pi\epsilon_o\, x} - \frac{z^2}{4\pi\epsilon_o\,(x-a)}$$

So for the 1D lattice of ions the periodic set of finite potential wells, for large array x may be assumed to go from, $x = \infty$ to $-\infty$ so as to give;

$$k = \frac{2\pi n}{N_a} \qquad n = 0, \pm 1, \pm 2, \ldots$$

Thus for a finite crystal k can only assume a set of discrete values, but as N is large therefore one has closely spaced discrete values. A simplified picture of the periodic potential discussed above is shown below.

It is quite similar to the finite potential analysis periodicity equal to

$$= a + b \;(-\infty < x > -\infty)$$

So that the expression;

$$\psi(x+a+b) = e^{ik(a+b)}\psi(x),$$

where, *k* is continuous takes care of the Bloch theorem.

To expound this relation, we first consider a case where: $0 < E < U_0$ where E is the total energy and U0 is the potential energy in a finite well. Inside the well then, the wave function becomes $\Psi_a(x)$ and from the outside of the well, the wave function takes $\Psi_b(x)$ respectively. Schrödinger equation inside the well in which the range is for $0 < x < a$ then can be expressed as:

$$\frac{d^2\psi_a}{dx^2} + \alpha^2\psi_a = 0 \quad \text{where} \quad \alpha^2 = \sqrt{\frac{2mE}{\hbar^2}}$$

Outside the well, the Schrödinger equation for $-b < x < 0$ will also be expressed as:

$$\frac{d^2\psi_b}{dx^2} + \beta^2\psi_b = 0 \quad \text{where} \quad \beta = i\beta_-, \quad \beta_- = \sqrt{\frac{2m(U_0 - E)}{\hbar^2}}$$

Now based on the two general solutions obtained above, the wave functions can then be expressed as;

$$\psi_a(x) = A_1 \sin \alpha x + B_1 \cos \alpha x$$
$$\psi_b(x) = A_2 \sin \beta x + B_2 \cos \beta x$$

This implies that the wave equation relates by;

$$\rightarrow \sin \beta x = \sin i\beta_- x = i \sinh \beta_- x$$

If the continuity condition on wave function is applied at this point and its derivative at $x = 0$ used, the wave function relates as shown in the curve below in Fig. 3.6 which is a continuity requirement.

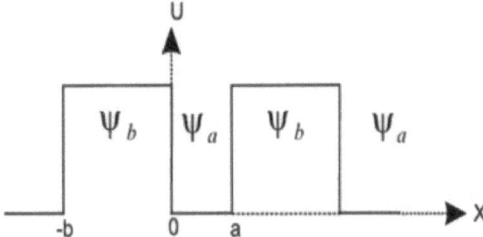

It will be noted at this point that at $x = -b$ is the same boundary as that of $x = a$, and the wave function and its derivative must then observe Bloch Theorem, i.e.

$$\rightarrow \psi_a(0) = \psi_b(0)$$

$$\frac{d\psi_a}{dx}\bigg|_0 = \frac{d\psi_b(0)}{dx}\bigg|_0$$

Where we have the relations at x = -b as periodicity requirements given as;

$$\psi(x+a+b) = e^{ik(a+b)}\psi(x) \quad \text{put } x = -b$$

$$\psi(a) = e^{ik(a+b)}\psi(-b)$$

$$\psi_a(a) = e^{ik(a+b)}\psi_b(-b)$$

Similarly,

$$\frac{d\psi}{dx}\bigg|_a = e^{ik(a+b)}\frac{d\psi_b}{dx}\bigg|_{-b}$$

Similarly,

$$\frac{d\psi}{dx}\bigg|_a = e^{ik(a+b)}\frac{d\psi_b}{dx}\bigg|_{-b}$$

At this point again, applying the boundary and periodicity conditions above will result into;

$$B_1 = B_2$$

$$\alpha A_1 = \beta A_2$$

$$A_1 \sin\alpha a + B_1 \cos\alpha a = e^{i(a+b)k}[-A_2 \sin\beta b + B_2 \cos\beta b]$$

$$\alpha A_1 \cos\alpha a - B_1 \sin\alpha a = e^{i(a+b)k}[\beta A_2 \cos\beta b + \beta B_2 \cos\beta b]$$

The constant values of A_2 and B_2 need to be eliminated. To eliminating the by using the first two equations in last two equations will give,

$$\rightarrow A_1\left[\sin\alpha a + \frac{\alpha}{\beta}e^{i(a+b)k}\sin\beta b\right] + B_1\left[\cos\alpha a - e^{i(a+b)k}\cos\beta b\right] = 0$$

$$\rightarrow A_1\left[\alpha\cos\alpha a - \alpha e^{i(a+b)k}\cos\beta b\right] + B_1\left[-\alpha\sin\alpha a - \beta e^{i(a+b)k}\sin\beta b\right] = 0$$

This results into an equation with two unknown constants. For nontrivial values of A_1 and A_2, the determinant formed from coefficient should be equal to zero.

$$\det\begin{bmatrix} \sin\alpha a + \frac{\alpha}{\beta}e^{i(a+b)k}\sin\beta b & \cos\alpha a - e^{i(a+b)k}\cos\beta b \\ \alpha\cos\alpha a - \alpha e^{i(a+b)k}\cos\beta b & -\alpha\sin\alpha a - \beta e^{i(a+b)k}\sin\beta b \end{bmatrix} = 0$$

Eigen Value Equation

When simplifying the above determinant we get the following Eigen value equation:

$$-\frac{\alpha^2 + \beta^2}{\alpha\beta}\sin\alpha a \sin\beta b + 2\cos\alpha a \cos\beta b = 2\cos k(a+b)$$

If the threshold frequency, f_0 is then taken for a constant system as:

$$f_0 = \sqrt{\frac{2mU_0}{\hbar^2}}$$

So that the normalized energy is given by;

$$\xi = \frac{E}{U_0}$$

Then for

$$E < U_0, \xi < 1$$

The constant β, will be given by;

$$\beta = i\beta_- = i\sqrt{\frac{2m(U_0 - E)}{\hbar^2}} = if_0\sqrt{1-\xi}$$

By taking, α, as;

$$\alpha = f_0\sqrt{\xi}$$

Then putting all the above values in the Eigen value equation, we form an expression given as;

$$\frac{1-2\xi}{2\sqrt{\xi(1-\xi)}}\sin\left(f_0 a\sqrt{\xi}\right)\sin\left(f_0 b\sqrt{1-\xi}\right) + \cos\left(f_0 a\sqrt{\xi}\right)\cosh\left(f_0 b\sqrt{1-\xi}\right) = \cos k(a+b)$$

If we consider the 1D crystal lattice earlier given by Fig 3.3, then $E<U_0$ does not necessarily mean that the electron is outside the crystal lattice or the potential well and therefore, for $E > U_0$, one will get;

$$\beta = \beta_+ = \sqrt{\frac{2m(E-U_0)}{\hbar^2}} = f_0\sqrt{\xi-1} \qquad \xi > 1$$

And substituting all these values in the Eigen value equation, we obtain

$$\frac{1-2\xi}{2\sqrt{\xi(\xi-1)}}\sin\left(f_o a\sqrt{\xi}\right)\sin\left(f_o b\sqrt{\xi-1}\right)+\cos\left(f_o a\sqrt{\xi}\right)\cos\left(f_o b\sqrt{\xi-1}\right)=\cos k(a+b)$$

where,

$$E > U_o \text{ in } \xi > 1$$

What if the lattice is infinitely long? Assume that we have an infinitely long lattice in the K-P model. In such a case the value of **k** can take any value and therefore the right hand side (RHS) of Eigen value equation, *cos k(a+b)* can take any value between +1 and -1. In the left hand side (LHS) the value of **S(ζ)** with value ζ ,between +1 and -1 result into allowed solutions for the values (ζ), (α) and (β) as it approaches (ψ) i.e. β→ψ. When a plot of **S(ζ)** as a function of ζ is shown below.

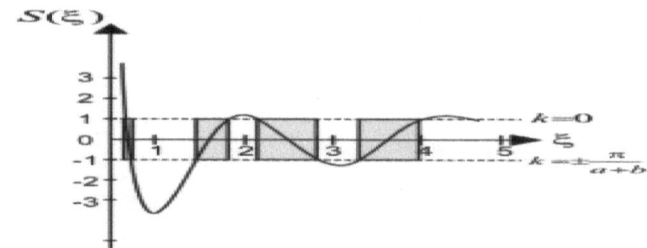

Note that the allowed values of ζ are shown by the shaded regions where, $S(\zeta) \leq \pm 1$. And thus they are allowed energy bands.

Bloch parameter, k

Taking a particle in a free space, the value k is equal to the wave number,

$$k = \frac{2\pi}{\lambda}$$

so that expected value of momentum will be given by;

$$\hbar k = <p>$$

For a particle bound to a periodic potential or crystal, **ℏk** = crystal momentum such that **ℏk** is not the actual momentum but the momentum related to the constant of motion which incorporates the crystal interaction. This crystal momentum parameter **k** is also periodic with a period of **2πn/(a+b)**. The **E-K** diagram is therefore the Energy versus crystal momentum characteristics of an electron in the crystal.

144

Energy Band Solution

Energy band solution indicates only the allowed energy and momentum states but not about time evolution of electron's position etc. given E, k gives possible values of position of finding the electron with a certain probability i.e, position is uncertain and this is demonstrated by the Heisenberg Uncertainty Principle summarized as;

$$\Delta E \Delta t \geq \hbar$$
$$\Delta p_x \Delta x \geq \hbar$$

If energy, E is known exactly, then the uncertainty in time is infinity and we cannot find anything about the electron's position. Therefore for a particle in motion, we need individual wave packets which forma an equivalent to the summation of constant wave function grouped about a peak energy.

$$\equiv \sum (\text{constant, E})$$

Therefore the probability of finding the represented particle in a given region of space is unity (P = 1) for some specified time. Thus the center of mass of a particle that is moving with a velocity υ in a classical idea that has a wave packet approach taken as also a mass in quantum mechanics (DM) approach. Therefore a packet of traveling wave with center frequency ω and center wave number **k**, describes a particle motion represented by this group velocity given by;

$$v_s = \frac{d\omega}{dk} = \frac{1}{\hbar}\frac{dE}{dk}$$

It is thus noted here then that **E** and **k** gives the center values of energy and crystal momentum.

Effect of External Force

The external force cannot be taken literally since it has an effect. An external force, F acting on the wave packet is taken as any force other than the crystalline force associated with periodic potential. This external force can arise from dopants present or an external electric field. Force F acting over a short distance, dx will do work on the wave packet and cause the wave packet energy to increase by;

$$dE = F\,dx = F v_g\,dt$$

From Newton's second law of motion, the rate of change of momentum is given by;

$$F = \frac{d(\hbar k)}{dt}$$

And this is equivalent to Mass x Acceleration given as

$$m^* \frac{dv_\varepsilon}{dt}$$

From this expression therefore, the effective mass will be given by;

$$m^* = \left(\frac{1}{\hbar^2}\frac{d^2E}{dk^2}\right)^{-1}$$

This finally results into a relationship given as;

$$m^* \propto \frac{1}{\text{curvature of } E-k \text{ plot}}$$

Based on this relation, at k = 0, the curvature of curve (b) band is more than the curvature of curve (a) as shown in figure Fig.4.1 below.

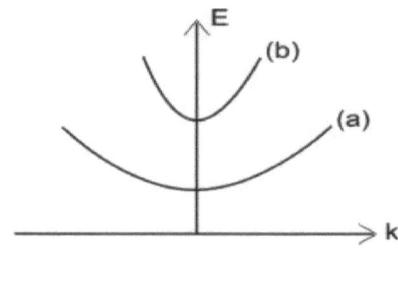

$$\left.\frac{d^2E}{dk^2}\right|_b > \left.\frac{d^2E}{dk^2}\right|_a$$

$$m_a^* > m_b^*$$

From this relation, we get that heavy mass causes slower movement resulting into a larger transit time and therefore mobility of a carrier will be proportional to $1/m^*$ which is also proportional to the curvature as seen in Fig.4.2 given by;

$$E - k \propto \frac{d^2E}{dk^2}$$

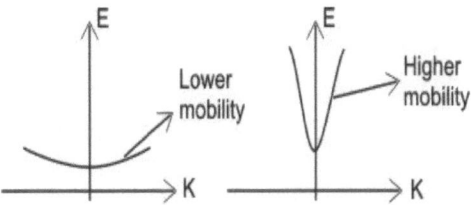

If the Kronig-Penny model is applied to the band segment below,

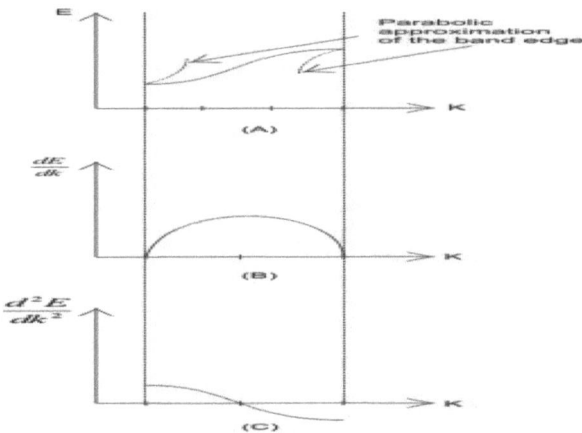

It is found that m* > 0 near the bottom or minimum of bands and for m* > 0 near the top part of each band. Therefore, for m* > 0 in response to applied force the particle or electron will accelerate in a direction opposite to than expected from purely classical calculation. Though the exact equation of the E-k diagram is not known, it is found that the top or bottom of the band edge or Brillouin zone edge of the E-k relation is approximately parabolic as shown in Fig. 4.3 above.

$$E - E_{edge} \cong A(k - k_{edge})^2$$

In the above equation, **A** is constant and using this expression,

$$\frac{d^2 E}{dk^2}$$

remains constant at the edge of the Brillouin zone or top or bottom of an energy band. Therefore m* ≡ constant that is near the edge of the Brillouin zones, or top, or bottom of an energy band is consistent.

Current Flow in Crystals

In crystals, atoms are well arranged. If there are N atoms in 1 dimensional crystal, it means that the distinct **k** values in each band will be equal to N spaced bands apart by;

$$\frac{2\pi}{N(a+b)}$$

Consider a case where each atom gives 2 electrons to the crystal approaching total of 2N free electron in the crystal. The 2N electrons will be distributed among the available energy states and at temperature of 0K, the 2N electrons will fit the 2N states in the lowest 2 bands which are the valence bands. This is shown in the figure below.

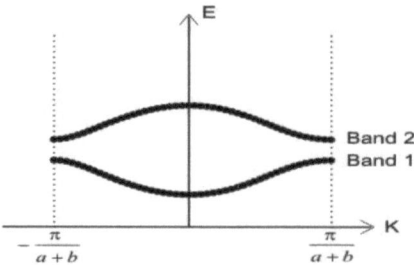

When temperature is increased to about room temperature levels, sufficient thermal energy is gained and a few electrons from the top of the 2nd band will move up to the bottom of the 3rd band as shown.

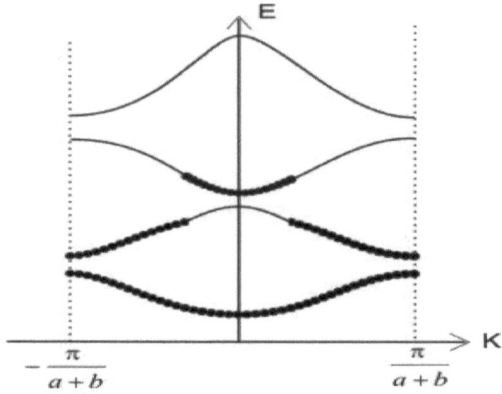

In such a situation, if a voltage applied to the crystal, a current will flow through the crystal. It will then be noted that the 4th band is empty and does not have electrons at room temperature and hence no current flows. These totally empty bands do not contribute to the charge transport process.

Concept of Holes

In the above figure 4.5, if we consider the 1st band where N states are fully filled by the N electrons at room temperature, when an external voltage is applied, the electrons will move to the other bands with velocity given by;

$$v_g = \frac{1}{\hbar}\frac{dE}{dk}$$

By symmetry, the band is symmetrically about $k = 0$ and for every electron with a given velocity, v_g in the same direction, there will be another electron with the same v_g in opposite – x direction. Thus first band produces no current and hence;

$$\sum v_i = 0$$

It is therefore confirmed that the totally filled energy bands do not contribute to the charge transport process and thus by asymmetry no applied electric field is introduced. But for nearly empty third band current, will be given by;

$$I_3 = -\frac{q}{L}\sum_{i(\text{filled sates})} v_i$$

where L ≡ length of 1D crystal would meant that the applied voltage v_i will be given by;

$$v_i = \frac{1}{\hbar}\frac{dE}{dK}|_i$$

The second nearly filled band current will equally be given by;

$$I_2 = -\frac{q}{L}\sum_{i(\text{filled states})} v_i$$

As the current desired increases, their summation becomes more difficult to evaluate as we have very large number of states filled. Thus if the band was completely filled, then we have;

$$\sum_{i(\text{all})} v_i = 0$$

This therefore simplifies the second current and can be expressed as;

$$I_2 = -\frac{q}{L} \sum_{i(\text{filled sates})} v_i$$

or

$$= -\frac{q}{L} \sum_{i(\text{all})} v_i + \frac{q}{L} \sum_{i(\text{filled states})} v_i$$

or

$$= \frac{q}{L} \sum_{i(\text{empty states})} v_i$$

From here, it is noted that the same current is the same as for a positively charge particle placed on the empty states when the remaining states are unoccupied. The overall motion of electrons in nearly filled energy band can then be described by the empty electronic states provided that the effective mass of the empty states is taken to be negative of m* as given below.

$$m^*_{\text{empty}} = -\left(\frac{1}{\hbar^2} d^2 E/dk^2\right)^{-1} \rightarrow = \left(\frac{1}{\hbar^2} \frac{d^2 E}{dk^2}\right)^{-1}$$

Explicitly, this expression shows that the empty energy states are at the top of the bands where the effective mass can be generalized as;

$$m^*_{\text{empty}} = +ve \text{ as } \frac{d^2 E}{dk^2} = -ve$$

The empty energy states at the top of a band tend to be with a positive charge and thus the m* are the equivalency of holes i.e.

$$m^* \equiv \text{Holes}$$

The electron or holes stay at the edge of a band in which the holes are at the top of Valence band while the electrons are in the bottom of conduction band and thus located where parabolic band approximation can be applied. In semiconductors, electrons and holes have constant effective mass and the classical treatment or behaviour forms a good parabolic band approximation.

Valence Band

In most materials, the valence band takes a maxima at zone center where k = 0, and thus the valence band is an approximation of three sub-band. Two of these bands are degenerate and thus have the same allowed energy at k = 0 while the third band has maximum at reduced energy k = 0. This shows that at k = 0, the shape and curvature are orientation of the bands is independent.

Conduction Band

Though the conduction behaves as the valance band, it has a minimum where electrons will gather and this minimum varies from material to material. Germanium carbon (GeC) has band minima at zone boundary. For SiC the band minima occurs at;

$$k \simeq = 0.8\left(\frac{2\pi}{a}\right)$$

which is measured from the zone center. On the other hand, GaAs has a band min at k = 0 though this semiconductor has a direct band-gap good for light emission and this makes them good in conserving momentum transition from Valance Band to the Conduction Band. These semiconductors have a structure appropriate for the design of a 3D tensor. The band gap energy for germanium at about 300K is usually used as reference for material processing, i.e.;

$$T = 300^0 K \; E_G \text{ for Ge} \simeq 0.663 eV$$

For silicon we have;

$$\simeq 1.125 eV$$

And for a ternary semiconductor and thus gallium arsenide has a band gap of GaAs;

$$\simeq 1.422 eV$$

Analysis show that as temperature, T, decreases a contraction of crystal lattice increases and the atomic bonds grow stronger. The consequence is that there is an increase in Band gap energy

$$E_G(T) = E_G(0) - \frac{\alpha T^2}{T + \beta}$$

E_{G0})		α	β
Ge	0.7437 eV	4.77×10^{-4}	235
Si	1.170 eV	4.73×10^{-4}	636
GaAs	1.519 eV	5.405×10^{-4}	204

Mobility

In solid state physics, mobility is considered to be a measure of ease of carrier motion within a semiconductor crystal. An electron in motion inside a crystal experiences collisions. The collision imped motion and as a result decrease electron or hole mobility. Since electron flow is regarded as current, the current density for hole flow is given as;

$$J_p = q\mu_p pE$$

while electron current density is given by;

$$J_p = q\mu_n nE$$

In these two expressions, **E** is the applied electric field and μ is the mobility, **q** the electronic charge and **n** & **p** are the respective carrier concentrations.

Lattice Vibrations

The term lattice can be defined as an arrangement of points or particles or objects in a regular periodic pattern in 2 or 3 dimensions. Consider the 1D lattice illustration shown in the figure below.

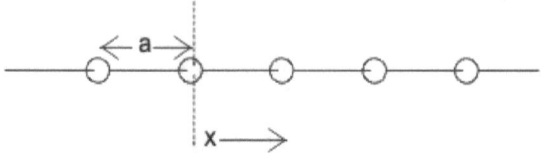

On the mechanical sense, Hooke's law is obeyed perfectly as the atoms vibrate. If one atom is displaced by **x** from equilibrium position, the potential energy of the atom is a function of distance from neighboring atoms and this can be summarized by the displaced position potential energy as it changes by the expression;

$$\Delta U = U(a+x) + U(a-x) - 2U(a)...$$

In this case U(x) is the potential energy when distance from neighbour is *x*. when Taylor series is used to expand the components;

$$U(a+x) \text{ \& } (a-x)$$

and keying only 1st significant term, it will give the change of potential energy as;

$$\Delta U = \left.\frac{\partial^2 U}{\partial x^2}\right|_{x \to a} = \frac{\gamma}{2}x^2$$

This expression forms a potential function of a harmonic oscillator however, this is only on the assumption that only one atom oscillates while other atoms are bond to each other. We expect that vibration of one atom in lattice will set other atom to vibration and vibrate according to the expression;

$$m\frac{d^2 x}{dt^2} = \gamma x$$

If we let (S_{l-1}), (S_l), and (S_{l+1}) be the displacement of $(l-1)^{th}$, l^{th} and $(l+1)^{th}$ atom from equilibrium as shown in the figure below then, force acting on l^{th} atom which is at a distance from adjacent atoms and for two adjustment atoms putting it to themselves where 1 is an integer, the force F on the lth atom will be given by;

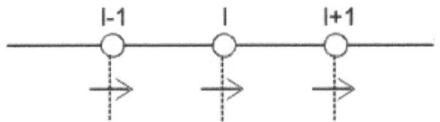

$$F = \frac{\gamma}{2}(S_{l+1} - S_l) - \frac{\gamma}{2}(S_l - S_{l-1})$$

$$= \frac{\gamma}{2}(S_{l+1} + S_{l-1} - 2S_l)$$

Applying this to the 2nd row to the lth atom will yield;

$$M\frac{d^2 S_l}{dt^2} = F = \frac{\gamma}{2}(S_{l+1} + S_{l-1} - 2S_l)$$

To give (S_{l+1}) as;

$$S_{l+1} = S_l + \left.\frac{\partial S_l}{\partial x}\right|_{x=0} a + \left.\frac{\partial^2 S_l}{\partial x^2}\right|_{x=0} \frac{a^2}{2} + \dots$$

And (S_{l-1}) will be given as;

$$S_{l-1} = S_l - \left.\frac{\partial S_l}{\partial x}\right|_{x=0} a + \left.\frac{\partial^2 S_l}{\partial x^2}\right|_{x=0} \frac{a^2}{2} - \dots$$

Then wave equation will now be simplified and expressed as;

$$M\frac{\partial^2 S_1}{\partial t^2} \cong \frac{\gamma}{2}a^2 \frac{\partial^2 S_1}{\partial x^2} \quad \ldots\ldots(ii)$$

Based on this final complex oscillation, a general wave equation that will result takes the form of a complex wave given by;

$$S_1 = A \exp j(\omega_q t - q\ell a)$$

Using this expression by taking regular frequency as, ω_q, where q is the wave number from equation labeled as (ii) above can be rewritten as;

$$-M\omega_q^2 = \frac{\gamma}{2}[\exp(jqa) + \exp(-jqa) - 2]$$

which when simplified further gives;

$$\rightarrow \omega_q = \pm\sqrt{\frac{2\gamma}{M}}\sin\left(\frac{qa}{2}\right)$$

Figure below shows a plot of ω_q as a function of the wave.

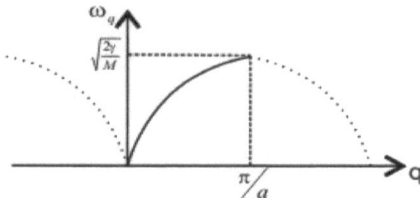

In a classical single harmonic oscillator approach, a single frequency is given by;

$$f = \sqrt{\frac{\gamma}{M}}.$$

In solid state physics, lattice vibrations are also characterized by a continuum of frequencies with a limiting maximum value given by;

$$f = \sqrt{\frac{\gamma}{M}}.$$

However, repetition occurs at $q > \pi/a$ or $q > -\pi/a$ to give a Brillouin zone concept. On this concept then let us consider a crystal of a finite member of atoms given by N. The boundary condition for the displacement of this lattice atom at the end of crystal may be taken periodic as to have no displacement at boundary conditions. Thus periodic boundary condition is given as;

$$q = \frac{2\pi n}{Na}$$

Lattice vibration is therefore applicable to semiconductor material like in which all atoms identical and one atom can be regarded as unit cell. Consider a material whereby 2 atoms form unit cell or 2 types of atoms form unit cell. Let this material be a semiconductors material and the unit cells be as illustrated below.

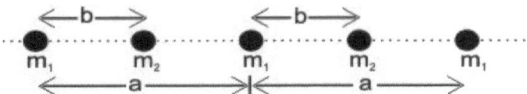

In such a case, the two equations are used to describe the vibrations in which 2 atom form a unit cell. Consider a condition where;

$$M_1 \rightarrow \text{odd no.}$$

and

$$M_2 \rightarrow \text{even no. position}$$

The using a simplified equation, we have for M_1 as;

$$M_2 \frac{d^2 S_{2l}}{dt^2} = \gamma_1 (S_{2l+1} - S_{2l}) - \gamma_2 (S_{2l} - S_{2l-1}) \cdots \cdots (1) \text{ position}$$

And M_2 given as;

$$M_1 \frac{d^2 S_{2l+1}}{dt^2} = \gamma_2 (S_{2l+2} - S_{2L+1}) - \gamma_1 (S_{2l+1} - S_{2l}) \cdots \cdots (2)$$

In the two above equations, the right hand side forms a wavelike expression and therefore we have an oscillatory frequency, ω_q with a wave vector, q|. If we assume that these expressions form a time harmonic solution then;

$$S_{2l} = A \exp[j(\omega_q t - q\ell a)]$$

$$S_{2l+1} = B \exp[j(\omega_q t - q(la + b))]$$

In the above expression, A and B are constants and substituting S_{2l} and S_{2l+1} in the expression for M_2 will form;

$$-\omega_q^2 M_2 A = -A(\gamma_1 + \gamma_2) + B \exp(-jqb)[\gamma_1 + \gamma_2 \exp(jqa)]$$

$$\omega_q^2 M_1 B = -B(\gamma_1 + \gamma_2) + A \exp(jqb)[\gamma_1 + \gamma_2 \exp(-jqa)]$$

Further analysis reduces the equations to;

$$\rightarrow p_1 A + p_2 B = 0$$

$$\rightarrow \gamma_1 A + \gamma_2 B = 0$$

These two equations have two unknowns that can be expressed as;

$$\det \begin{bmatrix} p_1 & p_2 \\ \gamma_1 & \gamma_2 \end{bmatrix} = 0$$

or as;

$$p_1 \gamma_2 - p_2 \gamma_3 = 0$$

which can be represented as;

$$\begin{bmatrix} p_1 & p_2 \\ \gamma_1 & \gamma_2 \end{bmatrix} \begin{bmatrix} A \\ B \end{bmatrix} = 0$$

Using a non-trivial approach to constants A and B above, we can represent it further as;

$$\rightarrow (\gamma_1 + \gamma_2 - \omega_q^2 M_2)(\gamma_1 + \gamma_2 - \omega_q^2 M_1)$$

which will result into;

$$= [\gamma_1 + \gamma_2 \exp(jqa)][\gamma_1 + \gamma_2 \exp(-jqa)]$$

The two solutions represent a dispersion relation between ω_q and q and solving for $\omega^2{}_q$ we get the expression given as;

$$\omega_q^2 = \left(\frac{1}{M_1} + \frac{1}{M_2}\right)\left(\frac{\gamma_1 + \gamma_2}{2}\right) \pm \left[\left(\frac{1}{M_1} + \frac{1}{M_2}\right)^2 \left(\frac{\gamma_1 + \gamma_2}{2}\right)^2 - \frac{4\gamma_1\gamma_2}{M_1 M_2} \sin^2 \frac{qa}{2}\right]^{\frac{1}{2}}$$

This expression can be simplified to form;

$$= \left(\frac{\gamma_a}{M_r}\right) \pm \left[\left(\frac{\gamma_a}{M_r}\right)^2 - \frac{4\gamma_1\gamma_2}{M_r(M_2 + M_1)\sin^2 \frac{qa}{2}}\right]^{\frac{1}{2}}$$

In which case;

$$\frac{1}{M_r} = \frac{1}{M_1} + \frac{1}{M_2}$$

And

$$\frac{\gamma_1 + \gamma_2}{2} = \gamma_a$$

A solution to the above expressions gives two values of ω^2_q for any wave number q.

$$\omega^2_q = \frac{\gamma_a}{M_r} - \frac{\gamma_a}{M_r}\left[1 - \frac{4\gamma_1\gamma_2 M_r}{\gamma^2_a(M_1 + M_2)}\sin^2\frac{qa}{2}\right]^{\frac{1}{2}}$$

And this implies that ω^2_q can be written as;

$$\omega^2_q = \frac{\gamma_a}{M_r} - \frac{\gamma_a}{M_r}\left[1 - \frac{4\gamma_1\gamma_2 M_r}{\gamma^2_a(M_1 + M_2)}\sin^2\frac{qa}{a}\right]^{\frac{1}{2}}$$

The values of **q** and those of ω^2_q are the obtained as;

$$q = 0, \quad \omega^2_{q_1} = 0, \quad \omega^2_{q_2} = \frac{2\gamma_a}{M_r}$$

The value of **q** will be;

$$q = \frac{\pi}{a}, \quad \omega^2_{q_1}$$

$$= \frac{\gamma_a}{M_r} - \frac{\gamma_a}{M_r}\left[1 - \frac{4\gamma_1\gamma_2 M_r}{\gamma^2_a(M_1 + M_2)}\right]^{\frac{1}{2}}$$

Likewise we also have;

$$\sin^2\frac{qa}{2} = 1$$

So that ω^2_q will be;

$$\omega^2_{q_2} = \frac{\gamma_a}{M_r} + \frac{\gamma_a}{M_r}\left[1 - \frac{4\gamma_1\gamma_2 M_r}{\gamma^2_a M_1 + M_2}\right]^{\frac{1}{2}} < \frac{2\gamma_a}{M_r}$$

A plot of ω_q as a function of the wave earlier can now be represented as shown in the figure below.

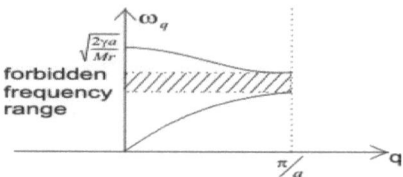

As earlier also, solving for S_{2l} and S_{2l+1} constant when q approaches zero, $q \to 0$ the ratio of amplitude of A to B becomes;

$$\frac{B}{A} = \left[1 - \frac{M_1 + M_2}{2M_1}(1 \pm 1)\right]$$

i.e $\frac{B}{A} = 1$ or $\frac{B}{A} = \left[1 - \frac{M_1 + M_2}{M_1}\right] = \frac{M_1 - M_1 - M_2}{M_1}$

This gives ratio A:B, as;

$$\left(\frac{B}{A}\right) = -\frac{M_2}{M_1}$$

When for lower branch A = B, then the ratio

$$\frac{B}{A} = 1 \text{ ie } B = A$$

And in this state, the vibrations of neighboring atoms are in phase. This is similar to those found when an acoustic wave propagation in the crystal to form an optical mode of a lower frequency. In a 3-D crystal, two distinct modes of transverse vibration exist or are formed along with optical or acoustic longitudinal mode.

Actual vibration

Consider an earlier expression obtained as;

$$S_\ell = \sum A_{qi} \cos(\omega_{qi} t - q_i la)$$

Solving this expression for q_i will yield;

$$\frac{\pi}{a} > q_i > 0$$

If we associate a vibration mode of wave number q quantum mechanical wave function (S_l) in the Schrödinger equation, then for a harmonic oscillator of each mode made up of vibration with a different energy state will give;

$$E_n = \left(n + \frac{1}{2}\right) hw_q$$

With this expression, the particles give a picture of lattice vibrations in which each mode of vibration relates with the number of particles each having an energy level given by;

$$\hbar \omega_q$$

At this condition, the state of vibration changes by creating or annihilating of such particles forming a Pseudo momentum because of periodicity which can be expressed as

$$q = q_1 - n q_n = q_1 - \frac{2}{a}$$

Thus the vibration of the lattice with a wave vector q having a collection of particles called phonons which associated by the energy, $\hbar \omega_q$ and pseudo momentum $\hbar q$ takes place at a particular temperature. Therefore lattice atoms execute random vibrations and phonons of various energies and characteristics would be visible and thus we can borrow the Hamiltonian of harmonic oscillator represented as;

$$H = \frac{x'^2}{2} + w^2 \frac{x'^2}{2}, \ x' = x\sqrt{M} \Rightarrow H = \frac{x'^2}{2} + \omega^2 \frac{x'^2}{2}, x' = x\sqrt{M}$$

Form this equation, the Hamiltonian of lattice system is given as;

$$H_1 = \sum_q \left(\frac{\dot{\alpha}_q \dot{\alpha}_q^*}{2} + w_q^2 \frac{\alpha_q \alpha_q^*}{2} \right)$$

By normalizing the expression for S_l, we obtain the equation;

$$= \frac{1}{\sqrt{NM}} \sum \alpha_q (t) e^{jqla} + \alpha_q^* (t) e^{-jqla}$$

This is the Kronig Penney model for stationary lattice atoms. This is however not entirely accurate as lattice vibration, Hamiltonian of vibrating lattice and electrons sum together to distort or introduce errors to this model. Individual electrons and lattice atoms sum up to give the expression;

$$= \sum_1 \frac{p_1^2}{2M} + V_1(S_1) + \sum_1 \frac{p^3}{2m} + V_e(\bar{r}, \bar{s}_1) + \sum \frac{q^2}{|\bar{r}_{ij}|}$$

The Eigen value equation resulting from this equation becomes truly Hamiltonian;

$$H_e \psi = E_e \psi$$

Ψ_e and E_e in this case are the modified energy instant Eigenvalues that resukt from the changing position of lattice atoms according to;

$$\psi \rightarrow \phi_k(\bar{r}, \bar{s}_1) \exp(j\bar{k}.\bar{r})$$

In situations where, *l* is the periodic component of the block function formed out of a perturbed lattice.

Lattice Scattering for Mobility

The position of the conduction band and valance bond extremes in a semiconductor depends on lattice spacing and these spacing changes during vibration and there is perturbation in the C_β and V_β.

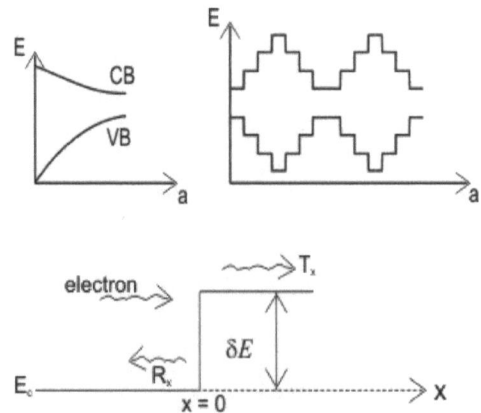

This effect by the lattice vibration and spacing is small for bound electrons with lower energy but the effect becomes more on the conduction electrons which are loosely bound to the atom or the nucleus and this result into Acoustic Vibration Compression & Expansion (AVCE) in lattice crystal. When these atoms are in a compressed position, the energy band is altered and causes the forbidden gap to increase. On the other hand, if it expands, the forbidden band gap width decreases. In calculating these two effects, the Potential Energy resulting out of the electron-electrons interaction is usually neglected as it is small under adiabatic condition. The incident waves formed takes the form of;

$$\psi_1(x) = Ae^{jk_ox}$$

And therefore the electron may be reflected at steps of δE or it may be transmitted as shown below

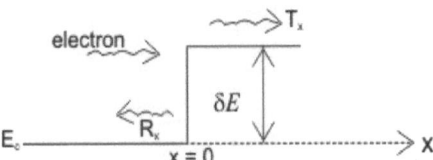

If the electron is reflected, the wave takes the form of;

$$\psi_2(x) = Be^{-jk_ox}$$

And this wave is for perfectly elastic collision case. If there exists a barrier on the transmission path, then the transmission over the barrier takes the form of;

$$\psi_3(x) = Ce^{jk_1n}$$

This lattice wave also causes δE to move with an acoustic velocity at which Doppler shift occurs which make the reflected momentum different from the incident momentum. This shift can be ignored assuming that the electron moved with a very high velocity resulting into the transmitted wave to lose energy and hence;

$$E_0 = \frac{\hbar^2 k_a^2}{2m_n^*} + E_c, \quad E_1 = E_0 - \delta E = \frac{\hbar^2 k_1^2}{2m_n^*}$$

So that;

$$x < 0, \quad \psi = Ae^{jk_ox} + Re^{-jk_ox}$$

$$x > 0 \quad \psi = ATe^{jk_1x}$$

Combining, the two expressions, we get;

$$\psi \,\&\, \frac{d\psi}{dx}$$

This form a continuous wave at $x = 0$,

$$\therefore R = \frac{k_o - k_1}{k_o + k_1}, \quad T = \frac{2k_o}{k_o + k_1}$$

We also have;

$$E_c + \frac{h^2 k_o^2}{2m_n^*} = E_c + \delta E + \frac{h^2 k_1^2}{2m_n^*} \quad \text{at } x = 0$$

$$\delta E = \frac{h^2 \left(k_o^2 - k_1^2\right)}{2m_n^*}$$

At $x = 0$

where ψ, is continuous.

$$\therefore A + AR = AT$$

or

$$A + B = C \rightarrow 1 + R = T,$$

As ψ is continuous we have;

$$k_o A - k_o B = k_1 C$$

$$k_o (A - AR) = k_1 AT \rightarrow 1 - R = \frac{k_1}{k} T$$

This causes the Deformation Potential and Mean Force Path to be given by;

$$|R| = \left| \frac{k_o - k_1}{k_o + k_1} \right|^2$$

And therefore, R is given by;

$$\therefore R = \left(\frac{m_\bullet^2 \, \delta E}{2\hbar^2 k_o^2} \right)^2$$

If the step height, δE, is to be small then;

$$k_1 \approx k_o \therefore 2k_o = k_o + k_1$$

This step hieght can be related to the occurring strain at that moment by;

$$\delta E = -X_c \left(\frac{\delta V}{V_o}\right)$$

Where χ_c is the deformation potential constant and is always equal to the shift of conduction per unit dilatational strain. Since thermal energy is the major cause of lattice vibrations, then as δp which is the maximum pressure that causes a volume change of δV, the C1 becomes constant and is given by the expression;

$$\frac{1}{2}\delta p \delta \gamma = C_1 KT.$$

From this expression, then compressibility can be given as;

$$\beta = \frac{1}{V_o}\frac{\delta V}{\delta p}.$$

$$\left(\frac{\delta V}{V_o}\right)^2 = \frac{2C_1 \beta KT}{V_o}.$$

$$\therefore K = \frac{(m_o^*)^2 C_1 \beta KT X_c^2}{2\hbar^4 K_o^{\ 4} V_o}.$$

The compressibility at a distance, δx, cause the probability of scattering or reflection to be given by $\frac{\partial x}{\lambda n}$, where λ_n is the mean free path, $\frac{l}{2}$ is the linear dimension of volume so that the Deformation Potential;

$$V_o \cdot \frac{\ell}{2\lambda_n} = |R|$$

$$\rightarrow \lambda_n = \frac{h^4}{8(m_n^*)^2 C_1 \beta X_c^2 KT}$$

Based on this approach, if an electron is in energy state E_1 at $t = 0$ state, then the probability that it will be in state E_2 after time $t > 0$ after a perturb potential, $\frac{\Delta V}{q}$ is applied will be given by;

$$|C_h \cdot (t)|^2 = \frac{1}{h^2} <\psi_2^* \Delta V \psi_1>^2 \left|\frac{\exp[j(E_2-E_1)t/h]-1}{j(E_2-E_1)/h}\right|^2$$

$$= \frac{2\pi}{\hbar^2}<\psi_2^* \Delta V \psi_1>^2 \sin\frac{(E_1-E_2)t/\hbar}{x(E_1-E_2)}$$

The expression above will give the probability of transition from E_1 to E_2 per unit time is called the transition probability and it is maximum only if $E_1 = E_2$. In this case, the collision involved show that energy is conserved. A real change of electron state from E_0 to E_1 occurs only if the energy associated with lattice vibration change by a creation or absorption of a phonon as the energy change is given by;

$$\delta E = -X_c \frac{\delta V}{V_o}$$

Acoustic phonon scattering

Strain in lattices is produced only by longitudinal vibration shear that is associated with transverse vibrations. However, this does not affect the energy Eigenvalues as it passes along the longitudinal classic wave or phonons which are equal to the coherent regions of compression or extension which are of the order of $l/2$ in linear extent where, l, is the length of disturbance. Based on this length l, the volume, V_0 of this linear extent is subject to a dilatorily stress producing a maximum pressure, δp and volume change, δV. Therefore, the stored Strain energy will be given by;

$$= -\frac{1}{2}\delta V$$

And if the source of strain energy is thermal, then the stored strain energy is proportional to kT so that the stored Strain energy will be given by;

$$\text{or} \quad -\frac{1}{2}\delta p \delta V = -C_1 kT \quad C_1 = \text{const.}$$

$$\beta = \frac{1}{V_o}\frac{\delta V}{\delta p} \quad \text{or} \quad \delta p = \frac{\delta V}{V_o \beta}$$

Therefore compressibility will then be expressed as;

$$\therefore \frac{1}{2}\delta p \delta V = \frac{(\delta V)^2}{2V_o \beta} = C_1 kT.$$

$$\therefore \left(\frac{\delta V}{V_o}\right)^2 = \frac{2C_1 \beta kT}{V_o}$$

With this analysis, the reflection coefficient will finally be expressed as;

$$|R|^2 = \frac{(m_n^* \delta E)^2}{(2\hbar^2 k_o^2)^2} = \frac{(m_n^*)^2 c_1 \beta kT X_o^c}{2\hbar^4 k_o^4 V_o}$$

Based on this equation for a distance, δx, the probability of scattering is $\delta x/\lambda_n$ where λ_n is the mean free path. Taking the linear dimension, $\frac{l}{2}$ the probability of reflection $\equiv |R|^2$ above, we can express $\equiv |R|^2$ as;

$$\therefore |R|^2 = \frac{1}{2\lambda_n} \rightarrow \lambda_n = \frac{h^4 k_o^4 IV_o}{(m_n^*)^2 C_1 \beta k T X_c^2}$$

$$\lambda_n = \frac{h^4}{8(m_n^*)^2 C_1 \beta k T X_c^2} \quad V_o = \frac{l^3}{8}$$

$$k_o = \frac{2\pi}{1}$$

If we let the component, λ_n, be independent of velocity so that the it tends towards the mean free time between scattering or collisions, then from Boltzmann's Theory of free particle in a gas, we get;

$$\tau_n = \frac{8}{3\pi} \frac{\lambda_n}{\bar{v}} = \frac{8}{3\pi} \frac{\lambda_n}{\sqrt{\frac{8kT}{\pi m_n^*}}}$$

$$= \frac{h^4}{(3\sqrt{8\pi})(m_n^*)^{3/2} C_1 \beta (kT)^{3/2} X_c^2}$$

Based on the above expression, Shockley show that from a full Quantum Mechanical treatment that, we can also express it as;

$$\bar{C}_n = \frac{\sqrt{8\pi}}{3} \frac{h^4 C_{11}}{(m_n^*)^{3/2} (kT)^{3/2} X_o^2}$$

where C_{11} is the elastic constant or the young's modulus for longitudinal extension in the crystal orientation of <110> so that the electron mobility can be calculated using the equation;

$$\mu_n = \frac{q\bar{\tau}_n}{m_n^*} = \frac{\sqrt{8\pi}}{3} \frac{qh^4 c_{11}}{(m_n^*)^{5/2} (kT)^{3/2} X_c^2}$$

So that the ratio of electron and hole mobility will be obtain from;

$$\frac{\mu_n}{\mu_p} = \left(\frac{mn*}{mp*}\right)^{-5/2} \left(\frac{X_c}{X_v}\right)^{-2}$$

Deformations sometimes hamper mobility and thus due to lattice Scattering, two constants are introduced in the above expression. These are χ_v which is the valence band deformation potential constant and χ_c which is the conduction band deformation potential constant. A more rigorous analysis can also be carried out with the perturbation theory. So far there are three possible models that can be used. These are:

- Deformable lattice ion model which considers the changes that the distribution gets distorted as the position of the atom remains the same.
- Rigid ion model which considers the ion if gets displaced as the potential gets displaced.
- Deformation potential model which considers the potential membrane getting distorted if the ion gets displaced.

CHAPTER FIVE
SOLID STATE THERMOELECTRIC FUNDAMENTALS

Thermoelectric Materials

A full understanding of thermoelectricity of solid state materials require an understanding many disciplines that include equilibrium thermodynamics, non-equilibrium thermodynamics, quantum mechanics, statistical mechanics, transport theory, crystallography and related solid state physics. The term *thermoelectric* imply that both thermal and electrical phenomena are involved. The term thermoelectric material refers to a material which exhibit a substantial thermoelectric effect. Every material exhibits *some* little thermoelectric effects although best electrical conductors are exhibit orders higher than 20 times of magnitude better than the best electrical insulators. Also, all materials conduct heat to some extent and that is why all materials generate a thermal EMF forces.

The Absolute Scale or Thermodynamic Kelvin, Temperature Scale

The term temperature is defined as a measure of the hotness though it can also be taken as the coldness of a system. For a measure to be rational and be useful between people of common interest, there must be an agreement on a scale of numerical values where the most familiar of the scales which is common is the Celsius or Centigrade Scale and is used in measuring devices for interpolating between the defining values of any measurable quantities. The only temperature scale with a real basis in nature is the Thermodynamic Kelvin Temperature Scale (TKTS), which can be deduced from the first and second laws of Thermodynamics. The low limit of the TKTS is absolute zero, or zero Kelvin, or 0K without the mark of degrees. Since it is linear by definition, only one non-zero reference point is needed to establish its slope. That reference point in case was chosen in the original TKTS as 273.15K, or 0°C. The 0°C is a temperature with which we all have a common experience. It is the temperature at which water freezes or coming from the other side, we say ice melts and at which water exists under ideal conditions as both a liquid and a solid under atmospheric pressure.

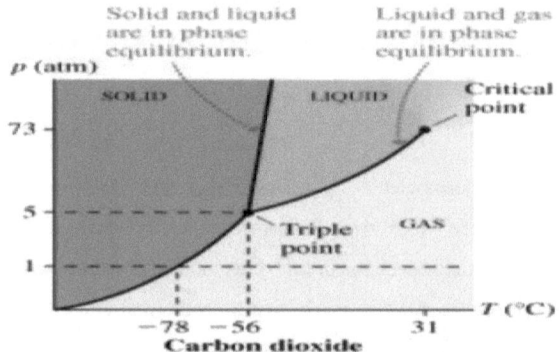

In 1954 the reference point was changed to a much more precisely reproducible point of 0.01°C. This is known as the triple point of water and is the temperature at which water exists simultaneously as a liquid and a solid under its own vapor pressure. The triple point of water will be the subject of extended discussion in a later article in this series of articles.

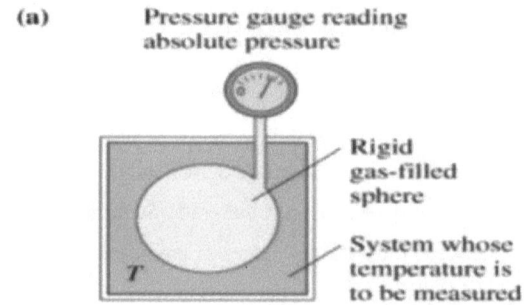

Pressure reading gauge

The unit of temperature of the TKTS is the Kelvin, abbreviated "K". The temperature interval °C is identically equal to the temperature interval K and °C or K may be used also to indicate a temperature interval. The difference between 1°C and 2°C is 1K or 1°C, but the temperature 1°C = the temperature 274.15K. Measurements of temperature employing the TKTS directly are hardly suitable for practicable thermometry. Most easily used thermometers are not based on functions of the First and Second Laws. The main purpose for the realization of the TKTS is to establish relationships between the Thermodynamic Scale of nature and the practical scales and thermometers of the laboratory or of industry, so that measurements made by non-thermodynamic means can be translated into terms of the TKTS, and rational temperature scales can be constructed on a basis related to realizable physical phenomena.

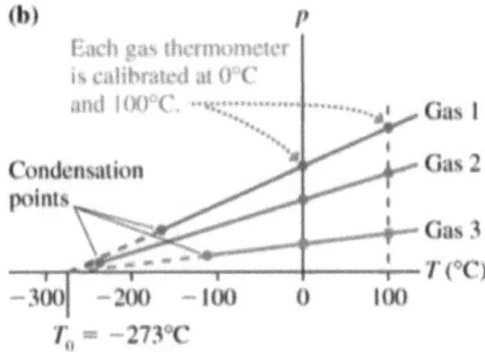

Gas behavoiur using the pressure gauge measurements

The variation due to pressure from the defined temperature of liquid-solid equilibrium is not large (but that does not mean that it is not to say that it may not be significant). The freezing point of water is reduced approximately 0.01K for an increase of pressure of 1 atmosphere. The variation due to pressure for liquid-vapor equilibrium is relatively very large. A three-phase equilibrium is represented by the triple point of water, the coexistence of liquid and solid water under its own vapor pressure, at 0.01°C. Because all three possible phases are determined by the physical state, it is generally possible to realize a triple point more accurately than a two-phase point. This may be obtained from the famous Phase Rule of Gibbs:

$$P - C - 2 = F$$

where P is an integer equal to the numiler of phases present, C is the number of kinds of molecule present and for an ideally pure material, $C = 1$ and F is an integer giving the number of degrees of freedom. Obviously, for the two-phase equilibrium there is one degree of freedom, pressure, and for the three-phase equilibrium $F = 0$; that is, the temperature is independent of any other factor. Figure below illustrates one, two and three-phase equilibria as1a, 2a, and 3a respectively.

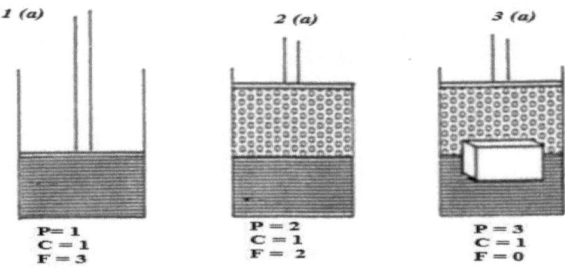

Figure: The Phase Rule of Gibbs

In the figure the parameters are defined as;

P = the number of phases present;

C = the number of components (1 for a pure material);

F = the degrees of freedom.

1a = is uncontrolled.

1b = is a melt or freeze point.

1c = is a triple point.

A rudimentary gas thermometer is shown in figure 2. Its operation will be illustrated by using it to show that the zero of the TKTS is 273.15K below the temperature of the normal freezing point of water, 0' on the Celsius Scale. Figure 2 shows a cylindrical bulb of constant volume, connected by tubing defined as constant-volume, to a U-tube manometer. A second connection to the manometer leads to a reservoir of mercury which contains a plunger, P, by means of which the column height of the manometer may be varied. The constant-volume bulb and tubing contain an ideal gas for purposes of accuracy. The bulb is first surrounded by an equilibrium mixture of ice and water (C, in Figure 2a). When the gas is in thermal equilibrium with the slurry in the bath, the pressure of the gas is adjusted by moving the plunger so that both columns of mercury in the U-tube are at the same height, corresponding to an index mark 1 as shown in the figure below. The parameters shown on the diagram are defined as;

A = helium gas.

B = mercury,

C = water + ice ($0\,^{\circ}C$),

D = water + steam ($100\,^{\circ}C$),

P = a plunger for adjusting mercury level.

In the next step, the ice bath is removed and replaced with a bath containing boiling water or more correctly, an equilibrium mixture of liquid and vapor water at a pressure of 1 standard atmosphere. As the manometric gas is heated by the boiling water it expands and the mercury in the manometer is displaced. The plunger is actuated to re-position the surface of the mercury in the left leg of the diagram at the index mark 1 restoring the criterion of constant volume in the closed gas system, a condition shown in Figure 2b. However, the mercury in the right, the open leg is not now at the index mark 1, but measurably higher than before. In fact, the difference in heights indicates that the pressure of the enclosed gas at the boiling point is 1.366099 times the pressure at the freezing point. We can then calculate:

$$[100 - 0)°C]/[(1.366099 - 1)] = 273.15$$

and from this we can understand the Celsius degree as 1/273.15 of the pressure ratio change between 0°C and 100°C. Thus if temperature is reduced by 273.15 Celsius or Kelvin intervals below 0'C, an absolute zero is reached which is the zero of the TKTS.

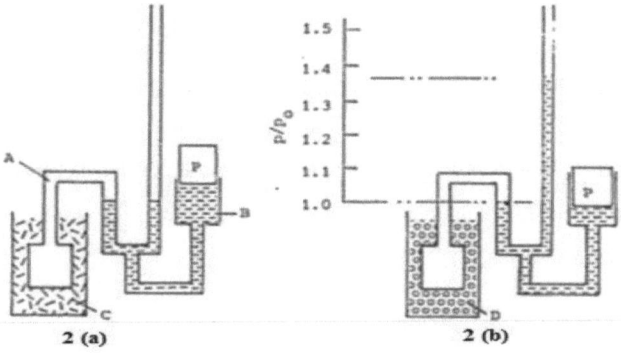

Figure: A rudimentary Gas Thermometer

The zero of the Celsius Scale is therefore 273.15K and any temperature value estimation can be found as;

$$T = t + 273.15$$

where **T** is temperature on the TKTS and **t** is temperature on the Celsius Scale. Note that the temperature interval and the zero of the TKTS have been defined without reference to the properties of any specific sub-stance. All constant volume gas thermometer measurements then can therefore be expressed in terms of the equation as;

$$P_1 / P_2 = (T_1 - T_0) / (T_2 - T_0)$$

where the T_s are temperatures on the TKTS which is the natural temperature Scale and the T_{0s} are the zero of that Scale and changes from 0 through 1 to 2 respectively as expressed in the equation above.

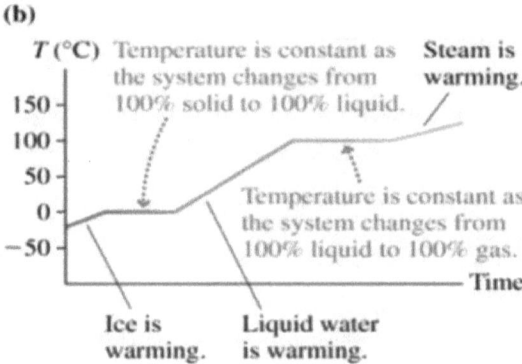

This relationship assumes an ideal gas. As you use and read this book you will have observed that the gas thermometer reflects **Charles'** or **Boyle's law** if the pressure or the volume of the gas, respectively, is only held constant. An ideal gas, is a gas whose behavior can be predicted exactly either from Boyle's or Charles' Law, which obeys it through all ranges of temperature or pressure, and where the relationship between concentration, $(\frac{n}{V})$, absolute temperature and absolute pressure is given as;

$$(\frac{n}{V})(\frac{T}{P}) = \frac{1}{R} = \text{constant}$$

more commonly written as;

$$PV = nRT$$

where R is the gas constant, identical for all ideal gases. R = 0.082053, and is known to about 30ppm.

Temperature-Dependent Effects

This expansion of air on heating became widely known in classical times, and was used in various dramatic devices. For example, Hero of Alexandria describes a small temple where a fire on the altar causes the doors to open. The altar is a large airtight box, with a pipe leading from it to another enclosed container filled with water. When the fire is set on top of the altar, the air in the box heats up and expands into a second container which is filled with water. This water is forced out through an overflow pipe into a bucket hung on a rope attached to the door hinges in such a way that as the bucket fills with water, it drops, turns the hinges, and opens the doors. The pipe into this bucket reaches almost to the bottom, so that when the altar fire goes out, the water is sucked back and the doors close again. (Presumably, once the fire is burning, the god behind the doors is ready to do business and the doors open…). Still, none of these ingenious devices is a *thermometer*. There was no attempt (at least none recorded) by Philo or his followers to make a *quantitative* measurement of how hot or cold the sphere was. And the "meter" in thermometer means measurement.

Fahrenheit's Excellent Thermometer

The first really good thermometer, using mercury expanding from a bulb into a capillary tube, was made by Fahrenheit in the early 1720's. He got the idea of using mercury from a colleague's comment that one should correct a *barometer* reading to allow for the variation of the density of mercury with temperature. The point that has to be borne in mind in constructing thermometers, and defining temperature scales, is that not all liquids expand at uniform rates on heating, water for example, at first contracts on heating from its freezing point, it then begins to expand at around forty degrees Fahrenheit, so a water thermometer wouldn't be very helpful on a cold day.

Amontons' Air Thermometer
Pressure Increases Linearly with Temperature

A little earlier in 1702, Amontons introduced an *air pressure thermometer*. He established that if air at atmospheric pressure (he states 30 inches of mercury) at the freezing point of water is enclosed then heated to the boiling point of water, but meanwhile kept at constant volume by increasing the pressure on it, the pressure goes up by about 10 inches of mercury. He also discovered that if he compressed the air in the first place, so that it was at a pressure of sixty inches of mercury at the temperature of melting ice, then if he raised its temperature to that of boiling water, at the same time adding mercury to the column to keep the volume of air constant, the pressure increased by 20 inches of mercury. In other words, he found that for a fixed amount of air kept in a container at constant volume, the pressure increased with temperature by about 33% from freezing to boiling, that percentage being *independent of the initial pressure*.

Thermal Equilibrium and the Zeroth Law of Thermodynamics

Once the thermometer came to be widely used, more precise observations of temperature and (as we shall see) *heat flow* became possible. Joseph Black, a professor at the University of Edinburgh in the 1700's, noticed that a collection of objects at different temperatures, if brought together, will all eventually reach the same temperature. We say nowadays that bodies in "thermal contact" eventually come into "thermal equilibrium" which means they finally attain the same temperature, after which no further heat flow takes place. This is equivalent to: **The Zeroth Law of Thermodynamics:** If two objects are in thermal equilibrium with a third, then they are in thermal equilibrium with each other. The "third body" in a practical situation is just the **thermometer**. It's perhaps worth pointing out that this trivial sounding statement certainly wasn't obvious before the invention of the thermometer. With only the sense of touch to go on, few people would agree that a piece of wool and a bar of metal, both at 0°C, were at the same temperature.

Measuring Heat Flow

The next obvious question is, can we get more *quantitative* about this "flow of heat" that takes place between bodies as they move towards thermal equilibrium? For example, suppose I reproduce one of Fahrenheit's experiments, by taking 100 ccs of water at 100°F, and 100ccs at 150°F, and mix them together in an insulated jug so little heat escapes. What is the final temperature of the mix? Of course, it's close to 125°F—not surprising, but *it does tell us something!* It tells us that the amount of heat required to raise the temperature of 100 cc of water from 100°F to 125°F is e*xactly the same* as the amount needed to raise it from 125°F to 150°F. A series of such experiments (done by Fahrenheit, Black and others) established that it *always* took the same amount of heat to raise the temperature of 1 cc of water by one degree. This makes it possible to define a **unit of heat**. Perhaps unfairly to Fahrenheit, **1 calorie is the heat required to raise the temperature of 1 gram of water by 1 degree Celsius.** Celsius also lived in the early 1700's. His scale has the freezing point of water as 0°C, the boiling point as 100°C. Fahrenheit's scale is no longer used in science, but lives on in engineering in the US, and in the British Thermal Unit, which is the heat required to raise the temperature of one pound of water by 1°F.

A second condition is that there must be the absence of any intermolecular forces acting on or within the gas and thus the internal energy, U, does not depend on the molecular distances, and $(dE/dU)_T = 0$. In reality though unfortunately there is no real ideal gas and the uncertainty of 30 ppm is a very large number. Though Helium gas tends to become closer to this behaviour, carbon dioxide varies most widely from ideality as observed in most experimental proves. However real gases approach ideality as their pressures are reduced and this reflect a reduction in density. Since it is not possible to measure the change in pressure of a gas at zero pressure or the change in volume of a gas be at zero volume, the requirement for an ideal gas is approached by making a number of measurements at a number of pressures and extrapolating to zero pressure values obtained. Such a system of measurements is shown in Figure below. Regardless of the nature of the gas, all gas thermometers at the same temperature approach the same reading as the pressure of the gas approaches zero values. One which is much used, that of Clausius, is the virial equation, which is a series expansion in terms of the density of the gas, and is written:

$$PV = nRT(1 + [nBv/V] + [2nCv/V^2] + [3nDv/V^3] + \ldots)$$

where the coefficients B, C, D, etc. are called the second, third, fourth, etc., volume virial coefficients and are constants for a given gas at a given temperature. Over the usual range of gas densities in gas thermometry, it is seldom necessary to go beyond the second virial coefficient.

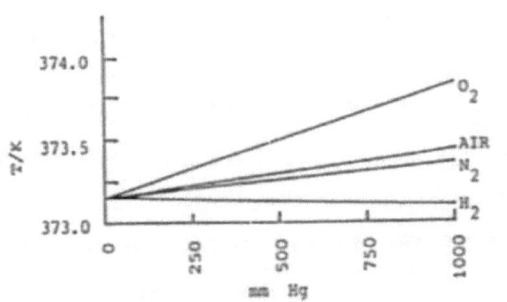

Figure : Pressure ratios of various gases at various pressures at the condensation point of steam.

The departure of real gases from ideality is only one of the problems of accurate gas thermometry. In real life situations we deal with real gases though in most experiments. A second is the purely mechanical matter of dead space. There must be a real connection to convey the pressure from the bulb to the manometer. It is inconvenient to locate the bulb and the manometer in the same thermo stated enclosure, and a common practice is to use two separate enclosures, each carefully thermos tatted. Figure below is a modification of Figure above to illustrate this configuration.

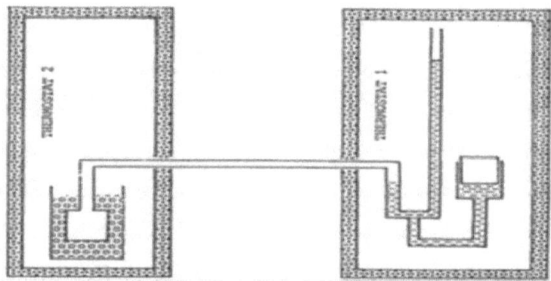

Figure: Thermostatted gas thermometer.

The U-tube manometer and the gas bulb are separately kept at constant temperature; the capillary is not. The mechanism price paid is that there is a capillary tube of generally of some length which goes through the wall of each of the two thermostats and whose temperature is essentially uncontrolled. The solutions are care and compromise. The bulb volume can be made as large as possible relative to the capillary volume. The temperature distribution along the capillary length can be measured at suitable intervals. The capillary volume can be kept small by providing a capillary of small diameter but not so small as to introduce thermo molecular pressures where the tube passes through a temperature gradient or conversely, a correction for thermo molecular pressure can be made. A third obvious problem is that of the thermal expansion of the materials of construction. An ideal constant-volume gas thermometer assumes that only the contained gas is subject to thermal expansion while in reality the whole system is subject to temperature changes which must be known or estimated and for which correction must be made.

A fourth correction required is for the hydrostatic head pressure of the gas in the system including that of the gas column itself. A fifth relates to the effects of sorption in which impurities in the gas or impurities remaining from a less than ideally clean system are absorbed on the walls of the bulb. Other parts of the thermometer system at the reference temperature, desorbed at a higher temperature, with the effect of elevating the measured gas pressure and then reabsorbed as the temperature approaches the reference temperature. Attention to the elimination of sorption has resulted in gas thermometry measurement of the normal boiling point of water as 99.975°C. Gas thermometry has claimed the attention of a number of fine experimenters for some generations, the most recent of whom have been Guildner and Edsinger, followed by Schooley, at the National Bureau of Standards.

This work forms much of the thrust to replace the International Practical Temperature Scale of 1968 with the International Temperature Scale of 1990 in an effort to more closely approximate thermodynamic temperatures in a practicable Scale. A concise account of the gas thermometer and gas thermometry at the NBS is given by Schooley, and should be consulted by anyone interested in experimental elegance. Schooley provides an example of the accuracy and precision of this work.

Table 1: Experimental gas thermometry

Fixed point °C	Gas therm °C	IPTS-68 °C	Uncertainty
Steam	99.975	100.000	±0.005
Tin	231.924	231.9681	±0.015
Zinc	419.514	419.58	±0.03

In the conclusion of work in preparation for ITS-90 by Schooley, gas thermometry is considered to be a finished matter at the NBS. The gas thermometer itself which should have been preserved as a **national shrine** or **monument** is now in the process of **dismembermen**t. As a generality, gas thermometry has led the development of **thermodynamic values** of the thermometric fixed points and a variety of other methods have been used largely to check its accuracy and consistency. These include acoustic thermometry, dielectric constant gas thermometry, noise thermometry and radiation thermometry, each appropriate to a portion of the range of the temperature Scale.

Thermodynamics Systems and Processes

Thermodynamics is the branch of physics that describes and correlates the physical properties of macroscopic systems of matter and energy. In other words, it is concerned with heat and related thermal phenomena. The various laws of thermodynamics enable us to accurately describe processes involved in heat energy. The formal study of thermodynamics has its roots in the industrial Original Objectives revolution, following the construction of the first successful atmospheric steam engines in England: first by Thomas Savery in 1697 and then by Thomas Newcomen in 1712. The word "thermodynamics" arises from the Greek terms representing heat (**therme**) and power (**dynamis**), which pretty well describes the efforts of the founders of this discipline. Their original objective was to gain the most power (*energy per unit time*) from a given amount of energy stored in some fuel which was typically turned into heat (*the first and the second law*). The classical study of thermodynamics primarily deals with the concepts of energy, energy transfer, and energy transformation within some material system as the macroscopic parameters of that system change. The formalism of thermodynamics arises primarily from the precise mathematical expression of two very fundamental observations. The first is the observation that energy, as defined in physics, is conserved - it cannot be created or destroyed - it is simply transformed from one form into another. For example, when a stone is dropped from some height, it loses gravitational potential energy as is falls, but the kinetic energy of the stone increases during this process is such a way that the total energy is conserved.

In many mechanical processes, however, we observe what appears to be "energy loss" due to friction, air resistance, etc. Careful observation of these situations usually shows that the mechanical energy which has been lost has simply been transformed into thermal energy, what we often call as heat energy. The study of thermodynamics has, to a large extent, grown out of our desire to understand the nature of this heat energy and to understand the transformation of mechanical energy into heat energy and vice versa. The second fundamental observation upon which the study of thermodynamics is based is the everyday observation that heat energy flows spontaneously from hot objects to cold objects, and never the other way around. These two observations, called the first and second laws of thermodynamics, respectively, were not stated in their present form until the 1850's.

Many of the early views of thermodynamics were based upon the false assumption that matter contained a substance called "caloric". This caloric was thought to be a massless substance which filled in the spaces between the "particle of matter" causing the expansion of a important only for polyatomic molecules. Unlike the translational kinetic energy, these energies are quantized, the rotational kinetic energy being more easily activated than the vibrational. A monatomic gas, however, can be characterized solely on the basis of its translational kinetic energy and the potential energy of interaction between the individual atoms. This internal potential energy is related to the average force of interaction between the individual molecules which constitute the gas. When these molecules are far apart (as in a gas at low pressure), their interaction is negligible.

When the molecules approach each other, they repel. In many cases, we can treat a gas as a large number of solid balls which do not interact unless the balls "hit". If the average distance between molecules is large in comparison to the size of the molecules, the potential energy of the gas can be neglected and the collisions between the particles are considered to be completely elastic collisions. These collisions will simply transfer kinetic energy from one molecule to another in such a way that the molecules will move in a random fashion. The energy associated with this random motion is what we think of as heat energy – unorganized energy. So the random kinetic energy of a gas must somehow be related to the temperature of the gas.

In the case of a solid, we can model the system after a series of balls held together by springs. Again, the individual particles possess both kinetic and potential energies. The potential energy stored in the springs is associated with the amount the springs are compressed or stretched from their equilibrium positions. The kinetic energy is just the energy of back and forth motion of the individual particles. We again associate the random kinetic energy of the individual particles with the temperature of the solid. If enough energy is added to a liquid or solid the binding energy or potential energy may be overcome and the individual molecules may become free to move around - no longer bound to the other molecules. This constitutes a change in of *phase* the system. The energy required to break these bonds and to initiate the phase change is called the *latent energy* (or *latent heat*) of the material.

Thermodynamic System Interactions

As we begin a thermodynamic analysis of any system, we must first clearly define the system under study and distinguish that system from its surrounding. Only then are we in a position to properly discuss the *interactions* between that system and its surroundings. Thermodynamic Systems A thermodynamic system is some pre-defined part of our universe which may interact with its surroundings *mechanically*, *chemically*, or *thermally*. It is that part of the universe in which we are interested. We may define the system in terms of *a quantity of matter* (a **closed** or **fixed mass** system [in engineering, a **control mass**]), or *a region of space* (an **open** or **fixed volume** system [in engineering, a **control volume**]) chosen for study. The region *outside* the system of interest is called the *surroundings*, and the real or imaginary surface that separates the system from its surroundings is called the *boundary*. The boundary may be *fixed* or *movable*.

One example of a thermodynamic system might be a gas contained inside of a cylinder which is closed off at the top by a movable piston (see Figure below). This gas is subject to the external atmospheric pressure and to the weight of the piston. If the atmospheric pressure were to increase, the volume of the gas would subsequently decrease, causing the pressure of the gas inside the cylinder to rise. The result of this process is a *mechanical* movement of the piston, and can be considered as work done on the gas to compress it. On the other hand, the gas may be heated by its contact with the outside air, by absorbing radiant energy from the sun, or by being heated with a Bunsen burner, causing the gas to expand. Another example of a thermodynamic system might be an iceberg floating in the ocean. This iceberg, which is floating due to the buoyant force of the ocean water on the less dense ice, may be slowly melting as the iceberg extracts heat from the surrounding air and water.

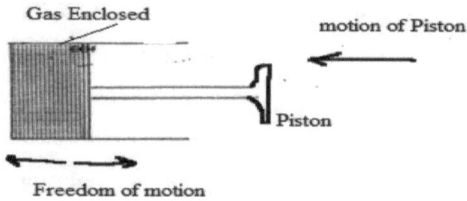

Figure : Piston showing mechanical work on a gas

This system will typically be idealized by assuming the piston is *massless and frictionless* and the fitting between the wall of the cylinder and the piston is *air tight*, so that no air can enter or leave the container. This system may be a little difficult to material when it became heated. The transfer of heat was simply viewed as the transfer of this caloric from one object to another. From this perspective, one might be able to say that an object "contained" a certain amount of heat. It was not until the mid to late 1700's that people began to notice inconsistencies in the caloric theory. One particular observation which indicated a problem with the caloric theory was the seemingly infinite amount of heat which could be produced when boring a canon. A team of horses was used to bore the hole in a cast iron canon. During this process, water was poured into the bore hole to keep the cast iron canon from overheating. This water would soon come to a boil and water would be continually added to the bore hole as the water boiled off. There seemed to be no limit to the amount of heat energy which could be derived from the boring process.

Statistical Thermodynamics and Internal Energy

Both the first and second laws deal with energy that is transferred from one system to another in the form of heat. This transfer of energy must be related to the energy that is stored in each system. We call this stored energy the internal energy. We now understand that the internal energy of a system is associated with the microscopic motion of the molecules that make up that system. The founding fathers of thermodynamics, however, did not have this model to build upon, and so classical thermodynamics is primarily a study of the changes in the *macroscopic* (large-scale, observable) properties of a system, with internal energy being one of these macroscopic quantities. Today, we have a somewhat easier job of understanding the basic concepts of thermodynamics, because we have a superior model upon which to build. We now understand that the internal energy of a system is comprised of two separate parts:

(1) the kinetic energy of translation, rotation, and vibration, and

(2) the potential energy of interaction between the various molecules.

For a gas, the rotational and vibrational kinetic energy is actually constructed, perhaps, but it is not conceptually difficult. Furthermore, the walls of the cylinder may be either (i.e., not allowing any heat to flow through the wall) or (heat conductive) *adiabatic diathermal* depending upon our particular needs. The external pressure acting on this system can be varied in a controlled fashion by adding to or removing from a pile of sand placed on top of the piston. Thus, this idealized system can interact *mechanically* with its surroundings and/or *thermally*.

Types of Interactions between Systems

A system may interact with its surroundings in many different ways. We use certain terms to precisely describe the type of interaction which may be taking place between a system and its surrounding. A system is said to be *isolated* if the system cannot interact with its surroundings *in any way* (i.e., if the system cannot gain or lose energy, particles, etc.). The universe as a whole may be considered as an isolated system. In fact, our solar system is very nearly isolated from the rest of our galaxy. If there is no *material* interchange between the system and its surroundings, the system is said to be **closed**. But, even in a closed system, energy may still be exchanged between the system and its surroundings by thermal conduction through the adjoining wall or through the movement of the adjoining wall as the gas of the system expands or contracts.

For a closed system, however, these processes must take place without a gain or loss of the particles which make up the system of interest. An **open** system, on the other hand, is one which may gain or lose material as it interacts with its surroundings. In practice, systems may be considered as open or closed depending upon whether a fixed *mass* or a fixed *volume* is chosen for study. A **closed** system (also known as a **control mass**) consists of a *fixed amount of mass*, and no mass can cross its boundary. An **open** system (or **control volume**) is usually a properly selected region of space which may, for example, enclose a device that involves mass flow, such as a compressor, turbine, or nozzle. A simple example of an open system is a hot water heater.

Suppose we want to know the amount of heat required to supply a steady stream of hot water at a given temperature. Since hot water leaves the water heater and is replaced by cold water entering the water heater, it is not convenient to concentrate of a fixed mass of water. Instead, we concentrate of the volume formed by the interior surface of the hot water heater. We will find that the applicable thermodynamic relationships differ, depending upon whether a system is open or closed. *It is, therefore, extremely important to properly recognize the type of system we have before we attempt to analyze it.* A system may have a **mechanical** contact with its surroundings. For example, if the piston separating region E from region in Figure used earlier were to move to the right (increasing the volume of the gas in region E, but decreasing the volume of the gas in region) the expansion of system E is an obvious example of the case where the system is *doing mechanical work on its surroundings* (i.e., on system).

Other forms of mechanical interaction are also possible. For example, a system may be composed of many charged particles arranged somewhat uniformly throughout the volume of the system. If an external electric field is present, the charges will respond to this field, and mechanical *work* will be done on the system which will store electrostatic energy in this system. Likewise, external magnetic forces, or gravity may cause the system to change its configuration is such a way as to store or release mechanical energy. A system may interact **chemically** with its surroundings if the boundary separating the two systems will allow a chemical interaction to take place. A common example of this phenomenon is seen when a calcite rock (such as limestone) is slowly dissolved as water (which contains dissolved carbon dioxide) flows through cracks in the rock and over its surface to produce caves and caverns. A system is said interact ***thermally*** with its surroundings when there is an energy change which takes place in the system without any change in the *macroscopic* properties of the system, such as the volume.

For example in Figure above, assume that the wall between system E and E*w* would prevent any particle exchange or chemical interaction, and that the wall remains fixed. Any increase or decrease in the energy of system E must arise due to a *thermal* interaction. In this case the boundary between the system and its surroundings is said to be *diathermal* (i.e., the wall permits heat to pass through). Typically, a thermal interaction between two systems can be associated with an increase or a decrease in temperature of the two systems - the hotter system cooling off, and the cooler system warming up. However, as we shall see later on, *not all heat transfers result in a change in temperature.*

Thermodynamic Equilibrium of state variables

One of the most important concepts of thermodynamics is the concept of thermodynamic equilibrium. This concept arises from our everyday experience. If we place a cup of hot coffee on the counter and come back much later, we find that the temperature of the coffee has changed - it has cooled off. If we leave the coffee on the counter long enough, the temperature of the coffee becomes equal to the temperature of its surroundings. More precisely, if a system is *isolated* for a long period of time, the different properties of the system (the temperature of our coffee; or the gas pressure, temperature, and volume for a ideal gas system) will eventually take on certain values. The of the system at that point in time *which will no longer change in time state* is *defined* to be a state of *thermodynamic equilibrium*. Once such a state is reached, the *state* properties or *macroscopic variables* will remain constant *unless and until the system is again allowed to interact with other parts of the universe.*

Thermal Equilibrium and Temperature

When we bring two systems together, we find that the hotter system will cool off, while the cooler system will heat up. This is according to experience, and is easily explained by saying that "heat" flows from the hotter system to the cooler system due to a thermal gradient, much like a fluid flows due to a pressure gradient or like electricity flows due to an electrostatic potential gradient. Since the flow of heat is often associated with a temperature change, we typically express the amount of heat which is added to or taken away from a system by the equation

$$dQ = CdT$$

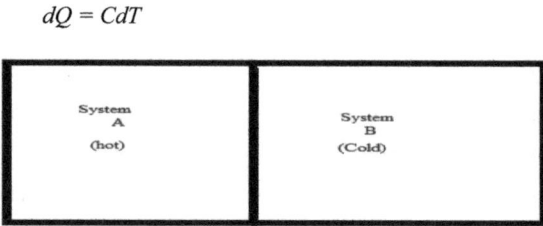

Figure: Two isolated gas systems A and B (hot and cold)

Where dQ is a differential amount of heat added to or removed from the system, dT is the differential change in temperature of the system, and C is the *heat capacity* of the system. By convention, we take dQ to be *positive* for heat added *to* the system. Thus,

$$dQ = CdT$$
$$= C(T_f - T_i)$$

The result is positive if the final temperature of the system T_f is greater than the initial temperature of the system T_i. The heat capacity C of a system is obviously proportional to the size of the system and to the composition of the system. And, although the heat capacity of a particular material is not the same in all temperature ranges, it is typically a slowly varying function of the temperature. Because the heat capacity as defined in this last equation depends upon the size of the system, we define the *specific* heat capacity, which is *independent* of the size of the system, by the relation;

$$c - C/m$$

giving;

$$dQ = mcdT$$

Historically, the heat capacity of all substances was related to the heat capacity of water. The amount of heat required to heat 1 gram of water from 4 °C to 5 °C was *defined* to be 1 calorie, giving the specific heat capacity of water as 1 cal/g-°C in this temperature range. Thus, when the temperature of the coffee in a coffee cup becomes equal to the temperature of the surroundings, we say that the coffee is in *thermal* equilibrium with its surroundings. The *temperature* of the cup of coffee is one of the macroscopic properties of our cup of coffee. We are familiar with the general concept of temperature and we make use of thermometers of various sorts (mercury thermometers, bimetallic coil thermometers, thermistors, thermocouples, etc.), but we need an *operational* definition of temperature. Just as we have precise definitions for the volume of a container or for the pressure exerted by a gas on the walls of that container, we must also have a precise definition for the *temperature* of the gas within the container.

The macroscopic parameter which we call *temperature* is, in fact, defined in terms of the concept of thermodynamic equilibrium. The temperature of a system is *that property of a system that determines whether it is in thermal equilibrium with other systems*. In order to understand this basic definition, consider again the definition of thermodynamic equilibrium. If two systems are connected by a diathermal wall that is one which permits the transfer of heat, *and neither system changes*, we say that the two systems are in thermodynamic equilibrium with each other, and that these two systems have the same temperature. An extension of this idea is the so-called *zeroth law* of thermodynamics which we shall discuss later: *two objects in thermodynamic equilibrium with a third must be in thermodynamic equilibrium with each other*. This is a practical way of defining how one can measure temperature. One simply finds a devise which has a measurable parameter (which we call a thermometric property) such as length, resistance, color, etc. which changes with temperature. We can then use this "thermometer" to determine if two systems are at the same temperature. If they are, then the thermometric property of the thermometer will not change.

Just as there are a number of different devices which can be used to measure temperature, there are a number of different temperature scales which arise from using different *defined* temperatures and predetermined temperature intervals. For example, if we choose the melting point of ice and the boiling point of water as 0° and 100°, respectively, we have defined the Centigrade (or, more properly, the Celcius) temperature scale. If, on the other hand, we choose 32° and 212° for these two states, we have introduced the Fahrenheit temperature scale. Thus in *mechanical equilibrium* there is no change in energy of the system due to changes in *external properties* (or macroscopic properties). For example, a system may be in mechanical equilibrium if there is no change in the size or shape of the system due to changes in external pressure.

Thermodynamic States and State Variables

State Variables

As was mentioned above, the *state* of an isolated system can be designated by giving the values of the system *parameters* or *variables* - those quantities which we can measure experimentally. The parameters or variables of interest are necessarily *macroscopic* quantities. We cannot in general measure the specific microscopic conditions which may produce the observed macroscopic conditions. Some of the variables or parameters which might be of interest in a given system are the mass, pressure, volume, temperature, magnetic moment, dielectric constant, surface tension and many more.

Intensive and Extensive Variables

To make our development of thermodynamics as simple a possible we are first going to look at the simplest possible system. We will consider a gas in which the macroscopic parameters of interest are simply the pressure, P, the mass (or, alternately, the number of moles 8 or molecules R of the gas), the temperature, T, and the volume, V. If you carefully select about the four parameters which we just listed you will realize that the volume, V of the system may be related in some way to the mass of the system through the density of the gas. We often find that some of the parameters of a system depend upon the *size* of the system. Thus, if you take a system in thermodynamic equilibrium and subdivide it into smaller and smaller pieces, some of the measurable quantities, such as the total energy, E, the mass, m, the volume, V , etc. are different for the smaller units than for the system as a whole and these depend upon the *extent*, or size, of the system. Some parameters, however, are independent of the size of the smaller sub-units and have the same value in each of the individual units as in the larger system. Those parameters or variables of a system which *do not* depend upon the size of the system are called *intensive* variables, while those that *do* depend upon the size of the system are called *extensive* variables.

An *extensive* variable, such as the volume V can be made *intensive* by dividing that variable by the mass (or the number of moles, or the number of atoms) of the system. Such a *reduced* variable is called the *specific* value of the parameter. For example, the *specific volume*, V, is given by

$$v = \frac{V}{m}$$

or

$$v = \frac{V}{n}$$

or

$$v = \frac{V}{N}$$

it is here noted that the specific volume of a gas is just the inverse of the density of that gas. In the above equations, the volume V of the gas in a container is an *extensive* variable, and is expressed by a capital letter, while the *specific volume* is an *intensive* variable, and is expressed with small letters. In most cases, extensive variables will be designated by upper-case letters, while intensive variables will be designated by lower case letters. Notable exceptions to this rule, however, are the *temperature* and *pressure* which are designated by capital letters, although they represent intensive variables.

Intensive variables are typically regarded as local variables. That is, their particular value may change from point to point within the system of interest or of analysis. For example, the density of a gas may vary from point to point within a gas. This density gradient, however, will create a diffusive flow of gas from high density regions to low density regions until the density gradients are essentially eliminated. In the same fashion, there may be temperature or pressure gradients within the gas which will cause thermal and mechanical transport within the gas until these gradients are eliminated. *When the system is left undisturbed for long enough* (i.e. related to the relaxation time for the different processes within the system), *any gradients in the local parameters will eventually go to zero, and the system is said to be in a state of thermodynamic equilibrium.* In such a state, the macroscopic properties of the system such as pressure and temperature have specific values *which are the same at all points within the system.*

The values of the macroscopic parameters of a system in a state of thermodynamic equilibrium are determined by the particular state (or condition) of the system and are called *state* variables. Thus, when the system is in a particular thermodynamic state, a specific value of all state variables is uniquely defined. This means we can write a relationship between the parameters of the system of the form;

$$f(P,T,v) = 0$$

This equation implies that there is some specific relationship between the parameters of the system when that system is in a specific equilibrium state. A knowledge of the specific value of all the state variables uniquely defines a particular thermodynamic state of the system. The number of variables required to completely specify the state of the system depends on the particular system involved. If the system is a gas with a definite number of similar molecules at moderate temperatures and pressures, the only parameters required to specify the state of the system are the pressure P, the volume V, the temperature T, and the number of molecules

R or moles 8 of the gas (a *gram-mole* is defined as the number of molecules of gas which have a mass in grams equivalent to the molecule's molecular weight and is given by 6.0225 x 10^{23} particles). Here, the volume V of the system is an *extensive* variable.

In many thermodynamics texts the equations are expressed in terms of the *specific volume*, *v*, so that the equations which are developed will not depend upon the size of the system. This also reduces the number of macroscopic parameters needed to completely specify the state of the system. Thus we might say that the parameters for our ideal system are the *specific volume v*, the pressure P, and the temperature T of the system, giving us only three distinct intensive variables which can be used to completely characterize the system, i.e.,

$$f(n, P, T, v) = f(P, v, T) = 0$$

This equation expresses the fact that the temperature is a function of the pressure and the (specific) volume; that the ressure is a function of the temperature and the specific volume; or that the specific volume is a function of the temperature and pressure, as expressed explicitly by the following equations:

$$T = T(v, P)$$
$$P = P(v, T)$$

and

$$v = v(P, T)$$

These three equations indicate that *the thermodynamic state of this simple gas is completely specified by two independent, intensive properties.* Although we are confident that such relationships exit between the observed properties of our gas, we do not often know in advance what these specific formulae are and on the other hand, knows quite well how the pressure is related to the temperature and the specific volume of a particular substance. If a system is suddenly brought into contact with another system at very different temperature, there will be thermal gradients set up inside the system and therefore the temperature will *not* be uniform throughout the system. There may also be pressure differentials, shock waves, turbulence, etc., established in the system. By making measurements of systems in equilibrium, then, we can determine the equilibrium values of the parameters of a system and determine a mathematical expression for the relationship among those variables.

Thermodynamic Processes for Pure Substances

A substance with a fixed chemical composition throughout is called a *pure* substance. A pure substance, however, does not have to be a single chemical compound. For example, air contained within a fixed volume at normal temperature and pressure is considered a pure substance since the composition of air is the same at all parts of the system. If the temperature of this system were to be decreased to the point where some of the components of air began to condense, however, this would no longer be a pure substance. Likewise, a mixture of oil and water would not be a pure substance since oil and water do not mix. Practically all pure substances can exist in different phases depending upon the temperature and pressure. The most common example is water. It can exist as a gas, a liquid, and a solid within the temperature and pressure ranges present on the surface of the earth. In the gaseous phase the molecules can move about freely within a fixed volume, bouncing off each other and the walls of the container, and completely filling the volume. That is according to Kinetic theory of gases.

In the liquid phase, the molecules are loosely attracted to each other and form groups of molecules that have a more or less fixed volume and density but which can still move about when subjected to external sheer forces. In a closed container the liquid and gaseous phase are typically in equilibrium. In the solid phase the molecules are more or less fixed in place but may oscillate about their equilibrium position. However, many solids exist in one of several different phases. Each phase is characterized by a specific molecular arrangement which is homogeneous within a given phase. Different phases may actually coexist within the solid and can be identified by an easily identifiable boundary surface whether real or imaginary boundary. For example, water ice may exist in different *seven* phases at high pressures.

Phase-Change Processes for Pure Substances

If we add heat energy to a pure substance, the temperature of that substance will rise until a critical point is reached where additional heat energy will begin to cause the inter-molecular bonds to break. As an example, consider an amount of water at 1 atm and 20 ^{0}C contained in a closed cylinder. We can assume the cylinder is capped by an ideal massless, frictionless, tight-fitting piston which rests on top of the water. The water in this phase is called a *compressed* or *sub-cooled liquid* (meaning that it is not about to evaporate). If we now add heat to the system, the water will expand slightly as the temperature rises and as result cause an increasing its specific volume (process 1 to 2 of Figure below). The pressure of the liquid will still remain constant at 1 atmosphere as the piston moves upward. This process will continue until the temperature of the water reaches 100 ^{0}C (point 2). Once the water reaches 100 ^{0}C any addition of heat will cause the intermolecular bonds to break and the water will begin to vapourize. A liquid which is about to vapourize is called a *saturated liquid*. As additional heat is added to the system the amount of saturated water will begin to decrease and the amount of water vapour will begin to increase, but the temperature of the mixture will remain constant at 100 ^{0}C.

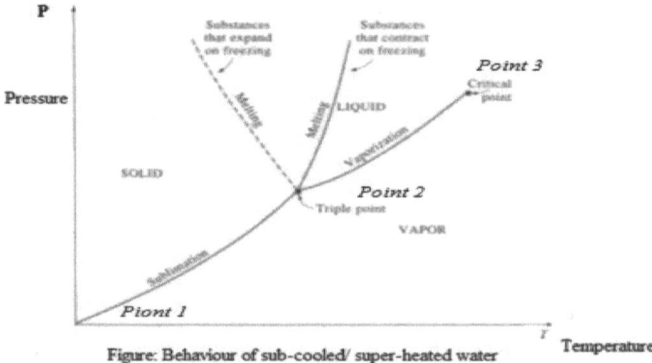

Figure: Behaviour of sub-cooled/ super-heated water

This process in point 2 point 3 above will continue as heat is added to the system until all the water is changed into vapour. At each point along the constant temperature curve the liquid and vapour coexist in equilibrium as a *saturated liquid-vapor mixture*. Along this curve any addition of heat energy increases the amount of water vapor and decreases the amount of liquid water, while any loss of heat energy decreases the amount of water vapor and increases the amount of liquid water. Once the phase-change process is completed we again have a single-phase substance and in this case water vapour, and the temperature of the vapour will begin to rise as we add additional heat (process point 4 to point 5). The water vapor in this last process which is no longer at the point of condensation is called a *superheated vapour*. Thus the processes we have described are reversible processes. If we begin with the superheated vapour and slowly remove heat from the system, the thermodynamic states can be retraced provided we maintain the same and constant external pressure.

Saturation Temperature and Pressure

If we increase the pressure on our sample of water, we find that we can raise the temperature beyond 100 ^{0}C before the water begins to vaporize or boil. In fact, we find that the temperature at which water begins to boil increases with pressure above 100 °C. At a given pressure, the *temperature* at which the liquid begins to boil is called the *saturation temperature*. Likewise, the *pressure* at which a liquid boils for a given temperature is called the *saturation pressure*. If we plot the saturation pressure vs. the saturation temperature, we obtain a *liquid-vapor saturation curve* such as that shown in figure below.

Obviously, the plot for water as it passes through a phase change must be modified if we change the pressure exerted on the water. To accomplish this, we can think of adding masses to the piston which sits on top of the water. Just as before, we can plot the temperature of the water varsus. the intensive volume as heat is added to the system. For each different external pressure, we will see a curve similar to the one we have in figure for superheated water except for the fact that the saturation temperature increases and the width of the saturated liquid-vapor region (the saturation line) grows narrower with increasing pressure as shown in Figure below. The width of the saturation line continues to grow narrower as the pressure increases until it disappears altogether. This occurs at the *critical point*, defined by a critical temperature, specific volume, and pressure, and is the points were the saturated liquid and vapor states are identical. At this point it is impossible to visually detect a liquid surface. Above the critical pressure there is no distinct phase-change process. The intensive volume increases continuously with increasing temperature and we can never tell when we pass from the condensed liquid to the superheated vapor.

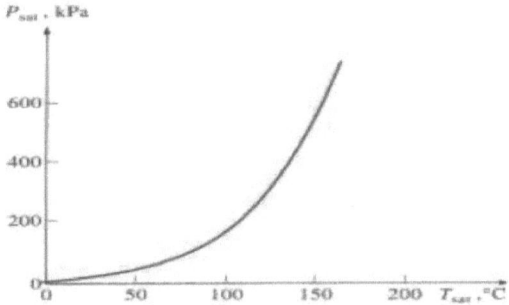

Figure: plot the saturation pressure vs. the saturation temperature

These same phase-change processes can also be plotted on a *P vs V* diagram. However, the *constant temperature* lines on the T vs. V graph have a downward trend, while the *constant pressure* lines on the X vs. V graph have an upward trend. This can be seen in Figures above. Three-Dimensional Representation of Process Diagrams Extended to Include the Solid Phase. The diagrams in Figures also represent three-dimensional T, V, P surfaces which include the solid phase. For each type of material, you will see a region on the T, V, P surface where the liquid and vapor phases can co-exist as well as a region where the solid and vapor phases can co-exist. These two regions are separated by the *triple line*. Along this line the substance can exist with all three phases in equilibrium. On a T vs. P graph often called the *phase diagram*, this triple *line* appears as a *triple point*. On this graph, all three phases of a substance are separated from each other by three lines, the sublimation line, the vaporization line, and the melting line which all intersect at the triple point.

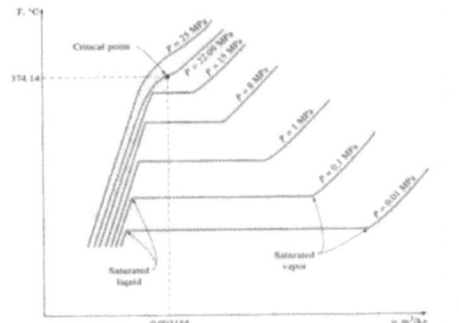

Figure: T-V Constant-pressure phase-change processes of a pure substance

Figure: T-V -diagram for a pure substance.

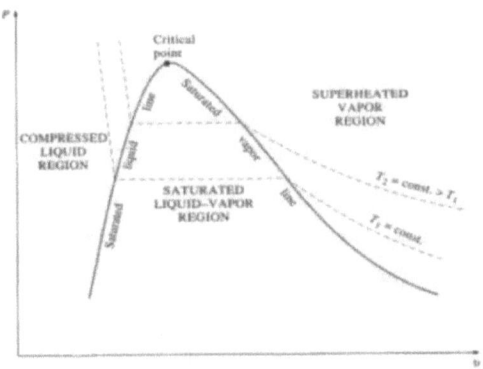

Figure: T-V diagram for a pure substance.

The three-dimensional T, V, P surfaces shown in Figures represents *all possible states of thermodynamic equilibrium* for that particular substance. This means that we can represent any *quasi-static process* (one which occurs slowly enough that the system is always essentially in equilibrium) for this substance by a line on this *P, V, T* surface. It would be nice to have a simple mathematical equation which would relate the various properties of any real substance (i.e., it would be nice to have a simple mathematical formula which would represent all the points of the T, V, P surface). Unfortunately, such a simple relationship, valid for all values of the properties of the system, does not generally exist. As a result, we often make use of tables of data which give relationships between the system properties and the critical temperature, pressure, and volumes, along with the saturation temperatures and pressures for various substances. For many *gases* at low pressures, however, there *is* a simple mathematical relationship between the observable properties of the gas, known as the *ideal gas law*.

Heat, temperature and thermal equilibrium are some of the major concepts most fundamental in solid state science. Any two isolated objects are said to be in *thermal equilibrium* if nothing happens when they are brought into contact. This intuitive result is called the Zeroth Law of Thermodynamics. This law forms the basis for the establishment the temperature scale and therefore objects in thermal equilibrium are said to be at the same temperature. Isolated objects at different temperatures brought into contact will exchange energy in an attempt to establish thermal equilibrium. Any work performed during this process is equal to the difference between the heat lost by one object and heat gained by the other object showing that energy is always conserved. This is the First Law of Thermodynamics.

Electrical Equilibrium

Material objects are composed of positive and negative charges and thus the concepts of electrical charge and electrical potential are fundamental. Opposite charges attract and like charges repel each other and therefore material objects are said to be in electrical equilibrium if there is no exchange of charge when they are brought into contact with each other. Such objects are said to have the same electrical potential. It also true that objects with different electrical potentials brought into contact exchange charge to establish electrical equilibrium. In nano-scale bulk materials have either zero or very small net electrical charge. As there occurs an exchange of charge there also occurs an exchange energy or *work*. An exchange of one Coulomb of electrical charge through a potential of one Volt results into an exchange of one Joule of work. An electrical current is the quantity of electrical charges that passes through a boundary each second and it is related to a change in electrical potential per unit of distance or the electrical gradient. A heat current is the quantity of heat which passes through a boundary each second and therefore by analogy a thermal force is related to the change in temperature per unit distance or the temperature gradient. Each thermal property has an analogous electrical property, as illustrated in table below.

	Thermal	*Electrical*	*Type*
Quantity	Heat	Charge	Reversible
Potential	Temperature	Potential	Reversible
Current Type	Heat Current	Electrical Current	Irreversible
Driving Force	Potential Difference	Temperature Difference	Irreversible

Equations of state

The first two laws of Thermodynamics establish *only* the general working frame for dealing with thermal phenomena. They simply affirm the *existence* of some state functions for a thermodynamic system: the internal energy and the entropy. The Third Law brings in a reference value for the entropy that is necessary for its complete definition. However, none of these fundamental laws allows the *effective construction* of the state functions. The specific values of the internal energy and of the entropy in a certain equilibrium state of a thermodynamic system remain characteristics of the considered system. For each particular macroscopic piece of matter some functional connections between the state parameters can always be established. Such relations are called *equations of state* in Thermodynamics. Only general properties of these equations can be inferred from the fundamental Laws of Thermodynamics. A small part of the equations of state can be established from arguments of theoretical Statistical Physics. This can be done only for quite simple systems. The value of theoretically established equations of state resides mainly in modeling and discussing more complicated situations. Nevertheless, most of the equations of state to be used in Thermodynamics are experimental results.

Caloric and thermal equations of state

It is usual to distinguish two types of equations of state, upon the experimental procedure for establishing them. Thus any thermodynamic system should have one *caloric equation of state* that is a relation of the form:

$$U = U(T,V,v)$$

whose construction needs *caloric* determinations (i.e. internal energy measurements by means of a calorimeter). Also, there should be one or several *thermal equations of state* that is relations of the form:

$$P = P(T,V,v)$$

which may be obtained through mechanical and temperature measurements only. The equations of the type:

$$\mu = \mu(T,V,v)$$

are usually included in the last category. Sometimes it may become useful to solve the above equation for the volume and for the other configuration parameters too are the following equations;

$$U = U(T,P,v),$$

and

$$V = V(T,P,v)$$

and

$$\mu = \mu(T,P,v).$$

Strictly speaking, these new relations *do not represent equations of state*. Nevertheless, they contain the same amount of information and will be treated on equal foot with the true equations of state thereafter. The caloric and thermal equations of state of a thermodynamic system contain all the information connected to the considered thermodynamic system. Nevertheless, as will be further demonstrated, the Laws of Thermodynamics induce a certain amount of redundancy in this information, which means that the equations of state are not strictly independent.

Functional relations between the equations of state TD system

Actually these which are also quite numerous, relations are consequences of the fundamental Laws of Thermodynamics. For example taking an equation previously discussed and written for a simple one-component system:

$$dS = \frac{1}{T} dU + \frac{P}{T} dV + \frac{\mu}{T} dv$$

If one replaces the internal energy, the pressure and the chemical potential by the corresponding equations of state, then, by The Second Law, the obtained differential form should be still integrable (in the new variables, T, V and v):

$$dS = \frac{1}{T}(\frac{\partial U}{\partial T})_{V,v} dT + \frac{1}{T}[(\frac{\partial U}{\partial T})_{V,v} + P] + \frac{1}{T}(\frac{\partial U}{\partial T})_{V,v} - \mu] dv$$

The integrability of the form in the above equation readily implies the following identities:

$$\frac{\partial}{\partial T}\{\frac{1}{T}[(\frac{\partial U}{\partial v})_{T,v} + P]\} = \frac{\partial}{\partial v}\{\frac{1}{T}(\frac{\partial U}{\partial T})_{V,v}\}$$

and

$$\frac{\partial}{\partial T}\{\frac{1}{T}[(\frac{\partial U}{\partial v})_{T,v} + P]\} = \frac{\partial}{\partial v}\{\frac{1}{T}(\frac{\partial U}{\partial T})_{T,v}\}$$

and

$$\frac{\partial}{\partial v}\{\frac{1}{T}[(\frac{\partial U}{\partial T})_{V,v}]\} = \frac{\partial}{\partial T}\{\frac{1}{T}(\frac{\partial U}{\partial T})_{V,v} - \mu]\}$$

Since the caloric equation of state that started in this section is supposed to be represented by a doubly differentiable function, one may write:

$$\frac{\partial^2 U}{\partial T \partial v} = \frac{\partial^2 U}{\partial v \partial T}$$

Thus, from an earlier equation it follows that:

$$\left(\frac{\partial U}{\partial v}\right)_{T,v} = \left[\left(\frac{\partial P}{\partial T}\right)_{T,v} - P\right]$$

Two other identities may be similarly obtained respectively:

$$\left(\frac{\partial \mu}{\partial v}\right)_{T,v} = -\left(\frac{\partial P}{\partial v}\right)_{T,v}$$

and

$$\left(\frac{\partial U}{\partial v}\right)_{T,v} = \mu - T\left(\frac{\partial \mu}{\partial T}\right)_{T,v}$$

The identities in the above two equations demonstrate the functional connections between the equations of state previously covered at the beginning of this section and the redundancy of the thermodynamic information contained therein. There is some practical importance of this kind of relations: they allow the theoretical construction of some hardly measurable functions like $T, v \left(\frac{\partial U}{\partial v}\right)_{T,v}$ through the determination of other quantities. By using equations of the form that we started with in this section one may obtain other functional connections of the same type as those in the above two equations). Passing to the new variables T, P and v, the fundamental relation reads:

$$dS = \frac{1}{T}\left[\left(\frac{\partial \mu}{\partial T}\right)_{T,v} + P\left(\frac{\partial v}{\partial T}\right)_{T,v}\right] dT$$
$$+ \frac{1}{T}\left[\left(\frac{\partial U}{\partial P}\right)_{T,v} + P\left(\frac{\partial v}{\partial P}\right)_{T,v}\right] dP$$
$$+ \frac{1}{T}\left(\frac{\partial U}{\partial v}\right)_{T,v} + P\left(\frac{\partial v}{\partial v}\right)_{T,v} - \mu] dv$$

The integrability conditions applied to the coefficients of dP and dT give (through the same arguments which produced above and below as:

$$\left(\frac{\partial U}{\partial P}\right)_{T,v} = -T\left(\frac{\partial v}{\partial T}\right)_{T,v} - P\left(\frac{\partial v}{\partial P}\right)_{T,v}$$

Another from this equation, one can obtain through the use of the general relation. Thus, one may write:

$$\left(\frac{\partial U}{\partial P}\right)_{T,v} = \left(\frac{\partial U}{\partial T}\right)_{T,v}\left(\frac{\partial v}{\partial P}\right)_{T,v}$$

which, by an earlier equation, gives:

$$\left(\frac{\partial U}{\partial P}\right)_{T,v} = -T\left(\frac{\partial v}{\partial T}\right)_{T,v} - P\left(\frac{\partial v}{\partial P}\right)_{T,v}$$

Ideal classical gas equations of state for thermodynamic systems

This is the thermodynamic system consisting in a collection of identical independent molecules. They are supposed to interact only with the walls of the recipient through instantaneous collisions. The total mass of the system is taken as constant. This is the simplest possible model and the corresponding equations of state are supposed to be known to the reader:

$$U = C_V T$$
$$P = \frac{vRT}{V}, \quad R = 8.314 \; J/mol/K$$

CV, the molar heat capacity of the gas, is a parameter depending on the structure of the gas molecules. For an ideal mixture of several perfect gases (that is a mixture whose components do not chemically interact) the equations of state become:

$$U = (v_1 C_{v1} + v_2 C_{v2} + v_3 C_{v3} ...)T$$

These last relations are direct consequences of the fundamental idea of a perfect gas: the molecules behave in the recipient without any reciprocal interaction and the total internal energy of the mixture should be the sum of the internal energies of the components). By similar arguments, the same addition property may be transferred to the partial pressures of the gas components.

The virial expansion

A thermal equation of state for a fluid with interactions among the molecules is known under this name. It is essentially represented by a series expansion of the fluid pressure upon the powers of the fluid density, v/V. The thermal equation of state for a perfect gas may actually be taken as a particular case of the virial expansion. In order to make a suitable comparison with this limit case, the factor vRT/V is forced out in front of this expansion:

$$P = \frac{vRT}{V}, [1 + \frac{v}{V} B(T) + \frac{v^2}{V} C(T), + ...]$$

where $B(T)$, $C(T)$, ... are the second, the third, ... virial coefficients, respectively. They are independent of the mass and of the volume of the fluid. The virial coefficients depend only on the fluid temperature and on its internal structure and may be computed through statistical mechanical considerations, starting from a suitable model of reciprocal interaction between molecules. By comparing such results with available experimental data, evaluations of the values of the intermolecular forces become possible. For a classical ideal gas we'll obviously have $B = 0$ and $C = 0$ and the density appears only to the first degree in the virial expansion. For high-density gases one may retain the second term of the parenthesis of the above equation and write:

$$P = \frac{vRT}{V}[1 + \frac{v}{V} B(T)]$$

At low temperatures the attractive (negative) potential energy between the molecules overcomes their (positive) kinetic energy. The total energy of a molecule will therefore be negative and its motion will be performed under the action of a global attraction towards the "interior" of the gas. Consequently, the gas pressure on the surrounding walls will decrease below the perfect gas case by a certain amount resulting from the dominant reciprocal attraction between the molecules. So, at low temperatures, $B(T)$ has to be negative.

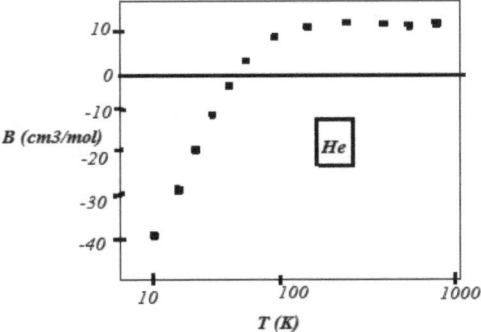

On the contrary, by heating up the gas, the average kinetic energy should overcome the potential energy and each molecule becomes repealed from the "interior" of the gas. The pressure on the walls will therefore increase to values above the perfect gas case and $B(T)$ becomes positive. These qualitative previsions are actually experimentally confirmed, as can be seen in the example of figure above where the measured values of the second virial coefficient of gaseous He are plotted as a function of temperature.

Van der Waals gas model of the thermal equation of state

The main form of this equation of state is the following:

$$[P + a\frac{v2}{V2}] \ (V - vb) = vRT$$

The purpose of the model is to correct the ideal gas thermal equation for the errors that may arise both when neglecting the intermolecular forces and when considering the molecules as volume-less points, that is when neglecting their own volume. Indeed, at usual distances between the molecules of a (even dense) gas, the forces are mainly attractive. The gas pressure on the walls, P, should thus be lower than the pressure in the "interior" of the gas by a certain amount ΔP. On the other hand, the "inside" pressure is the one that should enter the perfect gas equation. One may conclude that the desired correction for the intermolecular forces should consist in replacing P in the thermal equation of a perfect gas by $(P + \Delta P)$. The correction ΔP should be proportional to the intensity of the intermolecular interactions, that is to the average number of possible interactions per unit volume of the gas.

Given the total number of molecules of the gas, N, the number of distinct molecular couples that may be formed in a unit volume is essentially proportional to the squared molecular concentration N/V. We may thus write:

$$\Delta P = (\frac{N}{V})^2 = a(\frac{v}{V})^2$$

where $a\sim$ and a are positive constants depending on the nature of the gas molecules and of their reciprocal interactions. The correction concerning the volume occupied by each molecule at rest, the so-called **molecular volume**, should imply a decrease of the total volume (V) of the vessel. Thus, if the total molecular volume of the N molecules is ΔV, then the recipient volume available for free motion remains $(V - \Delta V)$. It is quite clear that the molecular volume ΔV should be proportional to the total number of molecules or, equivalently, to the number of moles of the gas: $(\Delta V = bv)$. By performing these corrections in the perfect gas equation of state (8.20), one immediately gets the Van der Waals equation. This equation is in fact another particular case of the virial expansion. Indeed, by solving the above equation with respect to the pressure, for low densities ($bv/V \ll 1$), one obtains:

$$P = \frac{vRT}{V}[1/\frac{1-b/\frac{v}{V}}{.} - \frac{av}{RTV}] = \frac{vRT}{V}[1 + \frac{v2}{V}(b - \frac{a}{RT})]$$

This equation is identical to an earlier equation if one takes:

$$B(T) = b - \frac{a}{RT}, \quad C(T) = 0$$

One may also observe that B is negative for low temperatures ($T < a/bR$) and that it becomes positive when the temperature increases ($T > a/bR$). So, even if $B(T)$ of the above equation shows no (slight) maximum value as is observed in the experimental curve of Figure above, the Van der Waals equation of state may still be used as a good approximation for a gas at intermediate densities. This equation has also the very important advantage of analytical simplicity which allows the use of the Van der Waals model as a test case for various statistical mechanical theories.

The thermal equation of state for a solid

The essential characteristic of a solid is its very weak compressibility and thermal expansivity. If one takes a reference volume V_1 for the considered piece of solid and if V is the volume resulting from a small temperature variation ($|T - T_1|/T_1 \ll 1$) and from a small pressure ($|P - P_1|/P_1 \ll 1$) then one may write the corresponding variation of the function $V(T,P)$ as a double power series expansion:

$$(V - V_1) = (\frac{\partial V}{\partial T})_{T,v} \times (T - T_1) \, (\frac{\partial V}{\partial P})_{T,v} \times (P - P_1) + \ldots\ldots\ldots$$

As it will be seen in the next chapters, there are specific notations for the volume derivatives with respect to the temperature and pressure:

$$(\frac{\partial V}{\partial T})_P = V\, a_p ; \quad (\frac{\partial V}{\partial P})_{T,v} = = V K_T$$

where a_P and K_T are the isobaric dilatation and the isothermal compressibility of the system, respectively. By limiting the expansion in the equation above to the first degree terms only and by gathering the constant terms in a unique constant V_0:

$$V - V_1 = (\frac{\partial V}{\partial T})_{P,v} \times (T - T_1) + (\frac{\partial V}{\partial P})_{T,v} \times P_1$$

one obtains the desired equation:

$$V = V_o\,(1 + a_p\, T - K_T\, P)$$

In a good approximation, the parameters aP and KT may be treated as constants.

The Equation of State for an Ideal Gas

Robert Boyle showed empirically that the pressure of a gas is inversely proportional to the volume of the gas when the gas is held at constant temperature, or

$$PV = \text{constant}$$

which is a function only of the amount of gas present only. Now for *different amounts* of the same gas, the constant of proportionality is different, but if the *specific* volume is used, we find that

$$\frac{PV}{n} = Pv = K$$

where K is a function *only* of the temperature and not of the size of the system (or number of molecules). In later experiments, Charles demonstrated that the product of the pressure and the specific volume when plotted against the temperature resulted in an approximate straight line for the case where the pressure was relatively low. The value of PV/n for this straight line was found to be 2.271×10^3 Joules per mole at the triple point of water (0 °C), and 3.102×10^3 Joules per mole at the boiling point of water (100 °C) for almost all gases at relatively low pressure. If we assume that this is a good approximation for *all* gases, then we can write the equation for this *ideal* gas as;

$$\frac{PV}{n} = a + RT$$

where a and R, are constants and where T is the temperature defined according to our chosen scale (or the so-called Celsius scale). If we plug in the values of PV/n mentioned above and solve for a and R we obtain:

$$a = 2.271 \times 10 \text{ Joules/mole}$$

and

$$R = 8.314 \text{ Joules/mole/°C}.$$

From our equation for TZ I8 we can therefore write:

$$PV = n(a + RT)$$

or, finally,

$$PV = na + nRT$$
$$PV = na + nR(T_0 + T) = nRT$$
$$\text{where To} = a/R = 273.2 \text{ °C}.$$

This last equation defines a new temperature scale which we call the thermodynamic or Kelvin scale with;

$$T(K) = T_0 + T$$

$$= 273.2\ ^0C + T$$

The present convention is for there to be a degree symbol associated with the Celsius temperature scale, but not with the Kelvin scale. The equation of state for an *ideal gas* is, therefore,

$$PV = nRT$$

where is the pressure, P is the *extensive* volume, n is the number of moles of the gas, R is a universal constant called the *ideal gas constant*, and T is the temperature in Kelvin. In physics we often find it more convenient to specify the number of particles R rather than the number of moles 8 and write the *ideal gas law* as;

$$PV = NkT$$

In this equation we introduce a new constant k, which is known as the Boltzmann constant. This constant is related to the ideal gas constant V by the relation;

$$nR = Nk$$

implying;

$$k = nR/N$$

We can find the value of k, (Boltzmann's constant) by setting $k = 1$ and $N = N_A$ (where N_A is Avagadro's number which give n by 6.0225×10^{23} molecules/gram-mole). We obtain:

$$k = 1.38 \times 10^{-23}\ \text{Joules/molecule-K}$$

In terms of *intensive variables* we can write the ideal gas law as;

$$Pv = kT$$

or

$$Pv = RT$$

where $v = V/N$, or $v = V/n$, respectively. The left side of the equations above, Pv, has units of energy per mole or energy per molecule. You should recall from modern physics that Boltzmann's constant is in terms of energy in Joules (or eV) per molecule per Kelvin, which means that the universal gas constant must be in terms of the energy in Joules (or eV) per mole per Kelvin. Based upon some fairly simple experiments, we have determined an equation of state which gives a functional relationship between the pressure, the specific volume, and the temperature of a gas at low pressures. But what do we mean by *low* pressures? Different substances are gases at different temperatures and pressures. For example, under "normal" conditions on the earth's surface, carbon dioxide is a gas, but at the low temperatures found on Mars, carbon dioxide often "freezes out" and solidifies into what we call dry ice. Such carbon dioxide is called "dry" ice because at normal atmospheric pressure on the surface of the earth solid carbon dioxide will not *melt*; carbon dioxide simply changes phase from a solid to a gas. This same phenomenon would also occur for water at pressures *below* the triple point pressure of 0.61 kPa.)

To determine just how well real gasses can be characterized by the ideal gas equation, we introduce the *compressibility factor*, Z, defined by;

$$Z = \frac{Pv}{RT} = \frac{v}{\frac{RT}{P}} = \frac{v \; actual}{v \; ideal}$$

where Z will be equal to unity for the case of an ideal gas.

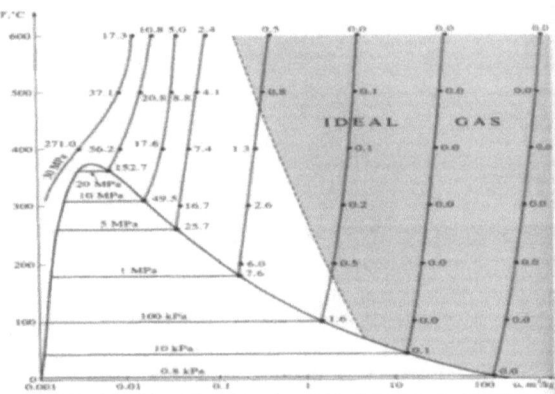

Figure: Percentage error involved in steam as an ideal gas

This compressibility factor obtained here is typically less than unity for real gasses and this is reasonable since the model for an ideal gas neglects the actual intermolecular attraction experienced by any real gas, and this intermolecular attraction would tend to effectively reduce the specific volume. The compressibility factor becomes *much less* than unity when the real gas is near its critical temperature and pressure. Under normal conditions, most familiar gases like air and its lighter constituents (oxygen, nitrogen, hydrogen, etc.) give errors of less than one percent when treated as ideal gases. Heavier gases, however, tend to deviate somewhat from the ideal gas equation. As an example, Figures above shows a - plot P-V for water vapor showing the percentage error involved in making the ideal gas approximation.

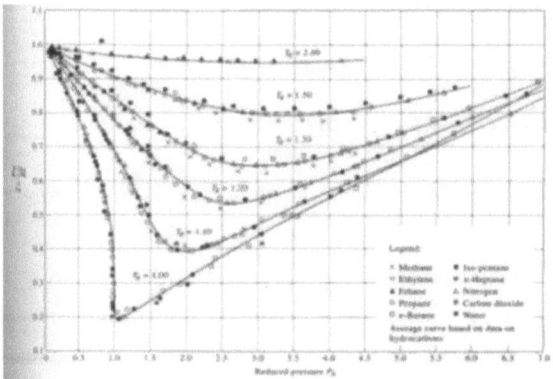

Figure. A comparison of the compressibility factor for various gases as a function of reduced parameters

The region where this error is less than 1% is designated as the ideal gas region. Thus, different gases may behave quite differently at the same temperature and pressure, one acting more like an ideal gas than the other. We find, however, that all gases behave very much the same at temperatures and pressures normalized relative to their critical temperature and pressure. Remember that the critical point of a substance is that temperature and pressure where the saturated vapor and saturated liquid coexist. If we define the reduced temperature and pressure by the equations;

$$T_R = T/T_{cr}$$

and

$$P_R = P/P_{cr}$$

we find that a plot of the compressibility factor as a function of these reduced quantities is essentially independent of the particular gas as shown in Figures above. From this figure we see that the ideal gas law works well (i.e., $Z \approx 1$). For any real gas:

1) for very low pressures ($P_R \gg 1$) regardless of the temperature, and

2) at high temperatures ($T_R \ll 2$) regardless of the pressure.

The greatest deviation from the ideal gas law occurs near the critical point where P_R and/or T_R = 1). Although the ideal gas law is somewhat limited in its application to real gases, we find it useful to postulate the existence of an *ideal gas* that obeys the ideal gas law *for all pressures and temperatures*. This is true primarily, because of the simple relationships between the properties of an ideal gas. We must remember, however, that the ideal gas law is just an approximation to reality.

The Equation of State for a Van Der Waal gas

It is argued that the equation of state for an ideal gas is a good approximation for real gases as long as we are dealing with relatively low reduced pressures. However, as the pressure of a gas is increased at constant temperature, the average distance between molecules decreases and intermolecular forces become more important. The equation of state for an ideal gas, however, implies that the specific volume of the gas would continue to decrease as the pressure increases. But even if the molecules could be modeled as hard spheres (i.e., where there is no interaction between molecules beyond the radius of a molecule), the volume could not decrease beyond a certain limit. Thus, if we considered each molecule to have a volume $\frac{4}{3}\pi a^3$, where 'a' is the radius of the molecule, then the volume of the gas could never become less than R times this volume. Using this argument, we let the *specific volume* occupied by the molecules themselves be given by;

$$b = \frac{V\,mol}{n} = [N \times \frac{4}{3}\pi a^3](\frac{1}{n})$$

which leads to a more accurate equation of state represented by the equation:

$$P(v-b) = RT$$

In this equation the size of the individual molecules is taken into account, and (v-b) represents the total specific volume accessible to all the molecules of the system (the volume of the container less the volume occupied by all the other molecules). In addition to this correction, we know that in most circumstances there really is an attraction between the molecules that make up a gas. This attraction is typically relatively weak (the induced dipole-dipole interaction being a $1/r^6$ force). A correction for this attraction can be incorporated by considering the pressure that the gas exerts on the wall of its container. When a molecule is completely surrounded by other molecules any attraction between the particles would average out to zero.

As a single molecule approaches the wall, however, there are more molecules on one side of this molecule (the side away from the wall) than there are on the other side of this molecule (the side toward the wall). This means that there is a net force on the molecule *away from the wall*. This net force effectively *reduces* the pressure which the gas would exert on the wall of the container. Thus, the pressure exerted on the walls of the container is actually than the pressure would *less* be if there were no intermolecular forces. This leads to the equation

$$P = \frac{RT}{(v-b)} - \delta P$$

where p is the actual pressure on the wall,

$$\frac{RT}{(v-b)}$$

is the pressure that would be exerted by the molecules on the wall if there were no interactions, and δP is the reduction in pressure due to molecular attraction. Now if we assume the interaction between molecules arises due to induced electric dipole moments, the electric force between these dipoles is of the order $1/r^6$ or $1/v^6$. Thus, we expect the change in pressure to be proportional to the inverse square of the specific volume. This means that the last equation can be written as;

$$P = \frac{RT}{(v-b)} - \frac{a}{(v^2)}$$

This form of the equation of state for a gas was first proposed by Van der Waal. In terms of *extensive* variables this equation can be written

$$[P + \frac{a}{(v^2)}](v - b) = RT$$

which is somewhat more difficult to remember but very often used. A diagram of the P_1, v_1, T surface for a Van der Waal gas was shown above. This type behavior is what we expect from a substance which changes phase and does resemble the P_1, v_1, T diagrams shown earlier for real substances. Van der Waal's equation should reduce to the ideal gas equation as the specific volume becomes large. To show this, we expand the equation in inverse powers of the specific volume:

$$P = \frac{RT}{v(1-b/v)} - \frac{a}{(v^2)}$$

$$= \frac{RT}{(v)}\left(1 - \frac{b}{v}\right) - 1 - \frac{a}{(v^2)}$$

or

$$P = \frac{RT}{v}(1 + \frac{b}{(v)} + [\frac{a}{(v)}]^2 + \ldots) - \frac{a}{(v^2)}$$

which can be written in the form

$$P = \frac{RT}{v_1}(1+\frac{RTb-a}{v_2} + [\frac{RTb^2}{v_3}]^2 + \ldots)$$

This last equation is in the form of a *virial expansion* of the equation of state. A virial expansion has the form;

$$P = \sum_i \frac{C_i}{v_i}$$

where the C_i is are the *virial coefficients*, and are typically functions of the temperature. The virial expansion often serves as a very good approximation to real gases, where the C_i's are chosen to give the best fit to the data. Obviously, as the specific volume of the system increases the first term is the most dominant, and, the virial expansion simply reduces to the ideal gas law.

Equations of State for Other Systems

Corresponding equations of state may be determined for liquids and solids. A simple one, valid for many solids at moderate temperatures and pressures is given by

$$V = V_0 (1 + \beta T - k P)$$

where β is the thermal coefficient of expansion and is the isothermal compressibility, respectively defined by the relations:

$$\beta = \frac{1}{V}(\frac{dV}{dT}) \text{ at constant pressure}$$

and

$$k = \frac{1}{V}(\frac{dV}{dP}) \text{ at constant temperature}$$

For solids, both β and k are small quantities $\beta \approx 10^{-6}$, $k \approx 10^{-12}$) but they are not quite constant, varying slightly with temperature and pressure. Similarly, we might be interested in the magnetic moment of a paramagnetic salt, and how this varies with the temperature (or volume, or pressure). A very good approximation for the magnetic moment of a paramagnetic salt valid for small fields and relatively low temperatures is given by the Curie law:

$$M = D(\frac{H}{T})$$

where M at constant temperature is the magnetic moment of the salt, D is the Curie constant, H, is the magnetic field, and T is the temperature. For such a salt, variations in pressure and volume are usually negligible.

System Parameters and Partial Derivatives

We have pointed out that the equation of state for both the ideal gas and for the Van der Waal gas can be represented by a three-dimensional surface upon which are located all the possible equilibrium states of the system in question. For a real system we may or may not know exactly what that surface is like (i.e., we may or may not know a correct mathematical expression which will properly describe *all* the equilibrium states of the system). But we may be able, even in the case where we do not know the equation of state exactly, to determine the characteristics of the P,V,T surface (i.e., to determine the equation of state) in some region around a particular set of values P_o, V_o, T_o. We can do this by determining the slope of the mathematical surface in the region around P_o, V_o, T_o, i.e. by determining the rate of change of one of the variables with respect to the other two in this region. For example, the differential change in the *state variable* T, as a function of differential changes in v and T can be expressed in the following way:

$$dp = (\frac{\partial P}{\partial v})dv + (\frac{\partial P}{\partial T})dP$$

at constant temperature and pressure respectively where the expression, $(\frac{\partial P}{\partial v})dv$ *at constant temperature* is the partial derivative of pressure with respect to volume with the temperature held;

$$dT = (\frac{\partial P}{\partial v})dv + (\frac{\partial P}{\partial T})dP$$

at constant volume and pressure respectively.

Likewise, we can write the change in temperature of the system in terms of changes in the volume and the pressure;

$$dP = (\frac{\partial P}{\partial v})dv + (\frac{\partial P}{\partial T})dP$$

at constant volume and pressure respectively and we can write the change in specific volume of the system in terms of the changes in pressure and temperature:

$$dv = (\frac{\partial P}{\partial v})dv + (\frac{\partial P}{\partial T})dP$$

at constant temperature and pressure respectively. If we look at the last equation carefully, we recognize a term which looks very similar to the thermal coefficient of volume expansion, β, which we defined earlier as;

$$\beta = \frac{1}{v}\left(\frac{dV}{dT}\right) \text{ at constant pressure}$$

. The thermal coefficient of volume expansion β, can be experimentally determined in the laboratory for different values of the pressure and temperature. Thus, even if we may not know a mathematical expression which is valid for all possible values of P, v, and T, we can measure β over different ranges of P, v, and T and form tables or graphs of β vs. P, v, and T, within certain regions of interest. The second partial derivative in the v-equation above is related to the isothermal compressibility, k, which is *also* an experimentally measurable quantity, defined by the equation;

$$k = \left(\frac{1}{v}\right)\left(\frac{\partial v}{\partial P}\right) \text{ at constant temperature.}$$

The minus sign in this equation is due to the fact that the quantity k, and β is inherently negative, and we wish to define, so that it is a positive quantity. Using these two experimentally measurable quantities, we can write

$$dv = -kvdP + \beta vdT$$

or, dividing by v,

$$dv/v = -kdP + \beta dT$$

which would allow us, if we knew the approximate values of k, and β (or their dependence upon P and T), to integrate this equation over a finite change in volume for a finite change in pressure and temperature. This would allow us to determine the approximate equation of state for this particular substance in the region where we know the values of k, and β. We can determinate of the Equation of State when k, and β are known. Let us assume that we find that the constants β and k, are given experimentally by;

$$\beta = 1/X$$

and

$$k = 1/P$$

respectively where β is a function of T only and k, is a function of P only, and these are inverse relationships. Then we can write;

$$dv/v = -kdP + \beta dT$$

which can be integrated between two different states A and B to give

$$\ln v_B - \ln v_A = -[\ln P_B - \ln P_A] + [\ln T_B - \ln T_A]$$

or

$$\ln \left(\frac{v_B}{v_A}\right) = -\ln \left(\frac{P_B}{P_A}\right) + \ln \left(\frac{T_B}{T_A}\right)$$

giving

$$\ln \left(\frac{v_B}{v_A}\right) = -\ln \left(\frac{P_A.T_B}{P_B.T_A}\right)$$

or

$$\frac{v_B}{v_A} = -\frac{P_A.T_B}{P_B.T_A}$$

or finally

$$(PvT)_A = (PvT) = \text{constant}$$

which is the ideal gas law. We have, therefore, shown that if we can determine a functional relationship for k and β we can determine the equation of state of the system.

The Partial Derivatives of State Variables

We showed earlier that changes in the state variables can be written:

$$dT = \left(\frac{\partial P}{\partial v}\right)dv + \left(\frac{\partial P}{\partial T}\right)dP$$

$$dP = \left(\frac{\partial P}{\partial v}\right)dv + \left(\frac{\partial P}{\partial T}\right)dT$$

and

$$dv = \left(\frac{\partial P}{\partial v}\right)dv + \left(\frac{\partial P}{\partial T}\right)dP$$

We will now show that once we determine any of the partials in the equations *two* above, we can determine any of the other four. This means that an experimental measurement of k, and β that are valid in the region of interest are all that is needed to determine the changes in pressure with volume and temperature, or changes in temperature with volume and pressure, etc. It is important to note here that these equations are valid *only* because the variables P v and T are *state* variables. This means that when a system is carried from state E to state F on a P v and T surface and then, along a different path, back to state E, the state variables *will return to their initial values*. That is, the pressure, volume, and temperature measured at state A (P_A, V_A, T_A) do not depend upon how the system came to state A. This statement is not generally true for *all* the quantities of interest in the study of thermodynamics, however.

For example, a gas may do work on its surrounding as it expands. The amount of mechanical work done during such a process is given by the integral

$$W_{A-B} = -\int_A^B PdV$$

But this integral depends upon the particular path taken from state A to state B. Thus, the work done in a given process depends upon the actual path taken, not just on the end points. This is also true for the amount of heat added to, or removed from, a system.

To help us distinguish between changes in state variables and changes in other variables, we will write the differential change in a state variable as dE or dP or dT, but the change in a variable which is *not* a state variable as dW or dQ. Now, let's demonstrate how the different partial derivatives are related to one another. We begin with the equation;

$$dv = (\frac{\partial P}{\partial v})dP + (\frac{\partial P}{\partial T})dT$$

at constant temperature and pressure. We can use the dT equation to substitute into this equation to give;

$$dv = (\frac{\partial P}{\partial v})dv + [(\frac{\partial P}{\partial T})dP]_T + [(\frac{\partial P}{\partial T})dT]_T] + [(\frac{\partial P}{\partial T})dT]_T$$

Since this equation is valid for *any* process on the P,v,T surface, it must be valid for an isothermal process in which $T = 0$. This would mean that the left hand side of this last equation must be identically zero, or;

$$1 - [(\frac{\partial v}{\partial T})_T + (\frac{\partial P}{\partial T})_T = 0$$

since the volume change in a given process cannot be zero at the same time that the temperature change is zero. If *both* v and T were zero, the state could not change at all. From the ideal gas equation, if T is held constant a change in state of the system means that both T and v must change by some amount. If either T or v is held constant in an isothermal process, the other must also remain constant, so that the state of the system does not change. This last equation gives us the so-called "reciprocal rule" for partial derivatives:

$$1 - [(\frac{\partial v}{\partial T})_T = 1/[(\frac{\partial P}{\partial T})_T]$$

Similarly, if we look back at Eq. and follow an isochoric process (i.e., one in which $dv = 0$), we have

$$(\frac{\partial v}{\partial T})_T + (\frac{\partial v}{\partial P})_T + (\frac{\partial p}{\partial T})_T = 0$$

or

$$(\frac{\partial v}{\partial T})_T + (\frac{\partial v}{\partial P})_T = - (\frac{\partial p}{\partial T})_T$$

and, using the reciprocal rule demonstrated above, the right hand term can be "inverted" and multiplied so that the equation becomes the so-called "cyclic rule" for partial derivatives:

$$(\frac{\partial v}{\partial T})_T + (\frac{\partial v}{\partial P})_T + (\frac{\partial p}{\partial T})_T = -1$$

In the equation above, we notice that the denominator must match the adjacent numerator in the terms on the left side of the equation and that *the actual ordering of the terms is irrelevant*. Using the two equations above, one can show that if we know just two of the partial derivatives, we can obtain all the other partial derivatives. We assume that we can determine the isothermal compressibility and the volume coefficient of thermal expansion:

$$\beta = \frac{1}{V}(\frac{dV}{dT}) \text{ at constant pressure}$$

and

$$k = \frac{1}{V}(\frac{dV}{dP}) \text{ at constant temperature}$$

If we want a general equation for dT in terms of the changes in temperature and volume, we can write:

$$dP = (\frac{\partial P}{\partial v})dv + (\frac{\partial P}{\partial T})dT$$

From the reciprocal rule, the first term on the right can be expressed as

$$(\frac{\partial P}{\partial v})dv = 1/(\frac{\partial v}{\partial P})dT$$
$$= -1/kv,$$

The second term on the right can be written in a slightly different form by using the cyclic relationship

$$(\frac{\partial v}{\partial P})_T + (\frac{\partial T}{\partial v})_P + (\frac{\partial p}{\partial T})_v = -1$$

which gives;

$$(\frac{\partial p}{\partial T})_v = - (\frac{\partial T}{\partial v})_P / (\frac{\partial v}{\partial P})_T$$

Thus, the partial of pressure with respect to temperature at constant volume is related to the isothermal compressibility, and to the thermal coefficient of expansion at constant pressure " by the equation;

$$(\frac{\partial p}{\partial T})_v = - \beta v / [-kv] = \beta/k$$

We can therefore express the change in pressure of a system in the following way:

213

$$dP = \frac{1}{v}dv + \frac{\beta}{k}dT$$

$$(P+\frac{a}{v^2})(v-b) = RT$$

from which we can calculate the change in pressure of the system in going from state A to state B if we know the difference in the specific volume and in the temperature of these two states, and if we know the functional dependence of " and , on the temperature and on the specific volume. Let's try to evaluate β for a Van der Waal's gas. (We showed earlier that β for an *ideal* gas was given by β = 1/T.) Again, the defining equation for β is;

$$\beta = \frac{1}{V}\left(\frac{dV}{dT}\right) \text{ at constant pressure}$$

Now the Van der Waal's equation is given by

$$(P+\frac{a}{v^2})(v-b) = RT$$

which is a cubic in v so that the evaluation is non-trivial. However, we can express the coefficient of thermal expansion, using the reciprocity and cyclic rules in terms of other partials which are more easily evaluated. Thus, we want to express v in T and using the cyclic relation v;

$$\left(\frac{\partial v}{\partial P}\right)_T \left(\frac{\partial T}{\partial v}\right)_P \left(\frac{\partial p}{\partial T}\right)_v = -1$$

we can write;

$$\left(\frac{\partial T}{\partial v}\right)_P = \left(\frac{\partial p}{\partial T}\right)_v / \left(\frac{\partial v}{\partial P}\right)_T$$

We can fairly easily solve Van der Waal's equation for the pressure T, to get

$$(P+\frac{a}{v^2})(v-b) = RT$$

$$P = \frac{RT}{(v-b)} - \frac{a}{v^2}$$

The partials of the pressure are

$$\left(\frac{\partial p}{\partial v}\right)_T = \frac{R}{(v-b)}$$

Dividing this last equation by the former one, we obtain, upon simplification,

$$\left(\frac{\partial p}{\partial v}\right)_T = \frac{RT}{(v-b)2} + \frac{2a}{(v3)}$$

for a Van der Waal's gas. You should verify that this expression reduces to the expected result as;

$$\beta = \frac{1}{V}\left(\frac{dV}{dT}\right) = \frac{Rv3\,(v-b)}{(RT\,v3-2a(v-b))2}$$

Thermodynamics

The term dynamics implies that motion or change with time and thus the word *thermodynamics* refers to the changes in properties with temperature. If an object changes from one thermal equilibrium state to another, thermodynamic principles can be applied to describe the change in the object's properties. Thermodynamics say nothing about the rate of change. Processes which occur at a non-zero rate between two objects not in thermal equilibrium are the subject of irreversible thermodynamics.

Electrical current as depicted by Ohm's law is proportional to the electrical force. The proportionality coefficient in this case is called the electrical conductivity which is a linear response. This means that one quantity changes linearly in response to a change of another quantity. Linear response approximation of the thermoelectric properties defines simple equations similar to Ohm's Law as shown in table below.

Thermoelectric Property	Definition	Under Condition	Type
Electrical Conductivity	$i = \sigma E$	$\nabla T = 0$	Direct
Thermal Conductivity	$Q = -\lambda \nabla T$	$i = 0$	Direct
Seebeck Coefficient	$E = \alpha \nabla T$	$i = 0$	Cross
Peltier Coefficient	$Q = \Pi i$	$\nabla T = 0$	Cross

Considering the parameters given in the table above, the first relation connects the electrical current to the electrical force while the second relation connects the thermal current to the thermal force. The electrical and thermal conductivities are therefore called direct effects. The electrical conductivity indicates how well a material conducts electricity and the thermal conductivity indicates how well a material conducts heat.

The Seebeck and Peltier coefficients are called cross effects. This is because they connect an electrical response to a thermal force or a thermal current to an electrical current. Cross effects form the basis for utilizing thermoelectric materials for energy conversion applications. On the other hand the Seebeck coefficient indicates how large a voltage generates a temperature gradient while the Peltier coefficient indicates how much heat passes through a material for a given current. Linear response coefficients could be defined under conditions of zero heat current or zero electrical gradients. In order to include the cross effects into the currents under arbitrary gradients, we need to add the effects together as;

$$i = \sigma(E - \alpha \nabla T)$$

$$Q = \Pi i - \lambda \nabla T$$

These expressions are perfectly correct and often convenient to use, but they are not symmetrical and in order to treat everything equally it may be preferred to re-write the expressions as:

$$i = \sigma E + \sigma \alpha (-\nabla T)$$
$$Q = \sigma \Pi E + \lambda (-\nabla T)$$

The expressions above very ideal represent a generalization of Ohm's Law. Lord Kelvin suggested that Peltier coefficient and Seebeck coefficient have a definite relationship given by;

$$S = PT$$

It is not till 1931 that Onsager derived a relationship using a technique based on thermal fluctuations and the result showed that Seebeck and Peltier effects are not independent effects, but manifestations of the same thing.

The Thompson Coefficient

A third effect called the Thompson effect asserts that when an electrical current flows through a material subject to a temperature gradient, heat is generated at a rate proportional to the electrical current which equally proportional to the temperature gradient, giving;

$$\dot{Q}_{Thompson} = \tau i (-\nabla T)$$

where is the Thompson coefficient. Therefore the total rate of heat generation within the material is then given by the sum of three terms:

$$i^2 \rho$$

which form the Joule heating effect.

$$-\nabla [\lambda (-\nabla T)]$$

which forms the rate that heat is conducted into the material.

$$\tau(-\nabla T)$$

which is the Thompson heat. The total heat is thus given as;

$$\dot{Q}_{Total} = i^2\rho - \nabla[\lambda(-\nabla T)] + \tau(-\nabla T)$$

Thompson showed that the coefficient was related to the temperature dependence of the Seebeck coefficient by;

$$\tau = T\frac{d\alpha}{dT}$$

Thompson effect represents the heat generated or absorbed due to the Peltier heat changes with temperature. In high speed semiconductor device design, analysis ignores temperature dependence of the Seebeck coefficient. Modern analysis techniques such as finite-element thermal models explicitly take into account temperature dependence of transport coefficients by adding the Thompson term to cause a double-counting effect in the heat balance.

Thermoelectric Theory of Solids

Virtually all materials exhibit currents which respond linearly to applied forces. For *superconducting* materials, electrical currents flow with no driving force at all and that is why applying Ohm's law fails. Superconducting materials are entirely beyond the scope of this book. A solid material is made up of a collection of atoms and ideal theories take only the geometrical arrangement and the type of atoms to predict all other important properties. Modern solid state theories are capable of doing exactly this. However, in some few special cases the first-principle calculations are used accurately, though these calculations are far too complex to perform reliably.

Lattice and Phonons

The main features of a lattice are well illustrated using a simple mass and spring model shown below. As from the figure below, the undisturbed lattices the atoms are regularly spaced corresponding to a unit cell repeat distances. The atoms vibrate about their equilibrium positions due to thermal agitation and practically, this motion is not entirely random since the movement of one atom stretches or compresses the springs connecting it to neighboring atoms. Vibrations propagate throughout the crystal and thus rather than describing the vibrations of the each individual atom, the sinusoidal disturbances of an entire groups of atoms, suggest middle and lower portions as having a sinusoidal disturbance is called a *phonon*. Phonon means particle of sound. Sound is an elastic wave of compression and extension which propagates through a solid. This is where the atoms are represented as point masses while bonding between the atoms are represented by very small flexible springs as;

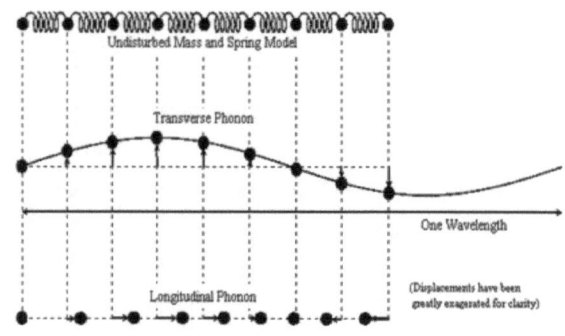

Mass and Spring Model for a One-Dimensional Crystal of Ions

A single phonon in it would exhibit regular patterns of atomic displacements and thus using mathematical techniques identical to ordinary Fourier analysis, any configuration of atomic displacements can be represented accurately by the summation of many sinusoidal disturbances. Since a collection of phonons can represent *any possible* configuration of disturbances from equilibrium, and since individual phonons represent particularly simple motions which actually do occur in solids, the phonon description has become the most common tool for describing the properties of solids.

In Quantum Mechanics a wave carries a momentum given by h/L, where h is Plank's constant as described by de Broglie. This momentum is not the same as that with a phonon velocity hence we have a phonon wavenumber defined by $2\pi/L$. Since there is a minimum allowed wavelength, there is also a maximum allowed momentum (h/a) and a maximum allowed wavenumber ($2\pi/a$). A crystal with a phonon in it must have a greater energy than a crystal without any phonons. When the bonds are being stretched and atoms moved, both kinetic and potential energy associated with each phonon vary such that the energy associated with a single phonon is very small, representing only a fraction of an electron volt. There may be several different types of atomic motion allowed for a given wavelength. Each type of allowed motion has a different amount of energy associated with it and travels through the crystal at a velocity characteristic of that type of phonon.

Phonon Dispersion

As shown in the earlier figure above, the masses and distances are very small. It is difficult to accurately calculate the strength of the springs which we regard as chemical bonds. The energies of the allowed phonons vary with the direction the phonon moves through the crystal and the momentum of the phonon. Individual phonons are studied using neutron scattering techniques. Experimental energy and momentum relationships are referred to as the phonon dispersion relation. The two most common models for how phonons move are the Debye Model and the Einstein Model. Debye model phonons have the same speed with an energy that is directly proportional to the wavenumber. It is also true that the speed of these phonons is the same as the speed of sound through a solid hence the term acoustic phonon is associated with this type of phonon to remind you that the acoustic (and elastic) properties of solids are associated with this type of vibration.

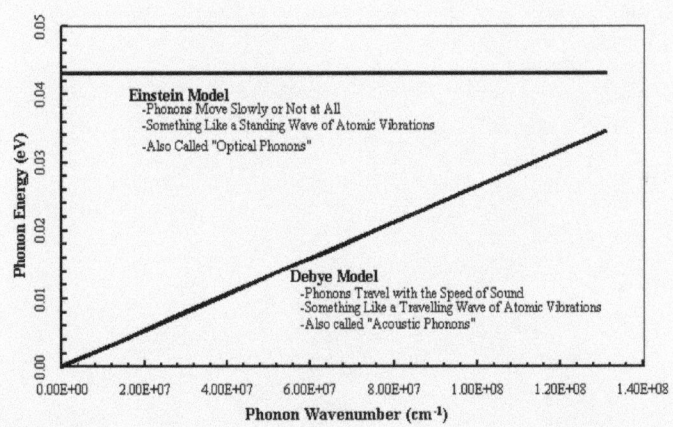

Einstein model considers phonons as standing still and only oscillates back and forth with the vibration wave crest immobile through the crystal. In ideal model behaviour, all phonons have exactly the same energy and such a phonon is called an optical phonon. This is because in ionic crystals the standing wave vibrations of charged ions represent an oscillating electrical dipole which can interact with electromagnetic radiation.

CHAPTER SIX
CARRIER SCATTERING

Phonon or lattice vibration scattering, Ionized or dopant or impurity scattering, Scattering by neutral impurity atoms and defects, Carrier-Carrier scattering and Piezo-electric scattering are the major concern of all semiconductor researchers. These are the common forms of scattering experienced in semiconductors. In all these types of scattering, Phonon or lattice vibration scattering and Ionized or dopant or impurity scattering predominates and therefore in most high speed devices where large carrier concentrations area not involved, these scattering must be minimized. When the mobility due to each scattering mechanism is taken as μ_i, the overall mobility that comprises Phonon or lattice vibration scattering and Ionized or dopant or impurity scattering is given as

$$\mu^{-1} = \sum_i \left(\mu_i^{-1}\right)$$

Individually, Phonon or lattice vibration scattering is given by Where C_{11} is the average longitudinal elastic constant of semiconductor, m* is the effectives mass of the carrier, E_{ds} is the displacement of edge of the band per unit dilation of lattice, and T is the absolute temperature. For ionized dopant or impurity density, we have;

$$\mu_I = \frac{64\sqrt{\pi}\epsilon_s^2 (2kT)^{3/2}}{N_I q^3 (m)^{3/2}} \left\{ \ln\left[1 + \left(\frac{12\pi \epsilon_s kT}{q^2 N_I^{1/3}}\right)\right]\right\}^{-1}$$

Where, N_1 is the ionized dopant or impurity density on scattering, ϵ_s is the permittivity and hence the therefore overall mobility for GaAs is given by;

$$\mu = \left(\frac{1}{\mu_L} + \frac{1}{\mu_I}\right)^{-1} \sim (m^*)^{-3/2} T^{1/2}$$

Mobility

Mobility then varies as;

$$\mu = \mu_{min} + \frac{\mu_0}{1 + \left(\frac{N}{N_{ref}}\right)^\alpha}$$

Considering Si as a semiconductor where N is the dopant concentration, then N_A or N_D i.e. at room temperature where, μ_1, N_{ref}, α and μ_0 are determined experimentally. Thus for parameters of Si, we have;

N_A or $N_D (cm^{-3})$ for Si	$\mu_n\ cm^2/(V-sec)$ at 300K	$\mu_p\ cm^2/(V-sec)$ at 300K
10^{14}	1358	461
5×10^{14}	1352	459
10^{15}	1345	458
5×10^{15}	1298	448
10^{16}	1248	437
5×10^{16}	986	378
10^{17}	801	331

Thus for parameters of GaAs we have:

$\frac{N_A}{N_D}$	μ_{min}	μ_0	α	N_{ref} (electrons)
0.4	1100	7100	0.542	5.09×10^{16}
0.8	200	8000	0.551	9.85×10^{15}
0.9	100	8100	0.594	4.02×10^{15}

Again from experiment we find μ_{min}, N_{ref}, α and μ_0 all have a temperature dependence of the form;

$$A = A_o \left(\frac{T}{300}\right)^\beta$$

For Si we have:

	Electrons	Holes	β
N_{ref} (cm^{-3})	1.3×10^{17}	2.35×10^{17}	2.4
μ_{min}	92	54.3	-0.57
μ_o	1268	406.9	-2.33(electrons), -2.23(holes)
α	0.91	0.88	0.146

When the values of Si are compared with GaAs, it is observed that GaAs behavior slightly different.

Field effect due to Scattering

In semiconductors, the drift velocity at is linear proportionality at low temperatures and is given by;

$$v_d = \mu$$

In such a condition, E field intensity greater than 1.5×10^3 V/cm v_d and E are no longer directly proportional and thus a nonlinear behavior is seen. Geometry is often at very small (submicron) and so even with 1 V across, 0.5μm dimension, $E = 1 + \frac{1}{2}$ μm = 2×10^4 V/cm. So mobility concept may fail in such high field zones. For very a very high E field, v_d saturates.

$$v_{d_{sat}} = \frac{v^o_{d_{sat}}}{1 + A e^{T/T_d}}$$

Inserting this formula to the Si at 300 K, gives $V_{d\ sat} = 10^7$ cm/s, at A = 0.8 and $T_d = 600$K for both electrons and holes. While for GaAs, V_d initially decreases with E after a critical field is 2×10^3 v/cm and then very slowly increases, In a region of high, E greater than 10^5 V/cm which can be considered as saturated velocity, $V_{d\ sat}$. However, in GaAs for holes the velocity saturates and J_p for drift = $qpvV_{dsat}$ Independent of E field. The variation of the drift velocity with electric field for holes and electrons are shown below.

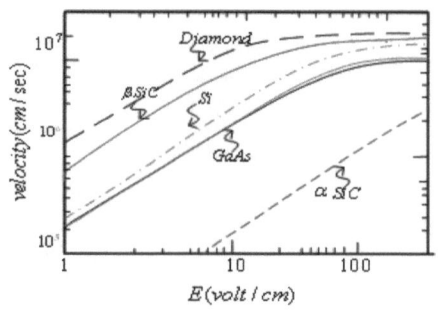

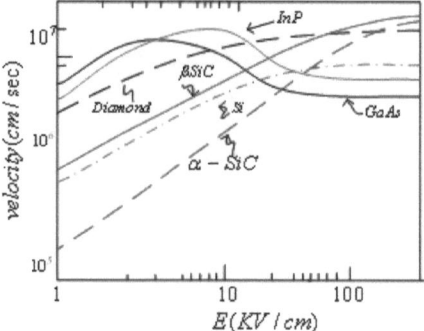

Electron Scattering

The conduction band minimum of GaAs is at **r** (k = 0) in which its secondary minimum of conduction band is at L (<111>). The L valley is sparsely populated at room temperature and with the E field, **r** valley electrons gain energy between scattering events reducing with temperature. Thus if these valley electron gain energy exceeds 0.29eV, the inter-valley transfer becomes possible and the population at L valley becomes enhanced at the expense of **r** valley and at the center of the **r** valley mass, $m^* \equiv 0.06m_0$ while at L valley mass, $m^* \equiv 0.06m_0$. It is thus noted that the effective mass increases in the L valley mobility while the drift velocity decreases in L valley as mobility varies as $\mu_n (m^*)^{-3/2}$. Given that the lower valley mobility, μ_1, population n_1; the upper valley mobility, n_2, then the total electron valley will be;

$$n = n_1 + n_2$$

In the E field that ranges from

E field $< E_{th}$, $n_1 \gg n_2$, also $\mu_1 > \mu_2$

This will give total conductivity, σ, as;

$$\sigma = q(\mu_1 n_1 + \mu_2 n_2)$$

If the above expressions could be arrange and re-arrange, the average mobility will be given by;

$$\bar{\mu} = \frac{\mu_1 \mu_1}{n_1 + n_2} = f_1$$

Where,

$$\nu = \bar{\mu} E$$

And

$$J = q \bar{\mu}_n E$$

are the strong function of E in the range of

$$0 < E < E_{th} \quad n_1 \cong n, \quad n_2 \cong 0$$

$$E > E_{th} \quad n_2 \cong n, \quad n_1 \cong 0$$

Carrie Density

It is difficult to express energies values E_1 and E_2 from the band theory. A good approximation can only be made from the band edges and these are the regions of bands normally populated by carriers. Between these energies E_1 and E_2 the number of allowed states available to electron or holes in the cited energy range per unit volume of the crystal can then be calculated.

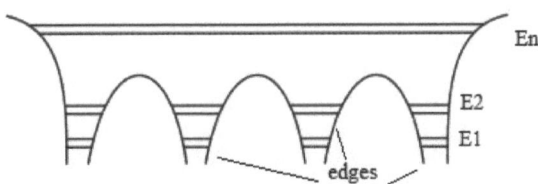

For electrons near the bottom of C Band the band forms a pseudo-potential well. The bottom lies at E c and termination of the band at the crystal surface forms the walls of the well. The energy of electrons is relatively small compared with surface barriers. One can think being in a 3 dimensional box. The density of states at the band edges ≅ density of states available to a particle of mass m* in a box with dimensions of the potential box.

Schrödinger equation

In a new approach, consider a particle of mass a *m* and total energy *E* in box of dimensions **a** by **b** by **c** as shown below and aligned in the x, y, z directions;

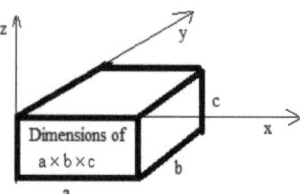

At the start of analysis, it is assumed that the potential in the box, U(x, y, z) is constant everywhere and equal to zero (U(x, y, z) = 0. Then the Time independent Schrödinger equation can be expressed as;

$$\frac{\partial^2 \psi}{\partial x^2} + \frac{\partial^2 \psi}{\partial y^2} + \frac{\partial^2 \psi}{\partial z^2} + k^2 \psi = 0$$

So that

$$k = \sqrt{\frac{2mE}{\hbar}} \quad \text{or} \quad E = \frac{\hbar^2 k^2}{2m}$$

When this expression is solved using the separation of variables approach as;

$$\psi(x, y, z) = \psi_x(x)\psi_y(y)\psi_z(z)$$

If this solution is used in the Time independent Schrödinger above, then;

$$\frac{1}{\psi_{x,y,z}} \left| \frac{\partial^2 \psi_x}{\partial x^2} + \frac{\partial^2 \psi_y}{\partial y^2} + \frac{\partial^2 \psi_z}{\partial z^2} \right| + k^2 = 0$$

Now since **k** is a constant in the above expression, then;

$$\frac{1}{\psi_x}\frac{d^2\psi_x}{dx^2} = -k_x^2$$

$$\frac{1}{\psi_y}\frac{d^2\psi_y}{dy^2} = -k_y^2$$

$$\frac{1}{\psi_z}\frac{d^2\psi_z}{dz^2} = -k_z^2$$

such that the value of k^2x, k^2y and k^2z to k^2 is expressed as;

$$k_x^2 + k_y^2 + k_z^2 = k^2$$

In one dimensional analysis we have;

$$\frac{d^2\psi}{dx^2} + k_x^2 = 0 \qquad 0 < x < a$$

And this gives the value of ψ_x as

$$\rightarrow \psi_x = A_x \sin k_x x$$

Hence, $\psi_x = 0$ for $x = 0$, $x = a$ and thus;

$$k_x = \frac{n_x \pi}{a} \qquad n_x = \pm 1, \pm 2, \ldots$$

In a similarly way, the values of ky, kz are found as;

$$k_y = \frac{n_y \pi}{b} \qquad k_z = \frac{n_z \pi}{c}, \quad n_y, n_z = \pm 1, \pm 1, \pm 3 \ldots$$

Where n is the number of modes in a μ wave resonator. Finally a generalized wave function will be given as;

$$\psi(x,y,z) = A \sin \frac{n_x \pi}{a} \sin \frac{n_y \pi}{b} \sin \frac{n_z \pi}{c}$$

And based on this wave function, we get it reduced to;

$$\frac{n_x^2}{a^2} + \frac{n_y^2}{b^2} + \frac{n_z^2}{c^2} = \frac{2mE}{\hbar^2 \pi^2}$$

Where, n_x, n_y and n_z area integers and therefore it shows that only a few discrete energy values are allowed inside.

Allowed energy levels

If the dimensions, **a, b** and **c** are very large then there are small increments from k_1 to k_2 and that is why there is a large number of states allowed in the **k** space and draw as shown.

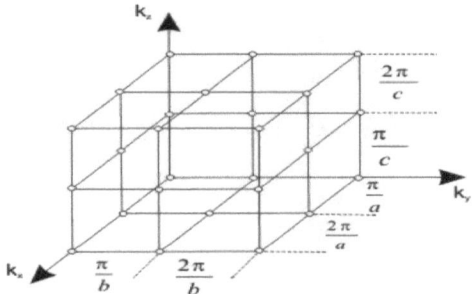

It therefore easy to calculate the k space vector as;

$$\bar{k} = \left(\frac{n_x \pi}{a}\right)\hat{u}_x + \left(\frac{n_y \pi}{b}\right)\hat{u}_y + \left(\frac{n_z \pi}{c}\right)\hat{u}_z$$

It is noted that the end points of all such vectors are dots and the k space unit cell of volume which contains one allowed solution is given by;

$$\frac{\pi^3}{abc}$$

If one counts all points on the k space it is thus necessary to divide by eight to obtain the number of independent solutions and thus given as;

$$\frac{No\ of\ allowed\ solutions}{Unit\ volume\ in\ k\ space} = \frac{abc}{8} \times 1\pi^3$$

From quantum mechanics and Bohr theory atomic, electrons have 2 spin states and therefore;

$$\frac{Allowed\ energy\ states}{unit\ volume\ in\ k\ space} = \frac{abc}{4\pi^3}$$

Now number of states with **k** value between arbitrary k and $k+dk$ is equal to;

$$4\pi k^2 (dk) \times \frac{abc}{4\pi^3} = \frac{k^2\ dk\ abc}{\pi^2}$$

So as to give the value of E as;

$$E = \frac{\hbar^2 k^2}{2m} \rightarrow k^2 = \frac{2mE}{\hbar^2}$$

The change in dE will be given by;

$$dE = \frac{\hbar^2 k \, dk}{m}$$

And will result into a **dk** value of;

$$dk = \frac{1}{\hbar}\sqrt{\frac{m}{2}} \frac{dE}{\sqrt{E}}$$

Finally when going from k to **E** space, the number of electron energy states with energy between **E** and **E+dE** will be given by;

$$\frac{abc}{\pi^2}\left(\frac{2mE}{\hbar^2}\right) \cdot \frac{1}{\hbar}\sqrt{\frac{m}{2}} \frac{dE}{\sqrt{E}}$$

$$= \frac{abc}{\pi^2}\left(\frac{m\sqrt{2mE}}{\hbar^3}\right) dE$$

And the density of energy states with energy difference between **E** and **E+dE** will be given by;

$$= \frac{\text{No. of states}}{(\text{volume})(dE\text{ range})} = g(E) = \frac{m\sqrt{2mE}}{\pi^2 \hbar^3}$$

The actual Conduction Band (CB) or Valence Band (VB) densities of state near band edges mass tends to be equal to effective mass, m* in the range when, E_c = min of CB energy and E_v = max VB energy and thus;

$$E > E_c \qquad g_c(E) = \frac{m_n^*\sqrt{2m_n^*(E - E_c)}}{\pi^2 \hbar^3} \qquad \text{CB} \qquad m_n^* \text{ for electrons}$$

And also,

$$E > E_v \qquad g_v(E) = \frac{m_p^*\sqrt{2m_p^*(E_v - E)}}{\pi^2 \hbar^3} \qquad \text{VB} \qquad m_p^* \text{ for holes}$$

When m^*_n is the average effective mass, (i.e. $m^*_n = m^*$) and for GaAs or compound semiconductors of Si or Ge m is complicated this calculation does not apply..

Effect of Fermi Dirac statistics in scattering

In Fermi Dirac statistics, it is assumed that not all allowed states are filled and electrons are indistinguishable in which according to Pauli Exclusion Principle, each state is occupied by one electron. This means that the total number of electrons is equal, fixed and given by;

$$N = \sum_i N_i$$

And this gives the total energy, $E_{total} = \sum_i N_I$ and is always constant energy. Therefore the electrons are viewed as indistinguishable "balls" which are placed in allowed state "boxes" in which each box behaves as single one ball. Hence the total energy of system is fixed and the balls mentioned above are grouped in rows according to their energy level. Thus the numbers of boxes in each energy level are related to the number of states of allowed electronic states at a given energy. Therefore the Fermi function is given by;

$$f(E_i) = \frac{N_i}{S_i} = = \frac{1}{1+e^{\alpha+\beta E_i}}$$

where $\alpha = \frac{E_F}{kT}$, $\beta = \frac{1}{kT}$, E_F = Fermi energy, k = Boltzmann's constant = 8.614 x 10^{-5} eV/K and therefore, for closely spaced energy levels, E_i becomes continuous variable E as shown in the figure 6.3 below.

$$f(E_i) = \frac{1}{1+exp[E-\frac{E_F}{kT}]}$$

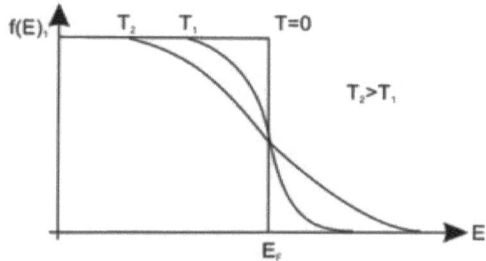

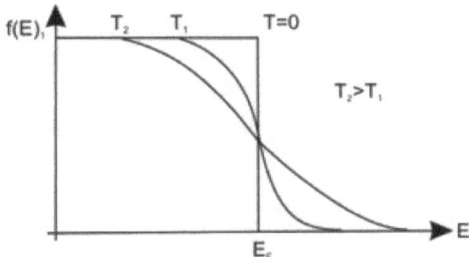

In this figure above, f(E) is the occupancy factor for electrons at energy E, g(E) is the density of states at energy E and 1-f(E) is the occupancy factor for holes at energy E.

Distribution of Electron due to scattering

The number of electrons is the CB with energy between **E** and **E+dE** is given by (E > E_C) and thus the distribution of electrons in conduction band, is given by;

$$= g_C(E)f(E)$$

And based on this distribution, the distribution of holes in the valence band is given by;

$$g_v(E)(1 - f(E))$$

The total carrier concentration in a band conduction will then be given by;

$$n = \int_{E_C}^{E_{top}} g_C(E)f(E)dE$$

The total no. of holes in valence band,

$$p = \int_{E_{bottom}}^{E_v} g_v(E)[1 - f(E)]dE$$

In a case where the E_{top} for the CB as it tends to infinity (∞) and E_{top} for the VB infinity ($-\infty$). For electronics, we have;

$$\rightarrow n = N_c \Im_{1/2}(\eta_c)$$

And for holes, we have;

$$\rightarrow p = N_v \Im_{1/2}(\eta_v)$$

Thus in the conduction and valence bands, the density of state will respectively be given by;

$$N_C = 2\left(\frac{2\pi m_n^* kT}{\hbar^2}\right)^{3/2} \quad N_v = 2\left(\frac{2\pi m_p^* kT}{\hbar^2}\right)^{3/2}$$

This is based on the Fermi Dirac integral of order ½ in which N_c is the effective density of conduction band states and N_v is the effective density of valence states as;

$$\Im_{1/2} = \frac{2F_{1/2}(\eta)}{\sqrt{\pi}} \quad ; \quad F_{1/2}(\eta) = \int_0^\infty \frac{\sqrt{\xi}d\xi}{1+\exp(\xi-\eta)}$$

And this reduces to give;

$$\eta_c = \frac{(E_F - E_C)}{kT} \quad \eta_v = \frac{(E_v - E_F)}{kT}$$

At room temperature (300k) we have:

Semiconductor	N_C (cm^{-3})	N_v (cm^{-3})
GaAs	4.21×10^{17}	9.52×10^{18}
Si	3.22×10^{19}	1.83×10^{19}
Ge	1.03×10^{19}	5.35×10^{18}

It can be noted also that the expression function below;

$$\Im_j(\eta) = \frac{1}{\Gamma(j+1)} \int_0^\infty \frac{\xi^j d\xi}{1+\exp(\xi-\eta)}$$

Forms the gamma function with significant conditions which include;

$$\Im_j(\eta) \rightarrow e^\eta \text{ as } \eta \rightarrow -\infty$$

which can be equalized as;

$$\frac{d\Im_j(\eta)}{d\eta} = \Im_{j-1}(\eta)$$

And hence;

$$\Im_{1/2}(\eta) \cong \left[e^{-\eta} + \xi(\eta)\right]^{-1}$$

So that the value can be obtained from the expression given by;

$$\xi(\eta) = 3\sqrt{\frac{\pi}{2}}\left[(\eta + 2.13) + \left(|\eta - 2.3|^{2.4} + 9.6\right)^{\frac{5}{12}}\right]^{-\frac{3}{2}}$$

This value has a max error of about 0.5 and therefore for n ≤ -3;

$$\Im_{1/2}(\eta) \cong e^{-\eta}$$

When further analysis is carried and closely approximated by exp (η) and thus

$$n \cong N_c e^{\frac{(E_F - E_c)}{kT}} \qquad E_c - E_F \geq 3kT \quad (\eta_c \leq -3)$$

$$p \cong N_r e^{\frac{E_v - E_F}{kT}} \qquad E_F - E_v \geq 3kT \quad (\eta_v \leq -3)$$

And at the Fermi level lie the band gap more than 3kT from both band edge and therefore the semiconductor is said to be non-degenerate as depicted in the figure below;.

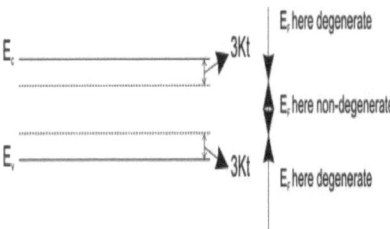

$$f(E) \simeq \exp\left[-\frac{(E - E_F)}{kT}\right]$$

Or it can be represented by the expression;

$$1 - f(E) \simeq \exp\left[+\frac{(E - E_F)}{kT}\right]$$

Maxwell-Boltzmann Approximation in carrier scattering

A simplified form obtained by further simplification gave rise to Maxwell-Boltzmann energy distribution (distribution at high temp for molecules in a low density gas). Thus in the intrinsic semiconductor, we have the holes and electrons relating as;

$$n = p = n_i, E_F = E_i$$

And thus the electronics and holes and their effects on the density of states in the conduction and valence bands can be given as;

$$n_i = N_c e^{(E_i - E_c)/kT}$$
$$= N_v e^{(E_v - E_i)/kT}$$
$$N_c = n_i e^{(E_c - E_i)/kT}$$
$$N_v = n_i e^{(E_i - E_v)/kT}$$

However, we can write the above expression for electronics and holes as;

$$n = n_i e^{(E_F - E_v)/kT}$$
$$p = n_i e^{(E_i - E_F)/kT}$$

And from these relations we have;

$$n_p = n_i^2$$

$$n_i = \sqrt{N_c N_v} e^{-E_g/2kT}$$

Where E_g is the Band-gap energy. It is the plot of allowed electron energy states as a function of position along a direction

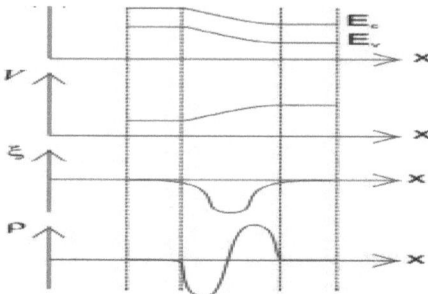

The carrier distribution against energy, E, for electrons can be graphed by multiplying g(E) and f(E) as shown below. From this graph, the carrier distribution against holes can be graphed by multiplying $g_v(E)$ and (1-f(E)). From the figure, when we have an electric field energy band diagram, bands bend with x so that;

$$E = \frac{1}{q}\frac{dE}{dx}$$

The potential energy will then be given by;

$$= -qV(x) = E_c - E_{ref}$$

Therefore,

$$V(x) = -\frac{1}{q}(E_c(x) - E_{ref})$$

$$E = -\frac{dv}{dx} \rightarrow \frac{1}{q}\frac{dE_c}{dx} = \frac{1}{q}\frac{dE_v}{dx}$$

Kinetic energy is therefore given as (E- E_c) in conduction band and (E_v –E) in valence band respectively.

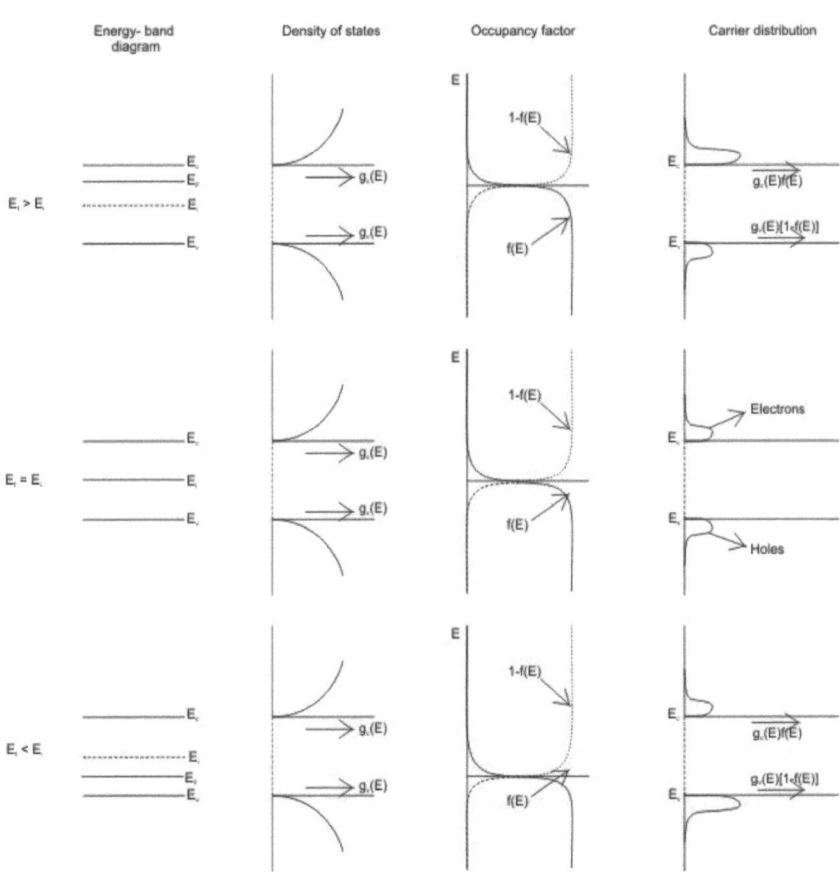

Change Neutrality Equation

Maxwell's equation where k_s is the semiconductor dielectric constant that is given as;

$$\nabla . E = \frac{\rho}{\epsilon \, o k_s}$$

Extrinsic Doping and Carrier Scattering

In material science and especially in thin films, impurities' are incorporated into a crystal to either contribute extra electrons or could accept electrons and this process is called doping. In the former, after the contribution of the electron or electrons the impurity itself becomes positively charged as it was neutral to start with and into concentration is called N_D^+. In the latter case the impurity after accepting electron/s becomes negatively charged and its concentration is called N_A^-. Assuming uniform doping under equilibrium, donors produce positive ions while acceptors produce negative ions as;

$$\rho = q(p - n + N_D^+ - N_A^-)$$
$$E = 0 \rightarrow \rho = 0$$
$$\rightarrow (p - n + N_D^+ - N_A^-) = 0,$$

This is the charge neutrality relation.

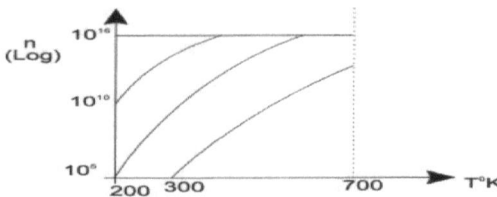

From the above expressions, the relationship for N_D^+ and N_A^+ are easy to express. If we let N_D to be the number of donor atoms while N_D^+ be the number of donor atoms that are ionized, then the ration of N_D^+ / N_D is the ratio of number of empty states to total number of states in donor energy and can therefore be expressed in a standard value by;

$$\frac{N_D^+}{N_D} = \frac{1}{1 + g_D \exp\left[\frac{E_F - E_D}{kT}\right]_g}$$

Where D, is taken to be equal to 2, D = 2. Similarly, if we let N_A to be the number of donor atoms while N^+_A be the number of donor atoms that are ionized, then the ration of N^+_A / N_D is the ratio of number of empty states to total number of states in donor energy and can therefore be expressed in a standard value by;

$$\frac{N^-_A}{N_A} = \frac{1}{1+g_A \exp\left[\frac{E_A - E_F}{kT}\right]}$$

Where A = 2. In the above expressions, g_D and g_A are the degeneracy factors. Thus from charge neutrality conditions, we have;

$$p - n + N^+_p - N^-_A = 0$$

And this ensures that the following expression is justified as;

$$\rightarrow N_v e^{(E_v - E_F)/kT} - N_c e^{(E_F - E_v)/kT} + \frac{N_D}{1+g_D e^{(E_F - E_D)/kT}} - \frac{N_A}{1+g_A e^{(E_A - E_F)/kT}} = 0$$

In the above equation, solving this equation so as to obtain the values of N_V, N_c, N_A, E_c, E_v, T, E_D, E_A then we can easily find the Fermi energy level, E_F and other parameters.

Free-out/Extrinsic, T

Consider a case where the values of N_D is much greater than that of N_A (i.e. $N_D >> N_A$) so as to have N_D much greater than *n* (i.e. $N_D >> n$). This will imply that the electron concentration is likewise much greater than the hole concentration. Thus as;

$$N^+_D \cong n$$

Will result into a computable constant at a given T as;

$$\rightarrow N^+_D = \frac{N_D}{1+g_D \exp((E_F - E_D)/kT)} = \frac{N_D}{1+g_D (n/N_C) \exp((E_C - E_D)/kT)} = \frac{N_D}{1+(n/N_\varsigma)}$$

Where;

$$N_\varsigma = \frac{N_C}{g_D} e^{-(E_C - E_D)/kT}$$

Therefore, the electron density will be given by;

$$n = \frac{N_D}{1+\frac{n}{N_\varsigma}} \rightarrow n^{2=} + N_\varsigma n - N_\varsigma N_D = 0$$

Solving this quadratic equation will give the number of electrons as;,

$$n = -\frac{N_\varsigma}{2} + \left[\left(\frac{N_\varsigma}{2}\right)^2 + N_\varsigma N_D\right]^{1/2}$$

The above expression applies only where the positive root chosen when $n \geq 0$ as;

$$n = -\frac{N_\varsigma}{2} + \left[\left(1+\frac{4N_D}{N_\varsigma}\right)^{1/2} - 1\right]$$

Similar result can be also be obtained for acceptor doped material using the above expressions. In most cases, the value of N_ς is typically much greater than N_D. For p-doped silicon we have $N_D = 10^{15}/cm^3$ at 300K so that;

$$E_C - E_D = 0.045 eV, g_D = 2$$

So that $N_c = 3.224 \times 10^{19}/cm^3$, $N_\varsigma = 2.826 \times 10^{18}/cm^3$ and this values result into n = 0.9996, so 99.96% of P atoms are ionized at room temperature.

Special case of High Temperature

When a semiconductor is kept at a high temperatures, most dopants area ionized and they may reach a point where;

$$N_D^+ = N_D, N_A^- = N_A$$
$$p - n + N_D - N_A = 0$$

However, from the mass action law,

$$np = n_i^2$$

Using this mass action law, we shall have Therefore,

$$\frac{n_i^2}{n} - n + N_D - N_A = 0$$

$$\rightarrow n^2 - (N_D - N_A)n - n_i^2 = 0$$

And the number of electrons will be given by'

$$\rightarrow n = \frac{N_D - N_A}{2} + \left[\left(\frac{N_D - N_A}{2}\right)^2 + n_i^2\right]^{1/2} \quad (n \geq 0)$$

And those of the holes will be obtained from;

$$p = \frac{n_i^2}{n} = \frac{N_A - N_D}{2} + \left[\left(\frac{N_D - N_A}{2}\right)^2 + n_i^2\right]^{1/2}$$

When a condition where;

$$n = N_D$$

Then;

$$p \approx n_i^2 / N_D$$

Is an indication that the material is donor doped and therefore;

$$(N_D \gg N_A, N_D \gg n_i)$$

And when anther condition arises such that;

$$p = N_A$$

Then

$$n = n_i^2 / N_A$$

And this is an indication that the material is acceptor doped and therefore;

$$(N_A \gg N_D, N_A \gg n_i)$$

Position of the intrinsic level, E_i

1) Exact position of E i

The term intrinsic refers to belonging to a thing by its very nature, Intrinsic semiconductor therefore have equal number of N_D and N_A such that when $N_D = 0$, N_A is also $N_A = 0$. And therefore in such a case, $n = p$, and $E_F = E_i$. This can be expressed as;

$$\rightarrow N_C e^{(E_f - E_c)/kT} = N_V e^{(E_V - E_F)/kT}$$

$$\rightarrow E_i = E_F = \frac{E_C + E_V}{2} + \frac{kT}{2} \ln\left(\frac{N_V}{N_C}\right)$$

$$= \frac{E_C + E_V}{2} + \frac{kT}{2} \ln\left[\left(\frac{m_p^*}{m_n^*}\right)^{3/2}\right]$$

$$E_F = \frac{E_C + E_V}{2} + \frac{3kT}{4} \ln\left[\left(\frac{m_p^*}{m_n^*}\right)\right]$$

It from here that we get that the E_i in Si is 0.0073 eV and this value is below the mid-gap while that of GaAs is 0.0403 eV which is above the mid-gap at 300 k.

2) Freeze out/Extrinsic T.

When the values of N_D is much greater than N_A, it implies that N_D is much greater than n_i and hence;

$$N_D \gg N_A, N_D \gg n_i$$

$$n = N_C e^{(E_F - E_C)/kT} = \left(\frac{N_C}{2}\right)\left[\left(1 + \frac{4N_D}{N_C}\right)^{1/2}\right]$$

$$E_F = E_C + kT \ln\left\{\left(\frac{N_C}{2N_C}\right)\left[\left(1 + \frac{4N_D}{N_C}\right)^{1/2} - 1\right]\right\}$$

The above analysis is useful for semiconductors low temperature calculations otherwise at high temperatures deviations start to arise.

3) Extrinsic

When impurities are introduced, though the mass action law is observed the Fermi level shifts and therefore $E_F - E_i$ is given by;

$$E_F - E_i = kT \ln(n/n_i) = -kT \ln(p/n_i)$$

$$E_F - E_i = kT \ln(N_D/n_i) \quad [N_D \gg N_A, N_D \gg n_i]$$

$$E_F - E_i = kT \ln(N_A/n_i) \quad [N_A \gg N_D, N_A \gg n_i]$$

Generation Recombination process

Semiconductors are very stable materials but when a semiconductor is perturbed from equilibrium state the carrier numbers of electrons or holes are modified. This may also cause combination or generation of carriers. Recombination-generation process result into ordering restoring mechanisms as the carrier excesses or deficits inside the semiconductor stabilized or removed. This perturbation is what causes optical excitation, electron bombardment and current injection from a contact. This is what causes drift and diffusion currents though they may result from other sources there along with process of recombination.

Band to Band Recombination

In the figure below, it shows a direct thermal recombination is occurring from the conduction band CB, to the valence band, VB. As it can be seen from the curve, there is direct annihilation of CB electron and a VB hole as the electron falls from an allowed CB state to hole in VB. The process of recombination in most cases is radiative with the production of a photon for each process.

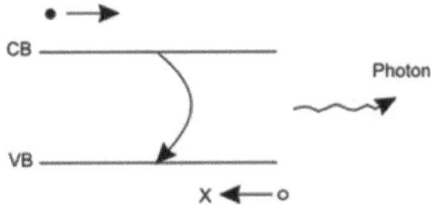

When we have band to band generation, either thermal or photon energy is absorbed as depicted below. However, when band to band recombination occurs, the photons are almost massless and therefore they have a very small momentum. The photon assisted transition is vertical on the *E-k* plot. In GaAs a semiconductor which is a direct band-gap semiconductor, there is little change in the momentum required for recombination process to proceed.

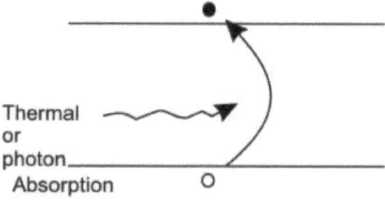

Thus the conservation of energy and momentum is met by the release of a photon.

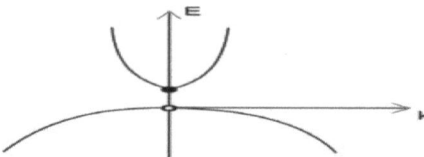

In an indirect band-gap semiconductor there is a change of crystal momentum associated with a recombination process. The emission of a photon will conserve energy but not momentum. In an indirect band-gap semiconductor, for a Band to Band recombination to proceed, a photon must be emitted or absorbed during the lattice vibration as shown below.

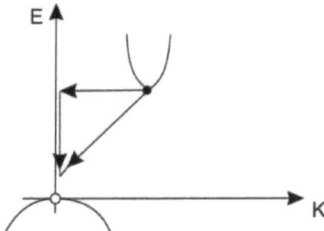

From the figure, it can be seen that the Band to Band process in indirect semiconductors is very complicated. This also imply that there is a diminished rate of band to band recombination occurring in indirect semiconductors and thus here in indirect semiconductors Recombination-Generation process, center recombination dominates. So, all R-G processes involve band to band radiative recombination in which the optical absorption coefficient as shown in figure.

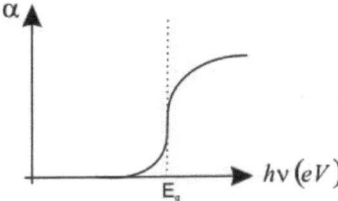

The Photon to electron-hole generation rate is given by;

$$G_r = \int_0^\infty c'\alpha(v)P(v)dv \rightarrow$$

This is based on Planck's radiation law in which the radiative recombination rate can be given by;

$$R_r = \alpha_r p n$$

Such that the thermal equilibrium is equal to generation rate .i.e.

$$G_r = R_r$$

The two expressions can therefore be combined to give,

$$G_r = \alpha_r p_o n_o = \alpha_r n_i^2$$

The excess carrier Radiative recombination rate U_r, can be obtained from;

$$U_r = R_r - G_r$$

$$= \alpha_r \left(pn - n_i^2\right) = \frac{G_r}{n_i^2}\left(pn - n_i^2\right)$$

Now if we let *p* and *n* be given as;

$$p = p_o + p_e$$
$$n = n_o + p_e$$

then the excess carrier Radiative recombination rate U_r, will be given by;

$$\rightarrow U_r = \frac{p_e}{\tau}$$

Combining the above expressions, the rate at which electrons and holes disappear will be obtained as follows;

$$\rightarrow U_r = \frac{G_r}{n_i^2}\left[(p_o + p_e)(n_o + p_e) - n_o p_o\right]$$

$$= \frac{G_r}{n_i^2}\left[p_o n_o + p_e n_o + p_e p_o + p_e^2 - n_o p_o\right]$$

$$= \frac{G_r}{n_i^2}[n_o + p_o + p_e]p_e$$

The component below has units of sec^{-1}

$$\alpha_r\left(p_n - n_i^2\right)$$

And therefore we can define τ, as;

$$\tau = \frac{1}{\alpha_r\left(p_n - n_i^2\right)}$$

While the minority carriers lifetime and Radiative recombination lifetime will be obtained from the expression below;

$$= \frac{n_i^2}{G_r(n_o + p_o + p_e)},$$

The expression is for an n-type semiconductor in which the Donor, N_d is approximated to;

$$n_o \approx N_d. \qquad p_o \approx 0$$

So as to get τ equal to;

$$\tau_r = \frac{n_i^2}{G_r N_d}$$

$$\frac{dp_e}{dt} = -\frac{p_e}{\tau} \rightarrow p_e(t) = p_e(0)e^{-t/\tau}$$

At high level injection levels where

$$p_e \gg n_o \text{ or } p_o$$

τ is equal or approximated to;

$$\tau_r \approx \frac{n_i^2}{G_r p_e}$$

Recombination-Generation Centers/Traps

The R-G process also involves deep energy level in which the impurity atoms approach the allowed energy levels in the mid-gap region. At these states, the crystal defects are also at deep level states in which electrons (e^{-1}) from conduction band, CB and hole from valence band, VB comes to the R-G center and gets annihilated. It can also be the considered based on the capture of having one electron from CB to deep state and then the same electron can jump to VB canceling a RG recombination and hence becoming non-radioactive and thus heat or lattice vibration is produced as shown in Fig. 8.6 and 8.7 respectively.

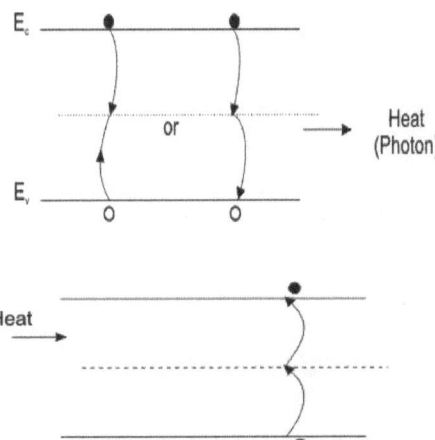

As it can be inferred is that the dominant mechanism in G- R in semiconductors involve the following;

$$\frac{\partial n}{\partial t} \rightarrow$$

This is the rate of change of electron concentration due to RG recombination and generation.

$$\frac{\partial p}{\partial t} \rightarrow$$

This is the rate of change of holes.

$$n_T \rightarrow$$

This is the number of R-G traps/cm³ filled with e^-

$$p_T \rightarrow$$

This is the number of R-G traps/cm³ empty.

$$N_T = n_T + p_T$$

This is the total number of traps or RG centers per cm³. This is a special case that involves trap mechanism in recombination or generation processes.

Electron capture

The number of electrons to be captured multiplied with the empty trap states can be expressed as;

$$\frac{\partial n}{\partial t} = -C_n N_T n.$$

As less captured of electrons occur, the C_n values tends to remain constant. The probability that the recombination-generation center is occupied by an electron will be given by;

$$f(E_T) = \frac{1}{1 + \exp\left(\frac{E_T - E_F}{kT}\right)} = f$$

So that;

$$\frac{\partial n}{\partial t} = -C_n N_T (1-f) n$$

The negative sign on C_n indicates that electron capture acts to reduce the number of electrons in CB and has units of cm³/sec. this now implies that thermal velocity × capture cross section will give;

$$C_n = v_{th} \sigma_n$$

CHAPTER SEVEN
SOLID STATE CONTINUITY EQUATIONS

Einstein and Continuity Equations

Diffusion is a process in which a particle tends to spread out a redistribute as a result of random thermal motion migrating from regions of high particle concentration to low particle concentration intending to produce uniform distribution of the diffusing elements. Electrons and holes are charged particles but they too so diffuse and the consequence is that they form the diffusion current. To obtain this diffuse current, derivation of Diffusion current make a number of assumptions which include among others; that the diffusing elements are on dimensional only, all carriers involved move with the same velocity υ, the distance moved by carriers between collisions is of a fixed length L in which, L is actually mean distance moved by carrier between collisions and always the randomness of thermal motion is equal to the number of particles moving in +x and –x directions as depicted bellow.

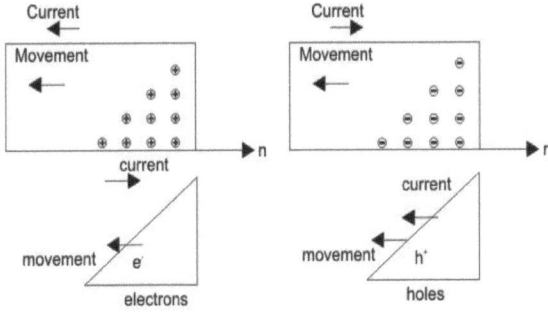

Within the section given by, Δx, which represents an equal outflow of particles per second from any interior section to neighbouring sections on the right and left we can derive an expression to relate diffusion current in a semiconductor. But because of concentration gradient the number of particles moving from right to left second is greater than number of particles moving from left to right, they form a gradient as shown below.

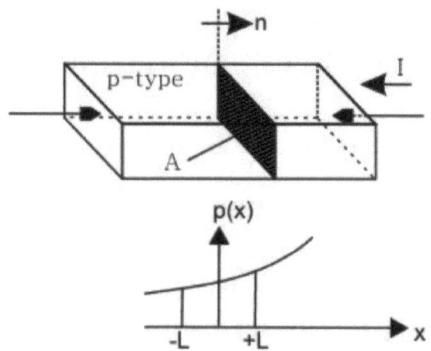

If we assume that half of the (1/2) of the holes in a volume *LA* on either side of x = 0 will move in the proper direction so as to cross x = 0 plane, then the holes (p) moving in +x (→) which cross x = 0 plane in time will be given by,

$$\frac{L}{v}$$

And it can therefore be expressed as;

$$= \frac{A}{2} \int_{-L}^{0} P(x)dx$$

While those moving in the negative direction (+x←) and crossing x = 0 plane in time, t will equally be given by;

$$\frac{L}{v}$$

$$= \frac{A}{2} \int_{0}^{-L} p(x)dx$$

But the holes in the range can be obtained by;

$$p(x) = p(0) + \left.\frac{dp}{dx}\right|_{x=0} x + \cdots \quad -L \leq x \leq L$$

And thus those in the positive +x direction will be given by;

$$\vec{p} = \frac{A}{2}\int_{-L}^{0}\left(p(0) + \frac{dp}{dx}\bigg|_{x=0} x\right)dx$$

$$= \frac{A}{2}\int_{-L}^{0}(p(0))dx + \frac{A}{2}\int_{-L}^{0}\frac{dp}{dx}\bigg|_{x=0} x\,dx$$

$$= \frac{A}{2}p(0)L + \frac{A}{2}\frac{dp}{dx}\bigg|_{0}\frac{x^2}{2}\bigg|_{-L}^{0}$$

Those in the negative –x direction will be given by;

$$\bar{p} = \frac{A}{2}p(0)L - \frac{A}{2}\left(\frac{dp}{dx}\bigg|_{0}\right)\left(\frac{L^2}{2}\right)$$

The net flow will be the difference between the +x and –x flow and can be expressed as;

$$\vec{p} - \bar{p} = -A\frac{dp}{dx}\bigg|_{x=0}\left(\frac{L^2}{2}\right)$$

So for a volume element flow, the number of +x directed holes with cross x = 0 plane in time will be given by;

$$I_{p|diff} = \frac{a(\vec{p} - \bar{p})}{\frac{L}{\bar{v}}}$$

Which can also be expressed as;

$$I_{p|diff} = \frac{-1}{2}qA\bar{v}L\frac{dp}{dx}$$

$$J_{p|diff} = -q\left(\frac{\bar{v}L}{2}\right)\frac{dp}{dx}$$

We can now therefore define the hole diffusion current by the expression;

$$D_p = \frac{\bar{v}L}{2} \rightarrow J_{p/diff} = -qD_p\frac{dp}{dx}$$

which can be generalized for holes and electrons as;

$$J_{p/diff} = -qD_p\nabla p$$

$$J_{n|diff} = -qD_n\nabla n$$

where D_p and D_n are constants within the cm² /sec dimensions.

When analysis of the Fermi level inside a material is done, it is found that it does not dependent on position of equilibrium and as such under equilibrium conditions, there is no current hence;

$$\frac{dE_F}{dx} = \frac{dE_F}{dy} = \frac{dE_F}{dz} = 0$$

Therefore a non-zero electric field is established inside a non-uniformly doped semiconductor under equilibrium and thus under equilibrium, the drift and diffusion currents balance out as;

$$J_{n|drift} + J_{n|diff} = q\mu_n n\xi + qD_n \frac{dn}{dx} = 0$$

This gives the electrons to be calculated from;

$$n = N_c \mathfrak{F}_{\frac{1}{2}}(\eta_c)$$

$$\eta_c = \frac{E_F - E_c}{kT}$$

And that is why the two contributions also balance as;

$$\mu_n n\xi + D_n \frac{dn}{dx} = 0$$

Or we can write;

$$\frac{dn}{dx} = \frac{dn}{d\eta_c}\frac{d\eta_c}{dx} = \frac{dn}{d\eta_c}\left(-\frac{1}{KT}\frac{dE_c}{dx}\right)$$

$$= -\frac{q}{kT}\frac{dn}{d\eta_c}\xi$$

Alternatively we can also express the relation as;;

$$q\xi\left(\mu_n n - \frac{q}{kT}\frac{dn}{d\eta_c}D_N\right) = 0$$

or

$$\frac{D_n}{\mu_n} = \frac{kT}{q}\left(\frac{n}{dn/d\eta_c}\right)$$

With this last expression, we arrive at the generalized form of Einstein relationship. When the number of electrons approaches energy gap levels, we get that;

$$n \to N_c \exp\left(\frac{E_F - E_c}{kT}\right), \quad \frac{n}{dn/d\eta_c} \to 1$$

Then from these relations, the generalized form of Einstein relationship can be rewritten as;

$$\frac{D_n}{\mu_n} = \frac{kT}{q}$$

Similarly, using this reduced generalized form of Einstein relationship at 300K, we obtain;

$$\frac{D_p}{\mu_p} = \frac{kT}{q} = 0.026V \text{ at } 300^0 \text{ k}$$

As discussed earlier, the current in semiconductor is found to be related as;

$$J = J_n + J_p = q\mu_n n\xi + qD_n \nabla n + q\mu_p p\xi - qD_p \nabla p$$

However in real materials, there exists AC and transients and under these AC and transit condition we need to add displacement current also to the above expression. In general we need to include carrier generation and recombination effects so as to have contributions from other processes that will give;

$$\frac{\partial n}{\partial t} = \frac{\partial n}{\partial t}\bigg|_{drift} + \frac{\partial n}{\partial t}\bigg|_{diff} + \frac{\partial n}{\partial t}\bigg|_{R-G} + \frac{\partial n}{\partial t}\bigg|_{other\ processes}$$

And this will imply that, all these other processes contribute a generalized current given by;

$$= \frac{1}{q}\nabla J_n - r_n + g_n$$

And this forms a continuity equation for electrons. The component, $\Delta.J_n$ shows that there will be a change in carrier concentration within a given small region if there is an imbalance between total carrier currents in and out of the region due to other processes. The component r_n and g_n refer to **R**ecombination and **G**eneration contributions. Therefore for holes we have the Continuity equations given by;

$$\frac{\partial p}{\partial t} = -\frac{1}{q}\nabla.J_p - r_p + g_p$$

Semiconductor Diodes

A diode can be defined as a thermionic tube having two electrodes; used as a rectifier or a semiconductor that consists of a p-n junction. There are many types of diodes which are designed for various applications. We use varactor diodes, GUNN diodes,, Tunnel diodes,, IMPATT diodes, Schottky diodes which are fabricated to have metal and a semiconductor and PIN diodes commonly used in microwave applications areas such as in heterodyning mixers for RF-IF conversion.

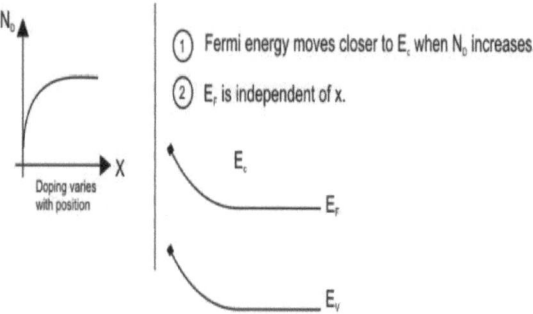

Diodes can also be found use in ddetectors as the one shown in which four diodes are used to balance the mixer. Some special switches also use diodes to operation the on and off states of an electronic device.

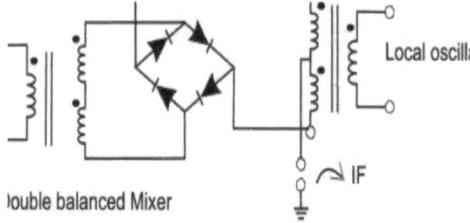

louble balanced Mixer

P-N junction Diodes

Before we analyze the P-N diodes, let us describe a PN Junction. We can begin with semi-conductor devices with the junction for three reasons; (1) The device finds application in many electronic systems, e.g. in adapters that charge the batteries of cell phones; (2) The p-*n* junction is among the simplest semiconductor devices, thus providing a good entry point into the study of the operation of such complex structures as diodes and transistors; (3) The *p-n* junction also serves as part of transistors and thus an abrupt p-n junction containing circuit.

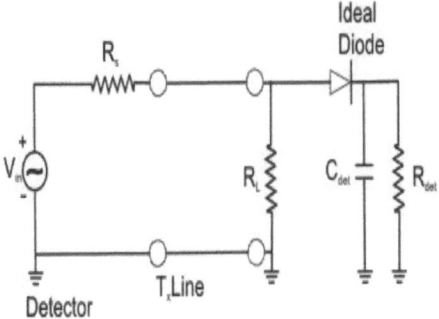

In a p-n junction, the built-in potential V_{bi} is given by;

$$qV_{bi} = E - (qV_n + qV_p)$$

However this buit-in potential can also be obtained from the expression given by;

$$= kT \ln\left(\frac{N_c N_v}{n_i^2}\right) - \left[kT \ln\left(\frac{N_c}{n_{no}}\right) + kT \ln\left(\frac{N_v}{P_{po}}\right)\right]$$

Which can be reduces to;

$$= kT \ln\left(\frac{n_{no} P_{pc}}{n_i^2}\right)$$

In terms of acceptor and donor densities, the buit-in potential can be expressed as;

$$= kT \ln\left(\frac{N_A N_D}{n_i^2}\right)$$

So that at equilibrium conditions of holes and electrons, the mass action law is maintained as;

$$n_{no} P_{no} = n_i^2 = n_{po} P_{po}$$

Thus in terms of holes and electrons, the buit-in potential is expressed as;

$$V_{bi} = \frac{kT}{q} \ln\left(\frac{P_{po}}{P_{no}}\right) = \frac{kT}{q} \ln\left(\frac{n_{no}}{n_{po}}\right)$$

This is the generalized expression used to obtain the holes and electrons as minority and majority carriers. The minority carrier in *n* side of the diode can be obtained as;

$$P_{no} = P_{po} \exp\left(-\frac{qV_{bi}}{kT}\right)$$

While the minority carrier in *p* side will be calculated from;

$$n_{po} = n_{no} \exp\left(-\frac{qV_{bi}}{kT}\right)$$

The following are some of the buit-in potentials of some semiconductors as obtained from the expression derived above;

V_{bi} in GaAs →1.1 to 1.4V

Si → 0.8 to 1V

Ge → 0.4 to 0.6V.

It should also be noted that at thermal equilibrium, the electric field, E in neutral regions are equal to zero or neutralized and thus the total negative is given by the charge in P side minus the total positive charge in n side. The figure below displays the Poissons approach to the relationship between charge and frequency in a crystal as

$$N_A x_p = N_D x_n$$

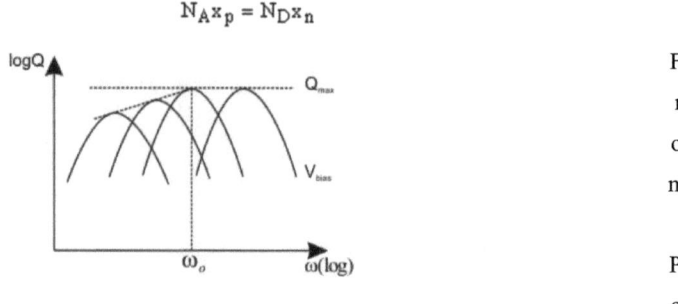

From Poissons Equation given by;

$$\frac{\partial^2 V}{\partial x^2} = \frac{d\xi}{dx} = \frac{\rho(x)}{\epsilon_s}$$

And using the concept that at thermal equilibrium, the electric field, E in neutral regions are equal to zero or neutralized and thus the total negative is given by the charge in P side minus the total positive charge in n side, we get that;

$$= \frac{q}{\epsilon_s}\left[p(x) - n(x) + N_D^+(x) - N_A^-(x)\right]$$

This means that in the *n* side which is in the range of $0 < x \leq x_n$ as;

$$\frac{d^2V}{dx^2} \cong \frac{q}{\epsilon_s} N_D$$

Likewise in the p side in the range of $-x_p \leq x < x_n$ will be given as;

$$\frac{d^2V}{dx^2} \cong \frac{q}{\epsilon_s} N_A$$

Integrating the above expression in the region $-x_p \leq x \leq 0$, we obtain;

$$\xi(x) = -\frac{qN_A}{\epsilon_s}(x + x_p)$$

Likewise integrating the expression in the region $0 \leq x \leq x_n$, we obtain;

$$\xi(x) = +\frac{qN_D}{\epsilon_s}(x - x_n)$$

Applying Maxwell's field at the junction region where $x = 0$ we obtain;

$$|\xi_m| = \frac{qN_D x_n}{\epsilon_s} = \frac{qN_A x_p}{\epsilon_s}$$

This reduces form also help us to obtain the in-built potential from;

$$V(x) = \xi_m\left(x - \frac{x^2}{2W}\right)$$

or

$$V_{bi} = \frac{1}{2}\xi_m(x_n + x_p)$$

In the above expression, W is the total depletion region width and thus eliminating the constant, ξ_m from the above three expression, we get W given as;

255

$$W = \sqrt{\frac{2\epsilon_s}{q}\left(\frac{N_A+N_D}{N_A N_D}\right)V_{bi}}$$

The above expression is only an approximation. A more accurate expression for depletion region width is obtained by introducing a correction factor to give the expression as;

$$W = \sqrt{\frac{2\epsilon_s}{q}\left(\frac{N_A+N_D}{N_A N_D}\right)\left(V_{bi}-\frac{2kT}{q}\right)}$$

W can be further be obtained by simplifying further to get it from;

$$= L_D\sqrt{2(\beta V_{bi}-2)}$$

Or it can be given through the Debye length as;

$$L_D = \sqrt{\frac{\epsilon_s kT}{q^2 N_B}} = \text{Debye Length}$$

In which the constant NB is a ratio obtained from the acceptor to donor density variations as;

$$N_B = \frac{N_A N_D}{N_A + N_D}$$

And if by chance one side is heavily doped the depletion region is in weakly doped place,

$$(\text{if } N_A \gg N_D, N_B = N_D \quad N_D \gg N_A, N_B = N_A)$$

Then the constant β, is given by;

$$\beta = \frac{q}{kT}$$

This will also imply that the depletion layer capacitance per unit area will be given by;

$$C = \frac{dQ_c}{dV}$$

Where, dQ_c, is the incremental increase in charge per unit area for any voltage increase.

$$dV = \frac{d(q_B W)}{d\left(\frac{qN_B}{2\epsilon_s}W^2\right)}$$

Combining the Debye length expression, the correction factor and the built-I potential, we obtain that the voltage increase can be expressed as;

$$= \frac{E_q}{w} = \sqrt{\frac{q\,\epsilon_s\,N_B}{2}}\left(V_{bc} \pm V - \frac{2kT}{q}\right)^{\frac{1}{2}} F/cm^2$$

Thus, inside depletion layer and assuming that the Boltzmann relation is just an approximation, there is an Abrupt Depletion Layer with a low injection in which injected minority carrier are less than the majority carrier and there is no generation current inside depletion layer, then the electron carriers are obtained from;

$$n = n_i \exp\left(\frac{E_F - E_i}{kT}\right) = n_i \exp\left[\frac{q(\psi - \phi)}{kT}\right]$$

While the holes carriers are obtained from;

$$p = n_i \exp\left(\frac{E_i - E_F}{kT}\right) = n_i \exp\left[\frac{q(\phi - \psi)}{kT}\right]$$

Where ψ and Φ respectively are given by,

$$\psi = \frac{-E_i}{q}, \quad \phi = \frac{-E_F}{q}$$

If voltage is applied externally, the minority carrier densities on both sides are changed but remain balanced according to mass action law i.e.

$$np \neq n_i^2$$

It can then be obtained that the majority carrier given as;

$$n = n_i \exp\left[\frac{(q\psi - \phi_n)}{kT}\right]$$

While those of hole carriers be given as;

$$p = n_i \exp\left[\frac{q(\phi_p - \psi)}{kT}\right]$$

and in both cases, Φ_p and Φ_n are quasi Fermi level for electrons and holes (E_{Fn} and E_{Fp}) under non-equilibrium condition as applied voltage or bias or optical field is maintained and they are separately. For Φ_n we have;

$$\phi_n = \psi - \frac{kT}{q} \ln\left(\frac{n}{n_i}\right)$$

And for Φ_p we have;

$$\phi_p = \psi + \frac{kT}{q} \ln\left(\frac{n}{n_i}\right)$$

Substituting these to obtain p_n so that we calculate the electron and hole currents, we get the expression;

$$p_n = n_i^2 \exp\left[\frac{q(\phi_p - \phi_n)}{kT}\right]$$

Here the forward bias and reversed bias are based respectively on the range;

$$\phi_p - \phi_n > 0 \quad pn > n_i^2 \quad \text{and} \quad \phi_p - \phi_n > 0 \quad p_n < n_i^2$$

In our earlier expression, taking $\xi = \Delta\psi$, the electron diffuse or carrier current J_n will be given by;

$$J_n = q\mu_n \left(n\xi + \frac{kT}{q} \nabla n\right)$$

Substituting J_n in the earlier expression, we get;

$$= q\mu_n n(-\nabla\psi) + q\mu_n \frac{kT}{q}\left[\frac{qn}{kT}(\nabla\psi - \nabla\phi_n)\right]$$

which is simplified to give J_n as;

$$= -q\mu_n n \nabla\phi_n$$

In a similar way J_p can be obtained to be given by;

$$J_p = -q\mu_p p \nabla\phi_p.$$

From these derivations, it is clear that the electron and hole current densities are proportional to the gradient of quasi Fermi level. Therefore if an applied voltage across the junction given by;

$$V = \phi_p - \phi_n \quad (B)$$

the boundary of depletion layer at *p* side in which $x = -x_p$ we get that;

$$n_p = \frac{n_i^2}{p_{no}} \exp\left(\frac{qV}{kT}\right) = n_{po} \exp\left(\frac{qV}{kT}\right)$$

where, n_{no} is the equilibrium electron density on the *p* side. Similarly

$$p_n = p_{no} \exp\left(\frac{qV}{kT}\right) \quad \text{at} \quad x = x_n$$

In which p_{no} is the equilibrium hole density on the *p* side. This will result on the **n** and **p** sides produce a continuity equation in steady state given by;

$$\frac{\partial n}{\partial t} = +\frac{1}{q}\nabla . J_n - r_n + g_n = 0$$

and

$$\frac{\partial p}{\partial t} = -\frac{1}{q}\nabla . J_p - r_p + g_p = 0$$

If we let the recombination rate be obtained from $g_p - r_p = g_n - r_p = \mu$ = net recombination, then in one dimension (1-D) we obtain the change in current density given by;

$$\nabla J_n = \mu_n E \frac{\partial n_p}{\partial x} + \mu_n n_p \frac{\partial E}{\partial x} + D_n \frac{\partial^2 n_p}{\partial x^2}$$

And therefore for **p** side steady state we have;

$$\therefore \mu_n \xi \frac{\partial n_p}{\partial x} + \mu_n n_p \frac{\partial E}{\partial x} + D_n \frac{\partial^2 n_p}{\partial x^2} - U = 0$$

And also for **n** side in the range of $0 < x \leq x_n$, we have;

$$\& -\mu_p \xi \frac{\partial p_n}{\partial X} - \mu_p p_n - D_p \frac{\partial^2 p_n}{\partial x^2} - U = 0$$

From these two expressions, we can confirm that charge neutrality holds approximately in the state where;

$$n_o - n_{n_o} \cong p_n - p_{n_o}$$

If we multiplying the first equation by $\mu_p p_n$ and second equation by $\mu_n p_p$ then using the Einstein relation as;

$$D = \left(\frac{kT}{q}\mu\right)$$

We can then express current density as;

$$\frac{-P_n - P_o}{\tau_a} + D_p \frac{\partial^2 P_n}{\partial x^2} - \left(\frac{\frac{n_p - P_n}{n_p} \frac{}{P_n}}{\frac{n_p}{\mu_p} + \frac{P_n}{\mu_n}}\right) E \frac{\partial p_n}{\partial x} = 0$$

Where D_p is given by;

$$D_p = \left(\frac{n_p - P_n}{\frac{n_{np}}{\mu_p} + \frac{P_n}{\mu_n}}\right)$$

This gives the lifetime τ_a as;

$$\tau_a = \frac{p_o - p_{no}}{r_p} = \frac{n_o - n_{po}}{r_n}$$

This is for n–type in a case where;

$$P_n \ll n_n \cong n_{n_o}$$

And at low injection we assume that;

$$\left(\mu_p P_n \frac{\partial \xi}{\partial x}\right) = 0$$

So that we obtain the a simplified expression given by;

$$\frac{-P_n - P_{no}}{r_p} - \mu_p \xi \frac{\partial^2 P_n}{\partial x^2} + D_p \frac{\partial^2 P_n}{\partial x^2} = 0.$$

In a case where the neutral region is given at $\xi = 0$, we get;

$$\frac{\partial^2 P_n}{\partial x^2} - \frac{P_n - P_{no}}{D_p \tau_p} = 0$$

Then injection occurs at a boundary condition given by;

$$\rightarrow P_n = P_{no} \exp\left(\frac{qV}{kT}\right) \text{at } x = x_n$$

$$p_n = p_{no} \text{ at } x = \infty$$

In real situations, the semiconductors have very large structure and thus for large structures we have;

$$p_n - p_{no} = p_{no}\left(e^{\frac{qV}{kT}} - 1\right) \exp\frac{-(x - x_n)}{L_p} \quad \ldots\ldots\ldots$$

Using this expression, the exponential decay of electron and hole current components in the depletion region are shown below.

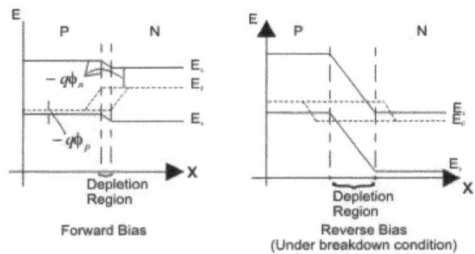

Forward Bias

Reverse Bias
(Under breakdown condition)

Diode Currents

A reverse-biased p-n junction carries a negligible current but exhibits a voltage dependent capacitance. Another interesting application of reverse-biased diodes is in digital cameras. If light of sufficient energy is applied to a p-n junction, electrons are dislodged from their covalent bonds and hence electron-hole pairs are created. With a reverse bias, the electrons are attracted to the positive battery terminal and the holes to the negative battery terminal. As a result, current flows through the diode that is proportional to the light intensity. We say the p-n junction operates as a photodiode. Thus a theoretical analysis proofs to be useful for high speed semiconductor devices. Consider a case in the *n* side where,

$$L_p = \sqrt{D_p \tau_p}$$

$$x = x_n, \; J_p = -qD_p \frac{\partial p_n}{\partial x}\bigg|_{x = x_n}$$

$$J_p = \frac{qD_p P_{no}}{L_p}\left(e^{\frac{qV}{kT}} - 1\right)$$

Similarly we obtain at the p side

$$J_n = qD_p \frac{\partial n_p}{\partial x}\bigg|_{x=-x_p} = \frac{qD_n n_{po}}{L_a}\left(e^{\frac{qV}{kT}} - 1\right)$$

Thus the total current will be given as;

$$J = J_p + J_n = J_s\left(e^{\frac{qV}{kT}} - 1\right) \quad \text{where } J_s = \frac{qD_p P_{no}}{L_p} + \frac{qD_n n_{po}}{L_n}$$

$$J = J_p + J_n = J_s\left(e^{\frac{qV}{kT}} - 1\right) \quad \text{where } J_s = \frac{qD_p P_{no}}{L_p} + \frac{qD_n n_{po}}{L_n}$$

This is shown in figure below;

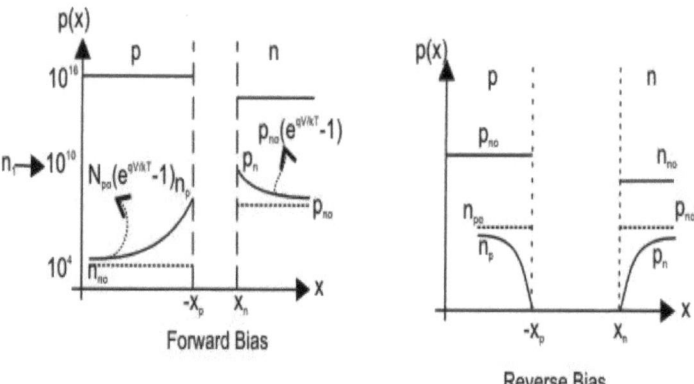

Forward Bias   Reverse Bias

When the abrupt junction with P^+ doping voltage drop is neglected, we get;

$$J_s \cong \frac{qD_p P_{no}}{L_p} \cong q\sqrt{\frac{D_p}{\tau_p}} \frac{n_i^2}{N_D}$$

This expression can be reduced to;

$$\cong \left[T^3 \exp\left(-\frac{Eg}{KT}\right)\right] T^{\frac{1}{2}}$$

The value V is constant for one sided abrupt junctions and based on the graph, its only Ge that forms an ideal equation that is valid while for GaAs and Si only qualitative agreement as shown in the figure below.

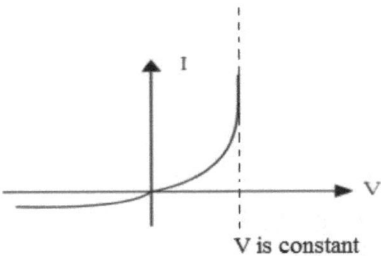

V is constant

However, this variation is due to the following reasons; there are surface effects; due to generation and recombination, due to tunneling of carriers between states in the band-gap, due to high injection conditions and also due to series resistance effect in the semiconductor. Under Reverse Bias the generation current, as considered for a generation rate of given by G, per unit volume is given as;

$$J_{gen} = \int_0^W qG dx = \frac{q n_i W}{\tau_e}$$

Therefore, for a one sided P-N diode in which theoretically ($N_A \gg N_D$) the reverse bias current will be given as;

$$J_R \cong q\sqrt{\frac{D_p}{\tau_p}} \frac{n_i^2}{N_D} + \frac{q n_i W}{\tau_g}, \quad \text{For Ge, } n_i \text{ is high}$$

From this expression, current J_R is dominated by the diffusion and therefore Shockley equation is strictly followed. It is also found that for Si and GaAs, n_i may be small and the generation term may be comparable or dominant. This shows that under forward bias condition the major Recombination-Generation process in the depletion region is a capture process. The Recombination current density is thus given by;

$$J_{rec} = \frac{qW}{2}\sigma v_{th} N_t n_i \exp\left(\frac{qv}{2KT}\right)$$

Hence obtained as;

$$= \frac{qn_i W}{2\tau_r}\exp\left(\frac{qV}{2KT}\right)$$

where, τ_r is the effective recombination lifetime with trap density, N_t.

N_t = Trap density

for

$p_{no} >> n_{po}$ and $V > \dfrac{3KT}{q}$

By substituting the above we finally obtain;

$$J_f = q\sqrt{\frac{D_p}{\tau_p}}\frac{n_i^2}{N_D}\exp\left(\frac{qV}{KT}\right) + J_{rec}$$

Diffusion Capacitance

The deflection layer capacitance-reverse junction when in forward biased is another contribution mad by rearranging the minority carrier density-Diffusion capacitance. Since the applied voltage is given by;

$$v(t) = \vartheta_o + \vartheta_1 \exp(jw_o t)$$

Then diffusion current can be given by;

$$J(t) = J_o + J_1 \exp(jw_o t)$$

where J_o is the forward Bias, J_1 is the Small signal amplitude hence for small signals as component of hole density in n-type side we have;

$$\widetilde{p}_n(x,t) = p_{n_i}\exp(j\omega_o t) + P_{no}\;; P_{ni} = f(x)$$

In a case where the voltage difference is great, i.e. $V_1 << V_o$ we have p_n given as;

$$p_n = P_{n_o}\exp\frac{\left(q(V_o + V_1 e^{j\omega t})\right)}{kT} \text{ at } x = x_n$$

$$P_n \cong P_{no} \exp\left(\frac{qV_o}{kT}\right) + \frac{P_{no}qV_1}{kT} \exp(j\omega_o t)$$

$$\exp(1+\delta) = e.e^\delta \approx e(1+\delta)$$

A similar expression can be obtained for the electron density in p side. Now for holes in the **n** side we know that continuity equation is given by;

$$\frac{\partial \tilde{p}_n}{\partial t} = \frac{-\tilde{p}_n - P_{no}}{\tau_p} - \tilde{p}_n \mu_p \frac{\partial \xi}{\partial x} - \mu_p \xi \frac{\partial \tilde{p}_n}{dX} + D_p \frac{\partial^2 \tilde{p}_n}{\partial x^2} \text{ at } x = x_n$$

And the equation can be further reduced to;

$$\frac{\partial \tilde{p}_n}{\partial t} = \left(\frac{\tilde{p}_n - p_{no}}{\tau_p}\right) + D_p \frac{\partial^2 \tilde{p}_n}{\partial x^2}, \; j\omega_o p_n \cong \frac{p_n}{\tau_p} + D_p \frac{\partial^2 p_n^2}{\partial x^2}$$

Or alternatively, the n side can be given a solution as;

$$\frac{\partial^2 \tilde{p}_n}{\partial x^2} - \frac{\tilde{p}_n}{\frac{D_p \tau_p}{1+jwr_p}} = 0$$

For a time varying case, the equation can be written as;

$$\frac{\partial^2 p_n}{\partial x^2} - \frac{p_n}{D_p \tau_p} = 0$$

Thus here,

$$\Gamma_p^* = \frac{\tau_p}{1+j\omega\tau_p}$$

The effect of frequency dependence on current signal when combined with the above complex equation will give Recombination current density a small signal current expressed as;

$$J_1 = \frac{qV_1}{kT} \left\{ \frac{qD_p P_{no}}{\frac{L_p}{\sqrt{1+j\omega\tau_p}}} + \frac{qD_n n p_o}{\frac{L_p}{\sqrt{1+j\omega\tau_n}}} \right\} \exp \frac{qV_o}{kT}$$

The current is thus admitted into the surface volume as diffusion current and thus the admittance due to diffusion only can only be given low frequency by;

$$Y = \frac{J_1}{V_1} = G_d + j\omega C_d \rightarrow$$

For the case where;

$$\omega\tau_p \omega\tau_n \ll 1$$

$$G_{do} = \frac{q}{kT}\left(\frac{qD_p p_{no}}{L_p} + \frac{qD_n n_{po}}{L_n}\right)\exp\left(\frac{qV_o}{kT}\right)$$

At low frequency also where G_d and C_d are function of bias frequency, we have

$$C_{do} = \frac{q}{kT}\left(\frac{qD_p p_{no}}{2} + \frac{qL_n n_{po}}{2}\right)\exp\left(\frac{qV_o}{kT}\right)$$

Only when $C_d \sim 1/\sqrt{W}$ then the deflection layer C_j capacitance will dominate at high frequencies.

Transient response of Diode

A diode will always respond to forward bias voltage or current. However the diode will not respond to the reverse voltage until excess minority carriers in neutral **n** and **p** regions have been withdrawn.

A Model diode

In a diode if we suddenly apply reverse bias, then the diode will pass a reverse current higher than reverse saturation current, I_a for a period of some time, t. The current then fall as the stored minority carriers are withdrawn rapidly and eventually reaching, I_a. One can use charge storage model to understand this transient behavior and thus consider a p^+_n junction-depletion region on the **n** side and an essential minority carrier is p only. Then we have;

$$p_c = \Delta p = p_c(x, t)$$

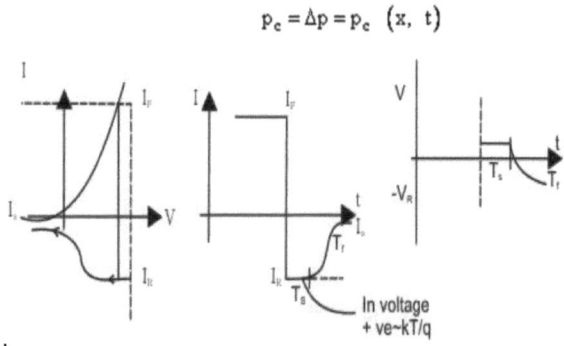

So that the hole distribution depicts in neutral **n**-region. Multiplying continuity equation by $qAdx$ and integrating it over the entire neutral **n** region from x_n to w_l, we get that;

$$-\int_{x_n}^{w_l} \frac{di_p}{dx} dx = \frac{qA}{\tau_p} \int_{x_n}^{w_l} P_c(x) dx + \frac{d}{dt} qA \int_{xn}^{w_l} P_c(x) dx$$

This expression can also be written as;

$$i_p(x_n, t) - i_p(w_1, t) = \frac{Q_p(t)}{\tau_p} + \frac{dQ_p}{dt}$$

In which the excess minority carrier stored in a n region at any time is given as;

$$Q_p(t) = qA \int_{x_n}^{w_1} P_c(x, t) dx \quad \text{at} \quad i_p(w_1, t)$$

The hole current at $x = w_1$ is related to the transit time and neutral n region of width w_1 and a time constant, τ_w is defined by the relation,

$$i_p(w_1, t) \cong \frac{Q_p(t)}{\tau_w}$$

In this case, τ_w is related to the passage time through the neutral **n**-region of width W_n, so we can write the hole current as;

$$i_p = \frac{Q_p(t)}{\tau_f} + \frac{dQ_p}{dt}$$

where

$$\frac{1}{\tau_f} = \frac{1}{\tau_p} + \frac{1}{\tau_w}$$

When this relation is plotted in a switching trajectory, the curves formed are shown below and from the curve; it can be shown that the edge of depletion layer is at x_n.

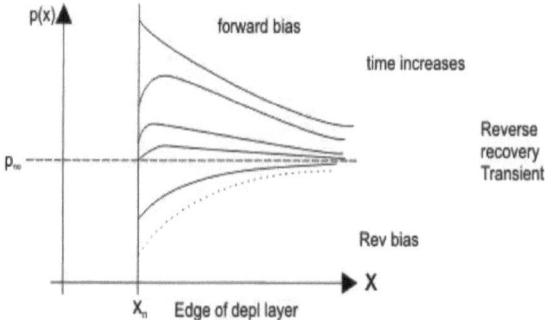

If this case we consider a storage phase, then we have;

$$i_p(x_n, t) = -I_R$$

Thus this almost constant value, - IR modifies the expression so that the equation is formed as;

$$\frac{dQ_p}{dt} + \frac{Q_p}{\tau_p} = -I_R$$

When the above expression is solved, the relation forms a solution given as;

$$Q_p(t) = C_1 \exp\left(-\frac{t}{\tau_F}\right) - I_R \tau_F$$

And in the equation C_1 is a constant and is determined by the condition that at time, t = 0, we have;

$$Q_p(0) = I_f \tau_f$$

This gives the constant, C_1 as;

$$C_1 = \tau_F(I_F + I_R)$$

$$Q_p(t) = (I_F + I_R)\tau_p \exp\left(\frac{-t}{\tau_p}\right) - I_R \tau_f$$

Now we assume a triangular hole distribution where the charge remains constant in the n-region at, $t = \tau_s$, as depicted,

268

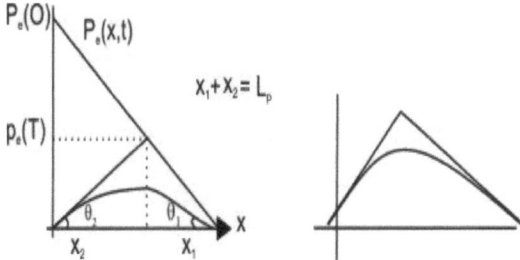

From the figure, we have for Θ_1;

$$\tan(-\Theta_1) \propto \left.\frac{-\partial P_c}{\partial x}\right|_{t=0} = \frac{I_F}{qAD_p}$$

And for Θ_2 we have the expression;

$$\tan\Theta_2 \propto \left.\frac{\partial P_c}{\partial x}\right|_{t=t_s} = \frac{-I_R}{qAD_p}$$

Where, τ_s is the time till the diode remains forward biased. Thus

$$\frac{P_c(\tau_s)}{P_c(0)} = \frac{1}{1+\frac{\tan\Theta_1}{\tan\Theta_2}} = \frac{1}{1+\frac{I_F}{I_R}}$$

Using the figure above, the stored charge is equal to the area of the triangle with a base given by;

$$(x_1 + x_2)$$

So that it can be expressed as;

$$\frac{Q_p(\tau_s)}{Q_p(0)} = \frac{P_c(\tau_s)}{P_c(0)} = \frac{1}{1+\frac{I_F}{I_R}}$$

At τ_s is equal to zero such that writing it in the form of;

$$Q_p(0) = I_F T_F$$

and then using it to solve for τ_s we get;

$$\tau_s = \tau_f \left(\ln\left(1+\frac{I_F}{I_R}\right) - \ln\left(1+\frac{I_F}{I_F+I_R}\right) \right)$$

However, a more accurate derivation of τ_s involves the solution of time-dependent continuity equation described earlier with appropriate boundary conditions and the solution obtained to be equal to;

$$\tau_s = \tau_p \left(\text{erfc} \left(\frac{I_F}{I_F + I_R} \right) \right)^2$$

Varactor Diodes

The word varactor is originally derived from the word or comes from a variable reactor. It is a device whose reactance can be varied in a controlled manner as the bias voltage is varied. That is why it is widely used in amplifier and harmonic generators, mixing or in mixers, in detection or in detectors, in variable voltage tuning and in many other applications. It is made from high speed semiconductor materials and therefore if we let the doping distribution be given as;

$$N_d(x) = N_0 x^m \text{ for } x > 0$$

then the abrupt doing profile is achieved by epitaxy or ion-implantation process where $m = 0$. In a varactor diode one side is heavily doped and on the other side the doping and the impurity concentration decreases with distance. This is what causes an abrupt doping profile as shown below.

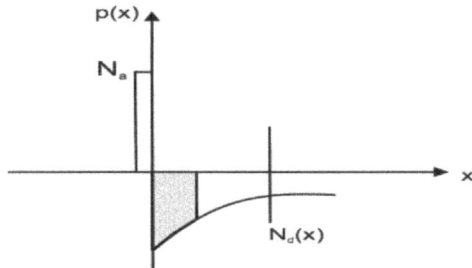

In this case, the value of m is found and m is a number less than zero i.e. < 0. These values take the form of -5/3, -3/2,-1 etc. as long as they are less than zero. Therefore the Poission's equation in the **n**-side is given by;

$$\frac{d^2V}{dx^2} = \frac{qN_d(x)}{\epsilon_s}$$

Applying the boundary conditions on this Poission's equation in the **n**-side, when the boundary conditions are;

$$V(x=0)=0, \quad V(x=W)=V, +V_{bi}$$

Which are the applied voltage and built-in voltage where W is the depletion width, then capacitance can be expressed as;

$$C_j = \frac{dq}{dv} = \left[\frac{qN_oA(\epsilon_s)^{m+1}}{(m+2)(V+V_{bi})} \right]^{\frac{1}{m+2}}$$

In this case, A is the area of the formed junction. This will then give capacitance as;

$$\log C_j = \text{constant} - \frac{1}{m+2}\log V$$

Considering the curve in figure above, the slope of this variation is given by;

$$S = \frac{d(\log C_j)}{d(\log V)} = \frac{1}{m+2}$$

The variation of capacitance, C, against voltage, V is shown below.

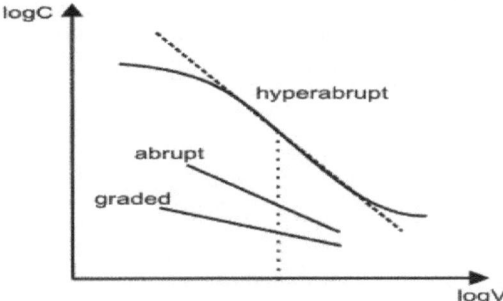

As depicted above, if m = - 5/3 then then S = 3 which indicates a Hyper-abrupt junction. When m = 0, then, S = 1/2 which forms abrupt junction and when m=0 then S=1/2 which forms a graded junction. It can be noted here that the larger the value of S, larger is the variation of C_j with biasing. The equivalent circuit of varactor diode is shown below.

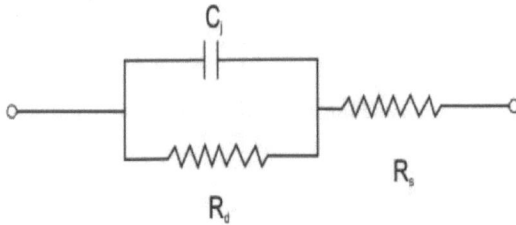

In the figure above, C_j is the junction capacitance, R_s is the series resistance, R_p is the parallel equivalent resistance of the generation-recombination current, diffusion current, and the surface leakage current. Both C_j and R_s decreases with reverse bias voltage and R_p increases with reverse bias voltage. Therefore the Quality factor Q of the varactor is given by;

$$Q = \frac{\text{ratio of energy stored}}{\text{energy dissipated per cycle}} = \frac{\omega C_j R_p}{1+\omega^2 C_j^2 R_p R_s}$$

The maximum Quality Factor therefore is;

$$Q_{max} \cong \left(\frac{R_p}{4R_s}\right)^{\frac{1}{2}}$$

Based on the above expression, the variation of Q with frequency is shown below. Therefore major considerations that made during the design for a varactor diodes are; (i) Capacitance, (ii) Voltage, (iii) Variation of capacitance with voltage, (iv) Maximum working voltage, (v) Leakage current and at high frequencies for a good varactor diode the equivalent circuit can be represented with lumped elements.

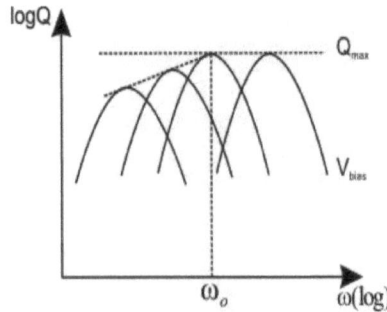

Varactor applications

Applications of the varactor are many and that is why they find applications in mixers, detectors and voltage tuning. In all these applications, the Varactor voltage is given by;

$$V(t) = R_s i(t) + \int \frac{1}{C_j(t)} i(t) dt$$

where $i_{(t)}$ is the current flowing which can also be independent expressed as;

$$i(t) = C_j(t) \frac{dV_c(t)}{dt}$$

The two expressions are difficult to solve. The only analysis done is based in their frequency domain where a set of coupled non-linear algebraic equation are solved and in this case, for voltage determination we have the expressions;

$$V(t) = \sum_{m=-\infty}^{\infty} \sum_{n=-\infty}^{\infty} V_{mn} \left[e^{j(2\pi m F_p + 2\pi n F_s)t} \right]$$

and for current determination we have the expression;

$$i(t) = \sum_{m=-\infty}^{\infty} \sum_{n=-\infty}^{\infty} I_{mn} \left[e^{j(2\pi m F_p + 2\pi n F_s)t} \right]$$

where *m* and *n* are harmonic numbers and $P_{m,n}$ is the average power flowing into non-linear harmonics n, given by f_p, and mf_p.

Let us consider an important application of pumped varactors. This parameter is the parametric amplifier. When the varactor is pumped at a frequency F_p and a signal is introduced as F_c then at F s the varactor behaves as impedance with negative real parts. Therefore the negative part can be used for amplification applications so that the series resistance limits the frequencies F_p and F_s and thus introduces noise. Therefore for losse in reactance, the power is given by;

$$\sum_{m=-\infty}^{\infty} \sum_{n=-\infty}^{\infty} P_{m,n} = 0$$

Therefore the Manley-Rowe frequency-power formulas for a case where we have lossless reactance are, the expression is given by;

$$\sum_{m=0}^{\infty} \sum_{n=-\infty}^{\infty} \frac{m P_{m,n}}{n F_p + m F_s} = 0$$

$$\sum_{m=-\infty}^{\infty}\sum_{n=0}^{\infty}\frac{np_{m,n}}{nF_p+mF_s}=0$$

Based on the above two expressions, the output voltage is given by;

$$V_{out}=\frac{R_{nonlinear}}{R_6+R_{nonlinear}}$$

Where the voltage is of analog nature given as;

$$(A\cos\omega t)\times\text{harmonic no.}=f(A\cos wt)$$

If the parametric amplifier is designed to only power can flow at input frequency f_p and output frequency is available at nF_p which is the frequency multiplier that is found in frequency dividers, rational fraction generators. Hence $p_1+p_n=0$ in which P_1 corresponds to power at f_p.

Parametric small signal amplifiers and frequency converters

If the RF signal at frequency, f_s is small compared to pump signal at frequency, f_s the power exchanged at the side band frequencies, nf_s and mf_s for, $m\neq 0$ is negligible. The corresponding Manley-Rowe equation now takes the form of;

$$\sum_{n=-\infty}^{\infty}\frac{P_n}{nf_p+f_s}=0$$

Under this Manley-Rowe equation conditions, the optimum gain takes the form of;

$$G_{max}=\left[\frac{\frac{m_1 f_c}{f_s}}{1+\sqrt{1+\frac{m_1^2 f_c^2}{f_s f_o}}}\right]^2$$

In this expression R_s being the series resistance and $S_{max}-S_{min}$ is the elastic swing, the modulation ratio m_1 is given by;

$$m_1=\frac{|S_1|}{S_{max}-S_{min}}$$

While the f_c the dynamic cut-off frequency is given by;

$$f_c=\frac{S_{max}-S_{min}}{2\pi R_s}$$

The P-I-N Diodes

In PIN diode, an *i* region present that is sandwiched between the *p* and *n* region as shown below. PIN diodes are fabricated by any of the following processes (i) epitaxial process, (ii) Diffusion of p and n in high resistibility substrate or by (iii) ion drift method.

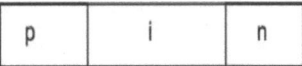

The i-region present in the PIN region is either a high resistivity a $p(\pi)$ or $n(\upsilon)$ layer such that in the PIN junction, the concentration, charge density and electric field profiles are shown. The PIN diodes are used widely in microwave wave circuits such as microwave switch with constant depletion layer and high power.

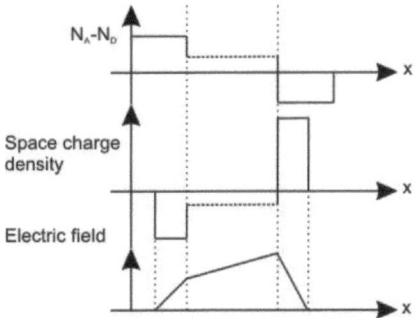

Since the PIN is a high speed switching junction, its switching speed is approximately equal to;

$$\cong \frac{W}{2v_s}$$

In the above expression, W, is the total depletion region width and υ_s is the saturation velocity across i region. In addition the PIN diode can be used as variable attenuator by varying device resistance that change approximately likely with forward current. It can be used also to modulate signals up to GHz range. The other use is in the photo detection of internal modulated light in reverse bias. Under Reverse Bias therefore, the junction capacitance is given by;

$$C_j = \frac{\epsilon_s A}{W}$$

Based on this, the series resistance is given as;

$$R = R_i + R_c (0.3\Omega)$$

Where R_i is the i-region resistance and R_c is the contact resistance so that the reverse bias current is given by;

$$I = -\frac{qAn_i W}{\tau_a}$$

Where in the above expression, τ_a is the ambipolar life time and its I-V characteristics is shown.

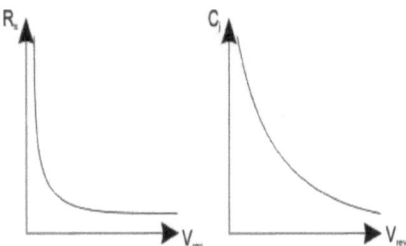

As mentioned above, PIN diodes find application in switching circuits in which the diode admittance (Y_r) in the reverse bias state and impedance (Z_f) in the forward bias can be expressed the diode admittance (Y_r) in the reverse bias state as;

$$Y_r = \frac{\left(\frac{f}{f_c}\right)^2}{R_s\left[1-\left(\frac{f}{f_r}\right)^2\right]} + j\omega\left\{\frac{C_j}{\left[1-\left(\frac{f}{f_r}\right)^2\right]} + C_p\right\}$$

And also the impedance (Z_f) will be expressed as;

$$Z_f = R_f + j\omega L_s$$

where the diode cut-off frequency is given as;

$$f_c = \frac{1}{2\pi R_s C_j}$$

and the reverse-bias series resonance frequency

$$f_r = \frac{1}{2\pi\sqrt{L_s C_j}}$$

In the above analysis, the Beam-lead PIN diodes are usually used in such frequency circuits.

Schottky diodes

When a metal-semiconductor junction is formed such that the carries see a barrier to flow from one terminal to the other, then the type of diode formed is called a schottky barries, as shown. When a Schottky diode is forward biased meaning that the negative is with respect to metal by a voltage, V_F, the barrier for electrons in semiconductor device decreases from qV_{bi} to $q(V_{bi}-V_F)$ and as a result, more electrons flows from the semiconductor to metal. Current I_{ms} therefore increases greatly as shown below.

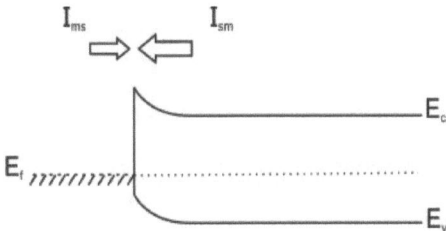

From the curve, it is noted that I_{ms} remains unchanged because there is no voltage drop across metal. Thus $q\Phi\beta$ remains unchanged. This also means that the reverse bias current drops across semiconductor and therefore increasing the barrier from qV_{bi} to $q(V_{bi}-V_R)$ where V_R is negative now, I_{ms} decreases more but I_{ms} remains almost unchanged. Small reverse current flows from Semiconductor to metal as shown.

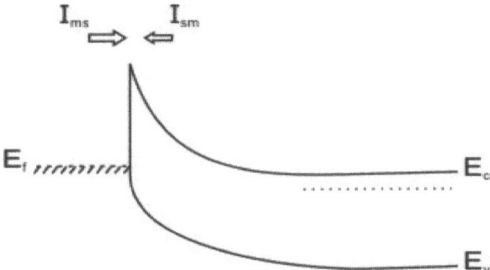

Now suppose we have a semiconductor diode whose parameter are such that $\Phi_{sc} > \Phi_m$ as shown. If you keenly analyze the curves above, there is no depletion layer formed in such a semiconductor because no barrier exists in semiconductor or in metal. Thus we have; (Metal) + (n-type) becomes rectifying for;

$$\phi_m > \phi_{sc}$$

(Metal) + (n-type) becomes non rectifying

$$\phi_m < \phi_{sc}$$

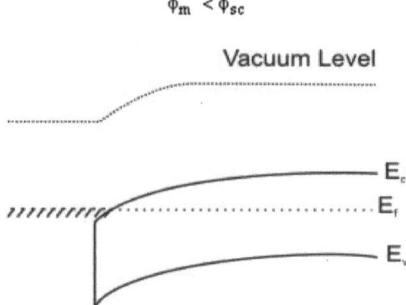

From the above expression, the opposite is true for p-type semiconductor such that $\Phi_{sc} > \Phi_m$ becomes rectifying with contact otherwise ohmic as shown below.

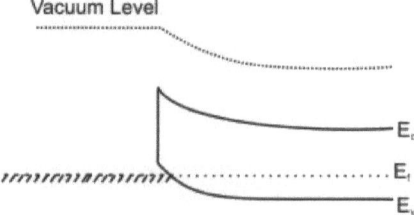

For now let us consider a rectifying contact. For a rectifying contact we have;

$$q\phi_B = q\chi + E_g - q\phi_m$$

$$qV_{bi} = q(\phi_{sc} - \phi_m)$$

$$W = \sqrt{\frac{2\epsilon_s}{qN_D}\left(V_{bi} - V - \frac{kT}{q}\right)}$$

$$|\xi(x)| = \frac{qN_D}{\epsilon_s}(W-x), V(x) = \frac{qN_D}{\epsilon_s}\left(Wx - \frac{1}{2}x^2\right) - \phi_B$$

From the above expressions, we find that the maximum, ξ, field occurs at;

$$x = 0 \quad \xi(x=0)$$

So that we can noe obtain Q_{sc} as;

278

$$\sqrt{\frac{2qN_D}{\epsilon_s}\left(V_{bi} - V - \frac{kT}{q}\right)} = \frac{2\left(V_{bi} - V - \frac{kT}{q}\right)}{W}$$

$$Q_{sc} = qN_D W = \sqrt{2q\epsilon_s N_D\left(V_{bi} - V - \frac{kT}{q}\right)}$$

In regard to Q_{sc} above, the space charge per unit area can then be determined form;

$$C = \left|\frac{\partial Q_{sc}}{\partial V}\right| = \frac{\epsilon_s}{W}\frac{1}{c^2} = \frac{2\left(V_i - V - \frac{kT}{q}\right)}{q\epsilon_s N_D}$$

To give a reduced for of Q_{sc} as in the expression below;

$$\frac{-d\left(\frac{1}{c^2}\right)}{dV} = \frac{2}{q\epsilon_s N_D}.$$

Thus if N_D is constant throughout then, E_0 does not vary much.

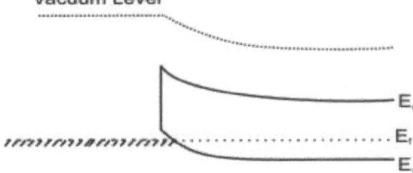

Schottky effect lowering of the barrier

When an electron is at a distance of x from metal surface, a positive charge is induced on metal surface. The force of attraction between electron and induced positive charge is equal to the force that will exist between electron and positive charge or image that was induced charge. The attractive force between the electron and the image charge will be given by;

$$F = \frac{-q^2}{4\pi\epsilon_o (2x)^2} = \frac{-q^2}{16\pi\epsilon_o x^2}$$

Therefore the work done on electron in the course of its transfer from infinity (∞) to point x will be given by;

$$E(x) = \int_{\infty}^{x} F(x)dx = \frac{-q^2}{16\pi \epsilon_o x}$$

This is equivalent to the potential energy of an electron at a distance x from the metal surface. This potential energy must be added to barrier energy is expressed as $-q\xi(x)$ to obtain total potential energy of electron shown;

$$-PE(x) = \frac{q^2}{16\pi\epsilon_o x} + q\xi(x)$$

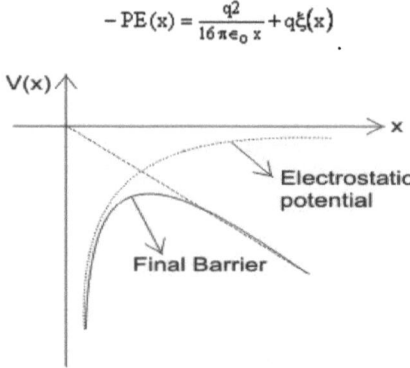

Combining the expression we get that in effect the barrier height varies with ξ field as;

$$\frac{d(P.E)}{dx} = 0 \rightarrow X_m = \sqrt{\frac{q}{16\pi\epsilon_o \delta}}$$

Current Transport in Diodes

Diodes of the Metal-Semiconductor type are known to be majority current flow devices and the mechanisms of current transport in these diodes involve;

- Transport of electrons from semiconductor over the potential barrier into the metal. It is the dominating factor in most moderately doped semiconductor with $N_D \leq 10^{17}/cm^3$ when operated at room temperature.
- Through the quantum mechanical tunnelling of electrons through the barrier. This is very important for heavily semiconductor since it is responsible for most of Ohmic contacts.
- Through electron-hole recombination in their depletion region.
- Through electron-hole recombination in the neutral semiconductor region.

In any diode, we do have a potential barrier and thus transport over potential barrier that involve high mobility materials can be explained through the thermionic emission theory while those involving low mobility materials can best be explained using the diffusion theory. Therefore, to comprehensively explain the transport of current in meta-semiconductor requires a combination of two processes. The thermionic emission theory considers the barrier height (q$\Phi\beta$ >> kT) and the thermal equilibrium of the diode. This is because net current flow does not effect this equilibrium since this currents flux is the one from the Metal to Semiconductor and the other is from the semiconductor to the metal. In this case, the shape of the barrier profile is not important. Current flow depends only on the component, q$\Phi\beta$ hence given as current density;

$$J_s \to M = \int_{E_f + q\phi_B}^{\infty} qV_n \, dn$$

In the above expression, the quantity, $E_f + q\Phi\beta$ correspond to the concentration of electrons with energy amount of $q\Phi\beta$ while $E_F + q\Phi\beta$ corresponds to the minimum energy needed to thermionically emit an electron into metal. V_x is the carrier velocity in x direction while dn is the electron density in energy range E and $E+dE$ given by;

$$dn = N(E)F(E)dE$$

$$= \frac{m_n^* \sqrt{2m_n^*(E - E_c)}}{\pi^2 \hbar^3} \frac{1}{1 + \exp\left[\dfrac{E - E_f}{kT}\right]} dE$$

The value of dn gives the density of states occupancy function that can be expressed as;

$$\frac{E - E_F}{kT} = -\frac{E - E_c + qV_n}{kT}$$

From this expression, if we assume that all energy in conduction band is equal to the Kinetic energy of the emitted electron, then this energy emitted can be expressed as;

$$E - E_c = \frac{1}{2} m_n^* v^2$$

So as to give the energy as;

$$dE = m_n^* v \, dv$$

Using the elemental energy emitted, we can thus express dn as;

$$dn = 2\left(\frac{m_n^*}{h}\right)^3 \exp\left(\frac{-qV_n}{kT}\right)\exp\left(-\frac{m_n^*v^2}{2kT}\right)\left(4\pi v^2 dv\right)$$

where the number of electrons per unit volume that have speed between v and (v + dv) overall direction is given as;

$$\exp\left(-\frac{m_n^*v^2}{2kT}\right)$$

This can be represented graphically as shown in Fig 12.9.

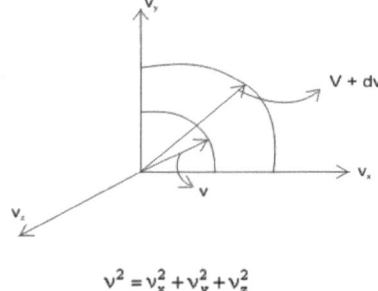

$$v^2 = v_x^2 + v_y^2 + v_z^2$$

From the graphical representation above, the volume of the elemental box can be given with rectangular coordinates dv_x, dv_y, dv_z to form a radial volume, **v**, as;

$$4\pi v^2 dv = dv_x dv_y dv_z$$

$$J_{s \to m} = 2q\left(\frac{m_n^*}{h}\right)^3 \exp\left(-\frac{qV_n}{kT}\right)$$

$$\int_{v_{0x}}^{\infty} v_x \exp\left(-\frac{m_n^*v^2}{2kT}\right) dv_x$$

$$\int_{-\infty}^{\infty} \exp\left(-\frac{m_n^*v_y^2}{2kT}\right) dv_y$$

$$\int_{-\infty}^{\infty} \exp\left(-\frac{m^*v_x^2}{2kT}\right) dv_z$$

From the above derivations, the component, V_{0x} is the minimum velocity required in the x direction so as to overcome and climb over the potential barrier. Thus considering that the potential barrier as discussed earlier is given by $qV_i - V$), then we have;

$$\frac{1}{2}m_n^* v_{ox}^2 = q(V_i - V)$$

$$J_{s \to m} = \left(\frac{4\pi q m_n^* k^2 T^2}{h^3}\right) \exp\left(-\frac{qV_n}{kT}\right) \exp\left(-\frac{m_n^* v_{ox}^2}{2kT}\right)$$

So as to give J_s current that has the potential barrier voltage as;

$$= \left(\frac{4\pi q m_n^* k^2 T^2}{h^3}\right) \exp\left(-\frac{qV_n}{kT}\right) \exp\left(-\frac{q(V_i - V)}{kT}\right)$$

$$= A^* T^2 \exp\left(\frac{q\phi_B}{kT}\right) \exp\left(\frac{qV_{appl}}{kT}\right) \qquad asq(V_n + V_i) = q\phi_B$$

In this expression, the effective Richardson Constant is included and is given by;

$$A^* = 4\pi q m_n^* k^2 T^2$$

Thus for free electrons, the Richardson constant, $A^* = 120 A/cm^2/k^2$. Based on this constant, materials have different constant as examples, those of n-type silicon and GaAS are;

$$A = 110 A/cm^2/k^2 \quad (n-Si)$$

And that of GaAs will be;

$$A = 8 \quad A/cm^2/k^2 \quad (n-GaAs)$$

Since GaAs, is an isotropic material where the rest mass is given as m_0, the lowest minimum of conductive band occurs at;

$$\frac{A^*}{A} = \frac{m_n^*}{m_0}$$

Multi-Valley semiconductor Diode Theory

If we take a_1, a_2, a_3 as directional cosines of the normal to the emitting plane relative to the principal axes of an ellipsoid in which m_x, m_y, m_z are the components of the effective mass tensor, the ratio of Richardson constant to those of the materials will be expressed as;

$$\frac{A^*}{A} = \frac{1}{m_o}\left(a_1 m_y^* m_z^* + a_2 m_x^* + a_3 m_x^3 m_y^*\right)^{\frac{1}{2}}$$

Taking silicon as an example, we have;

$$Si \rightarrow \frac{Si}{4\pi}\left(\frac{A^*}{A}\right) = \frac{6}{m_o}\left[\frac{\left(m_1^*\right)^2 + 2m_1^* m_1^*}{3}\right]^{\frac{1}{2}} = 2.2$$

where m^*_t transverse mass and longitudinal mass in both cases. Therefore the barrier height for electrons moving from metal to semiconductor remains the same and thus it is not unaffected by a voltage. This means therefore that it must be equal and opposite to current, $J_{s \rightarrow m}$ with V= 0 at room temperature. Therefore;

$$J_{m \rightarrow s} = -A^* T^2 \exp\left(-\frac{q\phi_B}{kT}\right)$$

And from this expression, the total current density for majority carriers will be given by;

$$J_n = A^* T^2 \exp\left(-\frac{q\phi_B}{kT}\right)\left[\exp\left(\frac{qV_F}{kT}\right) - 1\right]$$

Diode Diffusion Theory

Diffusion current is very important especially in Schottky diodes. It is actually the current that flow from the semiconductor to the metal in which it experiences a barrier height ht>> kT. The effects of electron collision within depletion region is therefore included so that the carrier conductance at x = 0 and x = w are unaffected by current flow. However impurity concentration of the semiconductor is non-degenerate resulting into a current in depletion region to depend on local field and concentration gradient. Thus it is expressed as;

$$J_x = J_n = q\left(n(x)\mu_n \xi + D_n \frac{\partial n}{\partial x}\right)$$

$$= qD_n\left[-\frac{qn(x)}{kT}\frac{\partial V(x)}{\partial x} + \frac{\partial n}{\partial x}\right]$$

In cases where we have the steady state, the current density is independent of x and therefore integrating both sides of the above current expression with an integrating factor given by;

$$= \exp\left(-\frac{qV(x)}{kT}\right)$$

We obtain J_n as;

$$J_n = \int_0^w \exp\left[-\frac{qV(x)}{kT}\right]dx = qD_n n(x) - \frac{qV(x)}{kT}\Big\}_0^w$$

If in this expression, we substitute the boundary conditions given below,

$$qV(0) = -q\phi_B$$

$$qV(W) = -qV_n - qV$$

The majority carriers at the start of flow will be given by;

$$n(0) = N_c \exp\left[-\frac{E_c(0)-E_F}{kT}\right] = N_c \exp\left(-\frac{q\phi_B}{kT}\right)$$

And at the w, point, majority carriers will be given as;

$$n(W) = n = N_c \exp\left(-\frac{qV_n}{kT}\right)$$

At this stage, we can substitute the above boundary conditions so as to obtain;

$$J_n = \frac{qN_c D_n\left[\exp\left(\frac{qV}{kT}\right)-1\right]}{\int_0^w \exp\left[-\frac{qV(x)}{kT}\right]}dx$$

This gives the barrier potential at point any point x as;

$$qV(x) = \frac{q^2 N_D}{\epsilon_s}\left[Wx - \frac{x^2}{2}\right] - q\phi_B$$

$$J_n = \left\{ \frac{q^2 D_n N_c}{kT} \left[\frac{2q(V_i - V)}{\epsilon_s} \right]^{\frac{1}{2}} \exp\left(-\frac{q\phi_B}{kT}\right) \right\} \left[\exp\left(\frac{qV}{kT}\right) - 1 \right]$$

It should be noted at this point that similar to thermoinic expressions obtained above, drift current also depends on Temperature, T, though in most cases explanations based on diffusion current is dominant if **n**-type semiconductor is heavily doped.

Tunneling current

The term tunneling can simply be taken as to move through by or as by digging. When a semiconductor is heavily doped, the depletion region becomes very narrow. This forces electrons to tunnel through it as they move. As a result, the tunneling current J_T can be obtained from;

$$J_{s \to m} = J_T f \exp\left(\frac{qV}{E_o}\right)$$

Using the expression of obtaining potential E_0,;

$$E_o = E_1 \coth\left(\frac{E_1}{kT}\right)$$

We get that the energy E_1 can be expressed by;

$$E_1 = \frac{qh}{4\pi} \sqrt{\frac{N_d}{m_n^* \epsilon_s}}$$

Substituting the expressions and making J_T the subject of the expression, we get;

$$J_T = A^* T \sqrt{\frac{\pi E_1 (q\phi_B - qV - E_c + E_{Fn})}{k \cosh(E_1/kT)}} \cdot \exp\left[\frac{-E_c - E_{fn}}{kT} - \frac{q\phi_B - E_c + E_{fn}}{E_o}\right]$$

CHAPTER EIGHT
HIGH SPEED FIELD EFFECT TRANSISTORS

A transistor is defined as a semiconductor device capable of amplification. It has the base, emitter and the collector terminals. A transistor in which most current flows in a channel whose effective resistance can be controlled by a transverse electric field is called the field-effect transistor (FET). A junction transistor having an n-type semiconductor between a p-type semiconductor that serves as an emitter and a p-type semiconductor that serves as a collector is called a *pnp* transistor. Metal semiconductor field effect transistor (MESFETs) are mostly made out of GaAs. Most of the MESFETs contenders in the market for GaAs IC have applications in the Analog and digital applications areas, communication technology satellite, fibre optics, cell phones, oscillator and 168 GHz for amplifiers.

Since MESFETs have many applications, new materials for MESFET that include SiC and GaN that are known to have a wide band-gap are being tried. Some of these materials (SiC, GaN) have shown a higher breakdown voltage 100kV, higher thermal conductivity and higher electron velocity. It is also noted that GaN has higher electron velocity than GaAs. Another semiconductor material that is in used is SiGe. MESFETs are suitable for compound semiconductor in which it is difficult to fabricate oxide semiconductors. Silicon dioxide materials are easy to obtain from Si elements. Therefore most MOSFETs are used in either Si or SiGe. Their n-channel is always very good. That is why it is used as faster electron transporter than a holes transporter in most materials.

Gate channel Metal-Semiconductor or Schottky diode normally in the "*ON*" MESFETs depletion mode with zero gate have a negative threshold voltage used while on the normal "*OFF*" MESFETs in which its enhancement mode have positive threshold voltage on the gate. During fabrication, ion implantation and electrode metallization are needed for the fabrication of MESFETs on a Semi-insulating GaAs substrate. However simpler technologies are available to fabricate MOSFETs or HEMTs.

FET operation

There are two types of operation, the Normally "ON" type and the Normally "OFF" type. This can be illustrated from the simplified diagram shown below.

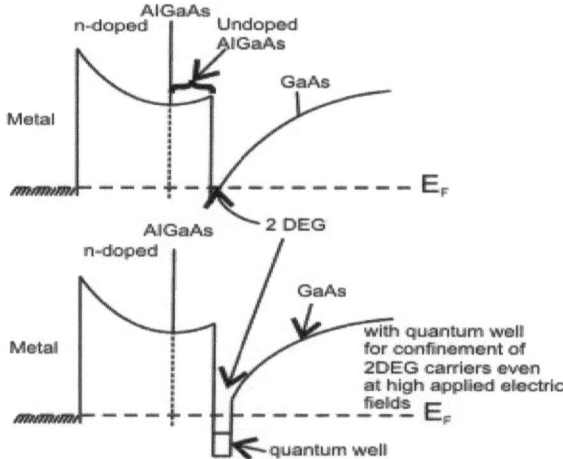

Taking L, as the length of the channel, Z as the width of the channel, then from the figure, we have the resistance of channel given as;

$$R = \frac{L}{q\mu_n N_d Z(a - W_o)}$$

Where W_o is taken as;

$$W_o = \sqrt{\frac{2\epsilon_s}{qN_D}}(V_i)$$

Where also V_i is taken as;

$$V_i = \phi_m - \phi_{SC}$$

In this condition, the channel acts like a constant resistor. If V_D in the expression increases the voltage distribution over channel changes as $V(x)$, causes the width of the depletion region to become;

$$W(x) = \sqrt{\frac{2\epsilon_s}{qN_D}\left[V_i + V(x)\right]}$$

Therefore, if $W(x)$ is given by the above expression then at W_{max} which occurs when source is grounded will be given by;

$$W_{max} = \sqrt{\frac{2\epsilon_s}{qN_D}(V_i + V_{DS})}$$

$$V_{DS} = V_D - V_S \cong V_D$$

This also implies that along the length of the channel resistance changes and therefore overall resistance of the channel becomes;

$$R = \frac{1}{q\mu_n N_D Z} \int_0^L \frac{dx}{a - W(x)}$$

It is noted also that as V_D increases it reaches a certain voltage and the current I_D saturates at;

$$V_{Dsat} = \frac{qN_D a^2}{2\epsilon_s} - V_i = V_p - V_i$$

From the above expression, V_P is called the pinch-off voltage and can be demonstrated as shown below.

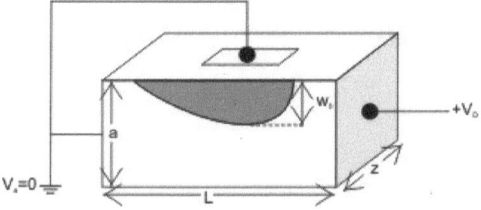

Also when $V_D > V_{Dsat}$, the additional voltage is found to appear across the depletion region so the depletion region widens further.

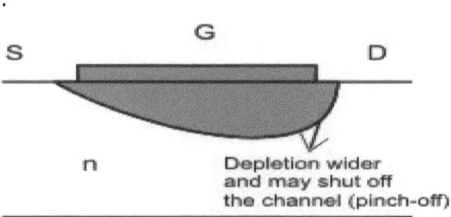

Based on the above conditions and for $W = a$, at $Vg<0$, we get V_{Dsat} as;

$$V_{Dsat} = \frac{qN_D a^2}{2\epsilon_s} - \left(V_i + |V_G|\right)$$

$$= V_p - \left(V_i + |V_G|\right)$$

Drain Current

From the figure above and in deriving the currents present in all the components, the following two assumptions are first made. That the will be a constant mobility ana the channel will be doped uniformly. Then considering the curve shown, we get that the drain current can be given as;

$$I_D = qN_D V_d(x) Z[a - W(x)]$$

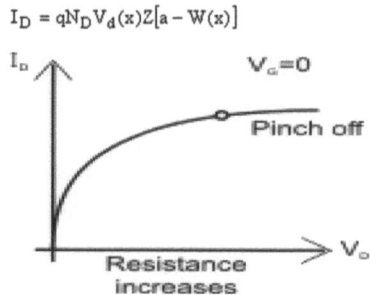

In the expression above, $V_d(x)$ is equal to electron drift velocity which as shown earlier is also given as;

$$= -\mu_n \xi_x$$

Or it can be expressed as;

$$= \mu_n \frac{dV}{dx}$$

$$W(x) = \sqrt{\frac{2\epsilon_s}{qN_D}[V_i - |V_G| + V(x)]} \quad V_G < 0$$

If this expression is substituted in the expression of drain current, I_D by putting $V_d(x)$ and $W(x)$ expression together and then integrating the combined expression from source, $x = 0$ to drain where $x = L$, then the drain current will be given by;

$$I_D = G_o \left\{ V_D - \frac{2}{3} V_p \left[\left(\frac{V_i + |V_G| + V_D}{V_p} \right)^{\frac{3}{2}} - \left(\frac{V_i + |V_G|}{V_p} \right)^{\frac{3}{2}} \right] \right\}$$

The constant here G_o is expressed as;

$$G_o = \frac{q\mu_n N_D Z}{L} \quad \text{for;} \quad 0 \leq V_D \leq V_{Dsat}$$

This is for only small drain voltages in the range of;

290

$$V_D \ll V_i - V_G$$

The expression for drain current, I_D, can best be expanded using the Taylor series and with the Retian 1st two terms to give;

$$\rightarrow I_D \simeq G_0 V_D \left[1 - \left(\frac{V_i + |V_G|}{V_p}\right)^{\frac{1}{2}}\right]$$

A plot of the I-V curve characteristics within the linear region given by V_D as V_G is made more and more negative causes I_D to decreases till finally the reduces to zero ($I_D = 0$) and the transistor to turn off or switches off. The turn off voltage or threshold voltage is therefore expressed as;

$$V_{th} = V_i - V_p$$

If g_m is taken as trans-conductance for small and then applied to obtain the drift voltage V_D in linear region as;

$$V_D = \frac{\partial I_n}{\partial V_G} = \frac{G_o V_D}{2\sqrt{V_p(V_i - |V_G|)}}.$$

This therefore gives the g_m value in saturation region as;

$$g_m = G_o \left[1 - \left(\frac{V_i + |V_G|}{V_p}\right)^{\frac{1}{2}}\right]$$

Using this expression, then the drain current in saturation region can also now be given as;

$$V_D = V_{Dsat} = V_p - V_i + |V_G|$$

So that the drain saturation current will finally be obtained as;

$$I_{Dsat} = G_0 V_p \left\{\frac{1}{3} - \left(\frac{|V_G| + V_i}{V_p}\right) + \frac{2}{3}\left(\frac{|V_G| + V_i}{V_p}\right)^{3/2}\right\}$$

If we let I_{Dsat} be independent of V_D, then the short channel MESFETs mobility μ, depends on the voltage, **E**. Therefore in the above equation, the value of V_G should be placed with sign.

Mobility in FETs

Mobility can be defined as the quality of moving freely. Thus from drain voltage, $V_d(x)$, we have expressed as;

$$V_d(x) = \frac{\mu_n |\mathcal{E}_x|}{1 + \frac{\mu_n |\mathcal{E}_x|}{V_D}}$$

Taking V_s as the electron saturation velocity, we can have E_s as;

$$E_s = \frac{dV}{dx}$$

$$I_D(x) = qN_D Z\mu_n a \left[1 - \sqrt{\frac{V_i - V_a + V(z)}{V_p}} \right] \frac{dV}{dx} \frac{1}{1 + \frac{\mu_n}{V_s}\frac{dV}{dx}}$$

Integrating the above expression from source to drain stages, we get;

$$I_D = \frac{G_o \left\{ V_D - \frac{2}{3}V_p \left[\frac{(V_i - V_a + V_D)^{\frac{3}{2}}}{V_p} - \left(\frac{V_t - V_G}{V_p}\right)^{\frac{2}{3}} \right] \right\}}{1 + \frac{\mu_n V_D}{V_s L}}$$

This last expression now can easily be used to plot a current against voltage (I-V) characteristics as shown below.

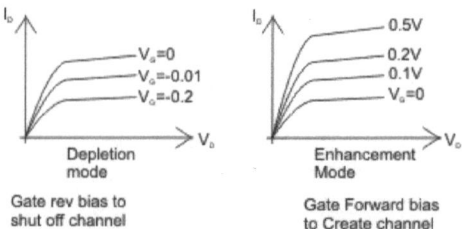

Depletion mode — Gate rev bias to shut off channel

Enhancement Mode — Gate Forward bias to Create channel

From the curves above, the reduction in I_D due to field dependent mobility show that G_o is same as before and the drain saturation voltage now ranges as;

$$V_{Dsat} < V_{pinch\ off}$$

In a reduced constant, gm can finally be given as;

$$g_m = \frac{\partial I_P}{\partial v_G}\bigg|_{V_a = const.}$$

Saturated velocity model

A model is an hypothetical description of a complex entity or process. The variations of velocity with electric field are shown respectively.

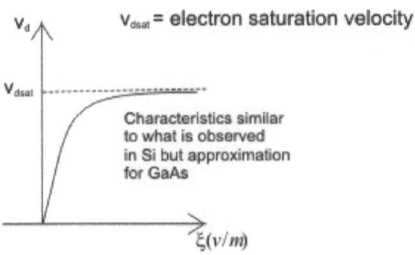

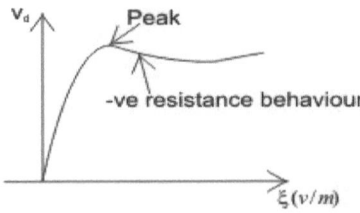

On this model, if we take it that at $V_d(x)$ is equal to V_s as;

$$v_d(x) = v_s$$

Then substituting this to obtain I_D independent of V_D, we get that drain current I_D can be given as;

$$I_D = qN_d . v_s Z[a - W] V_D$$

Then we get, W, as;

$$W = \sqrt{\frac{2\epsilon_s}{qN_D}(V_i - V_a - \alpha)}$$

When fabricating transistors and in practice, a combination of two expression below are used. These are;

$$v_d = \mu_n \xi_x$$

and

$$v_d = v_s$$

This analysis is better explained using experimental results in the V_D-I_D curve characters tics. The High Frequency Performance observed in transistors is due to small signal input current usually given as;

$$\tilde{i}_{in} = 2\pi f C_G \tilde{v}_G$$

In which the output current is given as;

$$\tilde{i}_{out} = g_m \tilde{v}_g$$

Based on these two equations, the cut-off frequency can be written in terms of g_m, N_D, mobility, μ and length, L as;

$$f_T = \frac{g_m}{2\pi C_G} \approx \frac{\mu_n q N_D a^2}{2\pi \epsilon_s L^2}$$

In the saturated velocity model as shown, we have the relation;

$$I_{Dsat} = Z(a - W)qN_D v_s$$

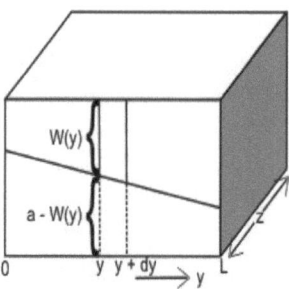

In this model, we obtain trans-conductance as;

$$g_m = \frac{Zv_s \epsilon_s}{W}$$

So as to give cut-off frequency as;

$$f_T = \frac{g_m}{2\pi C_G} = \frac{v_s}{2\pi L}$$

Hetero-junction in FETs

As in fast switching devices and for faster transit time in a RC circuit, the combined value of RC should be decreased. That is they are used in making HBTs/HEMTs. Two semiconductors with two different band-gaps can be grown one on top of the other or a material can be grown with variable band. When that is done and for a constant band-gap semiconductor, electrons and holes move in opposite direction once an electric field is applied. In a hetero-structure junction just shown, both can move same direction. Materials that have a higher Band Gap are denoted by **N** or **P** and those that have a lower Band Gap are denoted by **n** or **p**. Based on this argument we can have **pN, nP** or **p+ N** hetero-junction in which there are abrupt gradients. A few applications of hetero-structures are found in HBTs, diode lasers, LED, Photo-detectors, Quantum well devices and in solar cells.

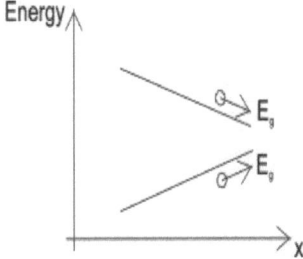

The materials that are grown together must be latticed matched. Few lattice matched compound semiconductors are;

$$Al_\alpha Ga_{1-\alpha}As \text{ and } GaAs, \; In_\alpha Al_{1-\alpha}As$$

and

$$In_\alpha Ga_{1-\alpha}As_{1-\beta}P_\beta$$

and

$$InP$$

and

$$In_\alpha Ga_{1-\alpha}As$$

All these are basically solid solution (or solution alloys of Al, As and GaAs) with no interface traps. To grow the above alloys, Molecular Beam Epitaxy (MBE) or Metal Organic Chemical

Vapour Deposition method (MOCVD) is used to growth these structures. Another method used is the Liquid phase epitaxy. When using Reflection High Energy Electron Diffraction method (RHEED) must be done calibration.it is noted here that for α < 0.45 the Band-gap minimum remains at, Γ, when keeping the alloy as direct band-gap and for α > 0.45 the band-gap minimum occurs at X making the material indirect band-gap.

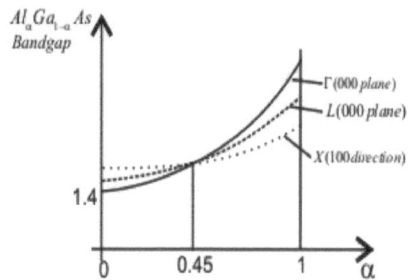

The energy diagram of the direct and indirect band-gap materials are shown below.

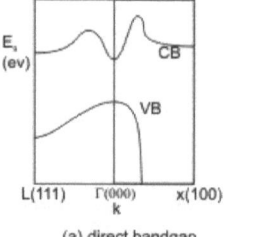
(a) direct bandgap

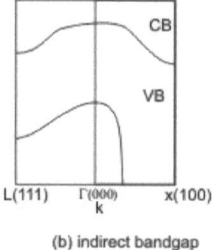
(b) indirect bandgap

We can use the above curves to analyze abrupt p-n junctions. Consider an abrupt p-n junction of **p** type GaAs and of **N** type AlGaAs such that it has more electron carriers than hole carriers. We can analyze it as follows. We first draw their energy band diagram side by side as shown. Then the two energy band diagrams obtained are brought in contact while keeping the E_F the same on both sides at this condition, electron flows from N side of **N** type AlGaAs to the p side and holes from p side to N side. The result is that a depletion region is formed.

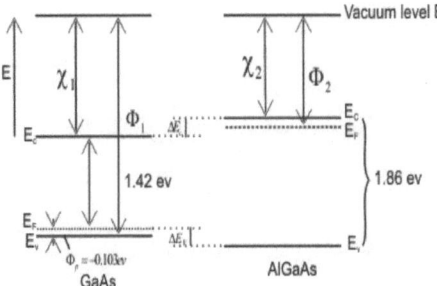

The flow continues till an equilibrium is attained and hence at thermal equilibrium, the Fermi level aligns itself on both sides as shown in figure 14.5. When this happens we get;

$$\Delta E_c = \chi_1 - \chi_2$$

$$\Delta E_v = (\chi_1 + E_{g1}) - (\chi_2 + E_{g2}) = \Delta E_c + E_{g1} - E_{g2}$$

The voltage, V_b becomes the sum of the contributions of V_{b1} and V_{b2}.

$$V_b = V_{b1} + V_{b2}$$

From this final expression, it is noted that the two sides of the abrupt junction created have two different permittivities and therefore applying the Poisson Equation for this condition we get the Poisson equation taking the form of;

$$\frac{d\xi}{dx} = -\frac{q}{\epsilon_p} N_A \quad \text{for} -x_p < x < 0$$

and

$$\frac{d\xi}{dx} = -\frac{q}{\epsilon_N} N_D \quad \text{for} \ 0 < x < x_N$$

When you critically analyze the two Poisson equations, we find that the E field is zero at $-x_p$ and x_N and therefore $\xi(x)$ can be obtained within the ranges from;

$$\xi(x) = -\frac{q}{\epsilon_p} N_A (x + x_p) \text{ for } -x_p < x < 0$$

and from

$$\xi(x) = -\frac{q}{\epsilon_N} N_D (x_N - x) \text{ for } 0 < x < x_N$$

The potential voltage profile can now be obtained by integrating $\xi(x)$ and assuming, $V(x_p) = 0$, so as to get potential voltage as;

297

$$V(x) = -\int \xi dx = \frac{q}{\epsilon_p} N_A \left(\frac{x^2}{2} + x_p x + \frac{x_p^2}{2} \right) \text{ for } -x_p < x < 0$$

From this equation, we can obtain the built-in potential voltage on the p-side as;

$$V_{bi} = \frac{q}{\epsilon_p} N_A x_p^2$$

While the built-in potential voltage on the N-side will be given by;

$$V_{b2} = \frac{q}{\epsilon_N} N_D x_N^2$$

From the two Poisson's equations, it is found that the electric flux density is continuous across the abrupt hetero-junction, hence we have;

$$\epsilon_p \xi(0^-) = \epsilon_N \xi(0^+)$$

And also this means that there is charge conservation. For charge conservation relationship to occur, we have;

$$N_A x_p = N_D x_N$$

It is therefore generally agreed that when $V \neq 0$ and by considering all the above expressions we can show that for applied X_p is given by;

$$X_p = \left[\frac{2N_D \epsilon_N \epsilon_p (V_{bi} - V)}{qN_A (\epsilon_n N_D + \epsilon_p N_A)} \right]^{\frac{1}{2}}$$

Further analysis also holds for the expression;

$$\frac{V_{b1} - V_1}{V_{b2} - V_2} = \frac{N_D \epsilon_N}{N_A \epsilon_p}$$

Substituting this expression to obtain X_p we get that at N, i.e. X_N;

$$X_N = \left[\frac{2N_A \epsilon_N \epsilon_p (V_{b2} - V)}{qN_D (\epsilon_n N_D + \epsilon_p N_A)} \right]^{\frac{1}{2}}$$

FET Current transport

It should be noted that in both the Junction diode and Metal-Semiconductor diode, there are many complex analysis required to fully find the minority and the majority currents that flow. Consider a p^{N+} junction and an N^{+n} junction shown below. If we assume this junction involve the Effective electron mass of;

$$Al_\alpha Ga_{1-\alpha} As$$

where

$$m_e^* = m_T^*$$

The mathematical calculation shows that in the range $\alpha < 0.45$, we get;

$$= (0.067 + 0.083\alpha)\, m_0 \quad \text{for} \quad \alpha < 0.45$$

$$m_e^* = \frac{1}{6}\left(2m_{tx}^* + 4\sqrt{m_{lx}^* m_{tz}^*}\right) \quad \text{for} \quad \alpha \geq 0.45$$

This is where the masses m^*_{tz} and m^*_{lx} are given as;

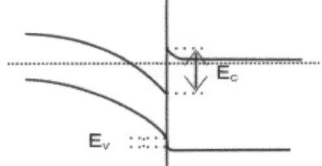

$$m_{tz}^* = (0.23 - 0.04\xi)m_0$$

$$m_{lx}^* = (1.3 - 0.2\xi)m_0$$

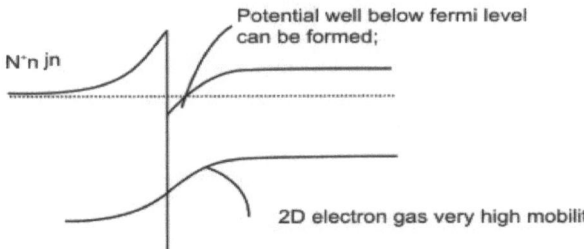

When these equations are combined, we get that the Effective hole mass will be given by;

$$m_h^* = \frac{m_{hh}^{*\frac{3}{2}} + m_{lh}^{*\frac{3}{2}}}{m_{hh}^{*\frac{1}{2}} + m_{lh}^{*\frac{1}{2}}} \quad 0 \le \alpha < 1$$

In this case the masses m^*_{lh} and m^*_{hh} are given by;

$$m_{lh}^* = 0.087 + 0.063\alpha$$

$$m_{hh}^* = 0.62 + 0.14\alpha$$

Using these expressions, show that the resulting Band-gap for $\alpha < 0.45$ is given by;

$$E_g(Al_\alpha Ga_{1-\alpha}As) = 1.424 + 1.247\alpha$$

and the change in potential will also be obtained as;

$$1.9 + 0.125\alpha + 0.143\alpha \quad \text{for} \quad \alpha \ge 0.45$$

$$\Delta E_v(\alpha) = 0.55\alpha$$

The potential voltage drop at the conduction band in will be given by;

$$\Delta E_c(\alpha) = 0.697\alpha \quad \text{for} \quad \alpha \le 0.45$$

and that of the valence band will be;

$$\Delta E_v(\alpha) = 0.476 - 0.425\alpha + 0.143\alpha^2 \quad \text{for} \quad \alpha > 0.43$$

The dielectric constant of this material is therefore obtained as;

$$\epsilon_s(\alpha) = (13.18 - 3.12\alpha)\epsilon_0 \quad F/cm.$$

While the electron affinity for the range $\alpha < 0.45$ is given by;

$$q\chi_e(\alpha) = 4.07 - 0.697\alpha$$

and in $(Al_\alpha Ga_{1-\alpha}As)$ the electron affinity for the range $\alpha < 0.45$ given by;

$$q\chi_e(\alpha) = 3.594 + 0.4253 - 0.143\alpha^2 \quad \text{for} \quad \alpha > 0.45$$

As a result the conduction band density of states is related to m^*_{de} by;

$$N_c = 2\left(\frac{2\pi kT}{h^2}\right)^{\frac{3}{2}} m_{de}^*$$

This is in the range $\alpha \le 0.45$, when;

$$m^*_{de} = m^*_\Gamma$$

and this means that the mass, m^*_{de} is given by;

$$m^*_{de} = 6^{\frac{2}{3}}\left[m^*_{\ell z} m^{*'}_{tx}\right]^{\frac{2}{3}} \quad \text{for } \alpha \geq 0.45$$

Combining all these expressions, the valence band density of states is given by;

$$N_v = 2\left(\frac{2\pi kT}{\hbar^2}\right) m^*_{dh}$$

Where the mass m^*_{dh} is expressed as;

$$m^*_{dh} = \left(m^{*\frac{3}{2}}_{lh} + m^{*\frac{3}{2}}_{hh}\right)^{2/3} \quad \text{for } 0 \leq \alpha \leq 1$$

If we take τ_f as the forward transit time which is also equal to the average time an electron spends in the Base region, then this time can be approximated to;

$$\tau_f \cong \frac{W_B^2}{2\eta D_n}$$

Where, W_B is the physical base width, η is the Fudge factor which is ≈ 1 and D_n is the electron diffusion constant but in practice W_B is kept below 100nm (W_B < 100nm).

BJT Models

First Order Model of BJT

This model is also referred to as the CE-forward active model. This is based on frequency and thus we can define a frequency, f_T as at short circuit current as;

$$f_T = \frac{g_m}{2\pi C_d} = \frac{1}{2\pi \tau_f}, \quad \text{which is the frequency}$$

At this frequency, the short circuit current gain is expressed as;

$$h_{fe} = \frac{I_c}{I_b} = \beta$$

In this case I_b is finite base current which can be obtained from;

$$I_e = I_b + I_c$$

From analogue electronics, we get that the emitter base current, I_e is given by;

$$= \frac{I_B}{\alpha}$$

This means that the collector current, I_c is given by;

$$\text{or} \quad I_c = \frac{I_b}{\alpha} \exp\left(\frac{qV_{be}}{kT}\right)$$

In this expression, β, is the low frequency current gain in the transistor and thus the circuit i show the first order model of the BJET.

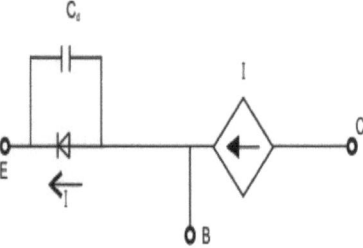

From this figure, the current gain is modeled as;

$$f_T/f = h_{fe}$$

So as to give collector current as;

$$I_c = I_s \exp\left[\frac{V_{be} - I_b R_B - I_e R_E}{V_T}\right]$$

Since in semiconductors, electrons diffuse, then the electrons diffusing across base are related to frequency as;

$$\tau_f = \tau_b$$

This gives the total transit time as;

$$\tau = \tau_f + \tau_e + \tau_{sc} + \tau_c$$

Making the cut-off frequency to be expressed as;

$$f_T = \frac{1}{2\pi\tau}$$

where τ_c is the time required to change the V_b by charging up the capacitances through base - junction resistance and hence expressed as;

$$\tau_c = \frac{V_T}{I_c}\left(C_{jBE} + C_{jBC}\right)$$

In capacitors, the space charge transit time, τ_{cs} is time required for an electron to drift through depletion region in BC junction. This space charge transit time where W_{dep} is the Base-Collector depletion region width is given by;

$$\tau_{sc} = \frac{W_{dep}}{2V_{act}}$$

This expression modifies the collector charging time τ_c to given as;

$$\tau_c = (R_E + R_c)C_{jBC}$$

In which R_E and R_c are emitter and collector resistances and this model is further expressed through the Gummel-Poon model shown, In this model we have the expression for $-g_m\tau_o$ as;

$$A_v = \frac{\partial V_o}{\partial V_i} = \frac{\partial I_c}{\partial V_i}\frac{\partial V_o}{\partial I_c} = -g_m \tau_0 = -\frac{V_o}{V_T}\omega\left(V_T = \frac{kT}{q}\right)$$

This effect can be expressed diagrammatically as depicted.

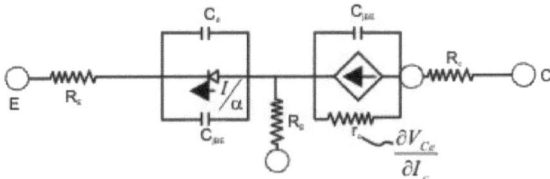

The BJT Current Model

In making very good quality BJT, there is need to always reduce the base width so that unwanted base current is reduce. The current model of BJT is that electrons are injected from the emitter and they diffuse across the Base. When there is a good design, then electron transfer from emitter to collector is maximized and I_b current which is due to the hole back injection is minimized. Electron diffusing across the base which is finally collected in the collector point gives rise to the collector current I_c given by;

$$I_c = \left(\frac{qA_{emit} D_{nbase} n^2_{ibase}}{W_{base} N_{base}}\right)\exp\left(\frac{qV_{BE}}{kT}\right)$$

In the expression, A_{emit} is the effective emitter area, D_{nbase} is the minority diffusion coefficient, W_{base} is the Base thickness, n_{ibase} is the intrinsic concentration in base and N_{base} is the base doping level. The base current due to hole back injection when derived from the above I_c will then be given as;

$$I_{Bback-inject} = \frac{qA_{emit}}{W_{emitter}} \cdot \frac{D_{Pemit} n_{iemit}^2}{N_{emit}} \exp\left(\frac{qV_{BE}}{kT}\right)$$

When these currents are compared and their ratios of desired to undesired current component are related, we get;

$$\frac{I_c}{I_{b_{in}}} = \frac{W_{emit}}{W_{base}} \cdot \frac{D_{nbase}}{D_{pemit}} \cdot \frac{N_{emit}}{N_{base}} \cdot \frac{n_{ibase}^2}{n_{iemit}^2}$$

And from these, the ratio given by;

$$\frac{W_{emit}}{W_{base}} \cdot \frac{D_{nbase}}{D_{pemit}}$$

is roughly of the order of 1 and by controlling or by controlled by doping, we get

$$\frac{N_{emit}}{N_{base}}$$

This means also that o make the ratio much higher than 1 then the base doping should be lower than the emitter doping such that;

$$\frac{N_{emit}}{N_{base}} \gg 1$$

This works for most transistors, however for high speed operation the base resistance R_B and the junction capacitance C_{BE} are required to be low. R_B decreases with increase in N_{base} and C_{BE} decreases with decreasing N_{emit}. Therefore the given by $N_{emit}/N_{base} \gg 1$ is very difficult to achieve. So high speed device performance, both (f_T) and high $\beta(I_c/I_B)$ ratio cannot be both achieved simultaneously. Thus optimization is required which gives recourse of keeping the N_{emit} low and N_{base} moderate. However, this increases the electron injection efficiency by using a hot electron injection and field assisted base transport. This is a concept used in HBT transistors.

Hetero-junction Bipolar Transistor (HBT)

When we consider the model above, it is clear that under normal homo junction BJT performance, the transistor obeys the ratio given by;

$$\frac{I_c}{I_{Bback}} = \frac{W_{emit}}{W_{base}} \frac{D_{nbase}}{D_{pemit}} \frac{N_{emit}}{N_{base}}$$

In the Hetero-bipolar transistor, the emitter is usually of wider band gap and can be represented as shown.

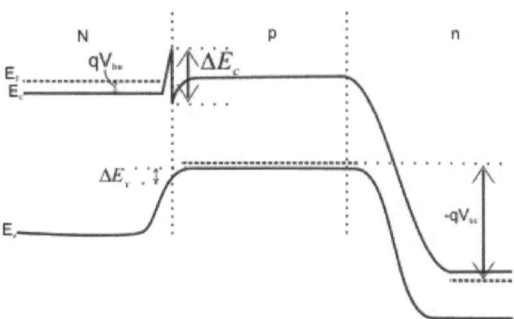

The electron can be easily injected from the emitter to the base but holes from p to the N-emitter experience a much larger energy barrier given by ΔE_v and hence the current I_b due to hole back injection is reduced. Thus the base region can be made highly doped which reduces the base resistance R_B. This is one of the advantages of HBT over BJT. Therefore in an abrupt HBT, we get;

$$\frac{I_c}{I_{BBack}} = \frac{W_{emit}}{W_{base}} \frac{D_{nbase}}{D_{pemit}} \frac{N_{emit}}{N_{base}} \exp\left(\frac{\Delta E_v}{kT}\right)$$

This shows that instead of using abrupt HBT transistor, we can also use graded HBT made of the materials from;

$$Al_\alpha Ga_{1-\alpha} As / GaAs$$

$$In_\alpha Ga_{1-\alpha} As / InP$$

Here the aluminum concentration is graded and hence called Graded Band Gap heterostructure. In a Graded Band Gap Hetero-junction the barrier for the hole can be made larger then electron barrier by ΔE_g and we get;

$$\frac{I_c}{I_{BBack}} = \frac{W_{emit}}{W_{base}} \frac{D_{nbase}}{D_{pemit}} \frac{N_{emit}}{N_{base}} \exp\left(\frac{\Delta E_g}{kT}\right)$$

Therefore the ratio, I_c/I_{BBack} can be made even large than abrupt Hetero-junction if design parameters are fine tuned.

Example:

Consider the following parameter obtained as shown below;

$$N_{emit} = 5 \times 10^{17} \text{ cm}^{-3} \quad W_{emit} = 1300 A^0$$

And also;

$$N_{base} = 4 \times 10^{19} \text{ cm}^{-3}$$

$$W_{base} = 1000 A^0$$

$$D_{nbase} = 20, \; D_{pemit} = 2 \text{cm}^2/V-s$$

Then for $Al_{0.3}Ga_{0.7}As/GaAs$ graded hetero junction we get;

$$\Delta E_g = 0.3 \times 1.247 = 0.374 eV$$

And the ratio of I_c/I_{BBack} is obtained as;

$$\frac{I_c}{I_{BBack}} = \frac{1300}{1000} \cdot \frac{205 \times 10^{17}}{24 \times 10^{19}} \exp\left(\frac{0.374}{0.0258}\right) = 2.5 \times 10^5$$

It is noted that when the ratio $I_c/I_{BBack} = 0.16$, then the Hetero-junction device is useless and when the $\Delta E_v = 0.55 \times 0.165$ and is an abrupt Hetero-junction of $Al_\alpha Ga_{1-\alpha} As/GaAs$, then it is a good material for Hetero-junction;

$$\frac{I_c}{I_{BBack}} = 0.16 \times \exp\left(\frac{0.165}{0.028}\right) = 97$$

Other good Hetero-junction materials include;

$$- Ga_{0.51}In_{0.49}P/GaAs$$

$$In_{0.50}Ga_{0.47}As/InP$$

$$In_{0.52}Al_{0.48}As/In_{0.53}Ga_{0.47}As$$

$$Si/Gi_{0.8}Ge_{0.2}$$

The structure of a HBT device is shown that is grown on **n** substrate. The various components of the Base currents are shown below.

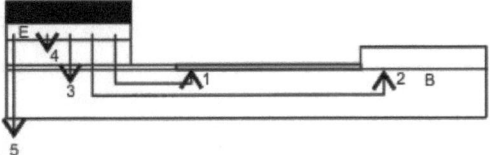

They include the following;

- Back injection of holes
- Extrinsic base surface recombination current
- Base contact surface recombination current
- Bulk recombination current in Base layer
- Depletion region recombination current in B-E depletion region.

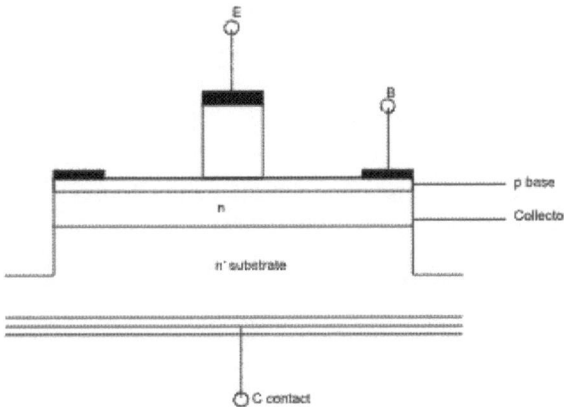

Bulk recombination current in the base region (I_{Bbulk}) is the dominant base current and the current gain is given by;

$$\beta = \frac{I_C}{I_{Bbulk}} = \frac{\tau_n}{\tau_b}$$

Where, τ_n is the minority electron recombination lifetime in Base and τ_b is the minority carrier transit time in Base. Thus when the Base is too thin, we have;

$$\tau_b = \frac{W_{base}^2}{2D_{rbase}}$$

τ_b can be decreased by an E field in base arising due to the linearly grading the bandgap in the Base region. A linearly graded Si/SiGe HBT is shown.

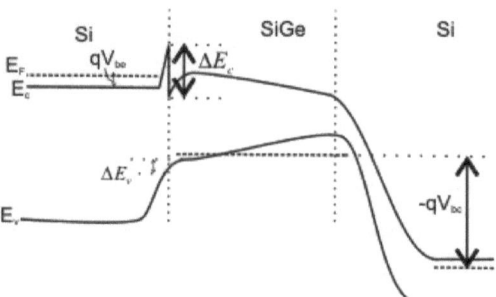

Thus the Poisson Equation is given by;

$$\frac{dE}{dx} = \frac{q}{\epsilon_s}(p - n + N_d - N_a)$$

Base-Collector junction where the Base doping is greater than the Collector Doping the depletion region is mostly inside the Collector, where the Electric field is high and electrons travel mostly with saturation velocity.

The Electron Carrier Concentration

The electron carrier concentration inside the collector is given by;

$$n(x) = \frac{J_c}{qV_{sat}} \quad \text{where } J_c$$

This is the collector current density and thus the Electric field in base - collector depletion region is given by;

$$\frac{dE}{dx} = \frac{q}{\epsilon_s}\left(-\frac{J_c}{qV_{sat}} + N_D\right) \qquad N_D = N_{collected}$$

When collector current density, J_c is small;

$$\frac{dE}{dx} = \frac{q}{\epsilon_s}N_C$$

where N_C is the Collector doping concentration. As J_c increases the ratio dE/dx slope becomes more negative as shown.

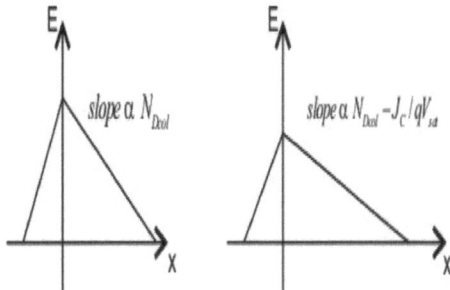

While the current density increases and the area under the field profile should remain constant the depletion region thickness would continue to increase until it reaches $x = X_c$ as shown below.

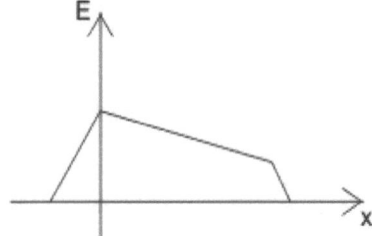

As the current density increases to a level such that $x = X_c$ the net charge inside the junction becomes zero and the field profile is constant as shown below.

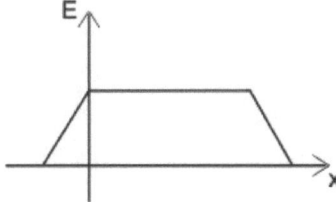

When J_c increases further such that, $x > X_c$ the net charge inside the junction becomes negative, the electric field takes negative values as shown below.

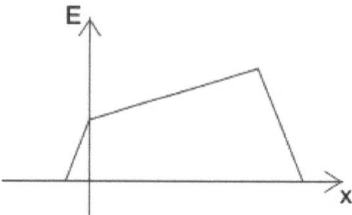

When there is no more field to prevent holes from spilling, base push-out or Kirk effect occurs as shown. The current gain decreases as the transit time associated with the thickened base layer increases;

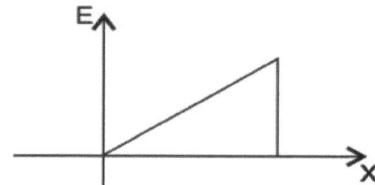

Emitter-Collector transit time

Emitter-Collector transit time is;

$$\tau_{ec} = \frac{1}{2\pi f_T} = \tau_e + \tau_b + \tau_{sce} + \tau_c$$

Where is gives τ_c as;

$$\tau_c = \frac{\eta KT}{qI_c}(C_{je} + C_{jc})$$

τ_c is the time required to change the base potential by charging the capacitances, $C_{je(BE)}$ (B-E junction capacitance) and $C_{je(BE)}$ (B-C junction capacitance) through the differential Base-Emitter junction resistance. Thus τ_b is the base transit time that can be expressed as;

$$\tau_b = \frac{W_{base}^2}{\eta D_{rbase}} \quad (\eta \sim 2)$$

As such the transit time through B-C depletion region will be given as;

$$\tau_{sc} = \frac{W_{Dep}^2}{2V_{sat}}$$

And the Collector charging time

$$\tau_c = (R_E + R_C) C_{jBC}$$

Hence the cut-off frequency is

$$f_T = \frac{1}{2\pi \tau_{ec}}$$

Hetero-junction FET

Hetero-junction FET are also called HEMT which stands for High Electron Mobility Transistor or TEGEFET which are also known to be two Dimensional Electron Gas (2DEG))or MODFET which stands for Modulation Doped Field Effect Transistor devices. In these devices, a potential is formed at junction of two dissimilar semiconductors (e.g. Al-GaAs/GaAs) where E_F is higher than the occupation levels of the electrons in the conduction band. The electrons are accumulated in this potential well and form a sheet of electron similar to the inversion layer in a MOS structure. The thickness of this formed sheet is $\leq$ 10nm, smaller than the De-Broglie wavelength of the electron in that very material. This sheet of free electrons then behaves like free atoms in a gas. That is why it is called electron gas. In this structure has been observed that they can have a mobility of about;

$$\mu_n \text{ of } 9 \times 10^6 \text{ cm}^2/v-s \text{ at 2k}$$

It can also be observed that this structure is similar to of the FET with 2DEG (Two Dimensional Electron Gas) as channel. Thus application of a bias voltage to gate modulates charges in the 2DEG and channel conducts current. This is similar to what is observed in FETs which are faster than most MESFET operations. The formation of 2DEG is shown.

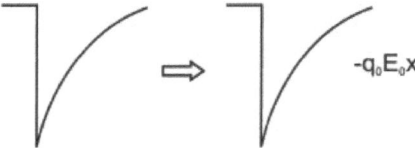

Quantum Well Picture

It can be seen that if the temperature is high or the applied field (V_{DS}) is high, the electrons are excited to high energies and may escape from the triangular well. It may also be scattered into conduction band of the barrier (spatial transfer) and the carriers would be lost from the channel. To avoid this, a quantum well may be introduced at the interface. This case we work at several possibilities of hetero-structures and thus charge transfer occurs leading to a conducting channel within single quantum well or multiple quantum well. The energies of the electrons are quantized in the step like density of states. Therefore many carriers can be put in the channel with a narrower dispersion of energies. Electron wave inside well is using the Kane model will be given by;

$$\psi = \sum e^{jkr} u_{ck}(r)\psi_n$$

In this expression, x is the growth direction and k is the transverse electron wave vector. Thus if the $u_{ck}(r)$ is the is the block wave form where ψ_n is the envelope wave function then the solution will be given as;

$$\left[-\frac{h^2}{2m^*(x)}\frac{\partial^2}{\partial x^2} + V_c(x)\right]\psi_n(x) = \varepsilon_n \psi_n(x)$$

Where m*(z) is the effective mass, $V_c(z)$ is the potential and ε_n is the confinement energy of carriers. Since the boundary conditions should be continuous at the interfaces, then they must be given as;

$$x_n(x) \quad \text{and} \quad \frac{1}{m^*(x)}\frac{\partial x^n}{\partial x}$$

The triangular potential well is shown and therefore the triangular quantum well Potential given as V(X) is linear for $x > 0$ and ∞ at $x = 0$. This causes the Schrödinger equation to be modified as;

$$-\frac{h^2}{2m}\frac{d^2\psi_n(x)}{dx^2} + qE_o x \psi_n(x) = E_n \psi_n(x)$$

In the above expression the boundary condition is given as;

$$\psi_n(0) = 0$$

Using the boundary conditions, the above equation has two independent solutions. One solution that is non-singular found at $x \to \infty$ and it is the **AIRY function** (A_i). Thus the resulting wave function as shown will be given as;

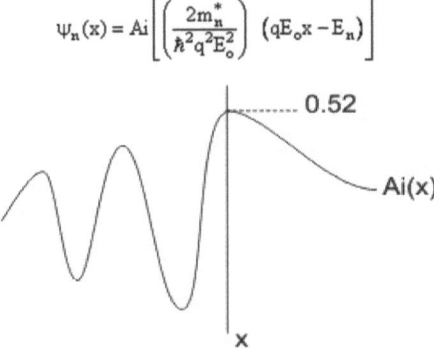

In summary therefore the quantized energy levels will be given by;

$$E_n = -\left(\frac{\hbar^2 q^2 E_o^2}{2m_n^*}\right)^{\frac{1}{3}} a_n$$

An expression where a_n is the nth zero of $A_i(x)$, hence the value of a_n can be obtained as;

$$a_n \cong -\left[\frac{3\pi}{2}\right]\left(n+\frac{3}{4}\right)^{\frac{2}{3}} \qquad n = 0,1,2\ldots\ldots\ldots$$

So that at n^{th} potential voltage, the energy can be given as;

$$E_n \cong \left(\frac{\hbar^2}{2m}\right)^{\frac{1}{3}}\left[\frac{3\pi q E_o}{2}\left(n+\frac{3}{4}\right)\right]^{\frac{2}{3}}$$

The basic idea of HFET or HEMT also known as MODFET- Modulation Doped Field Effect Transistor is that at equilibrium charge transfer occurs at the hetero-junction to equalize the Fermi level on both side. Doping the N side gives wide base. Electrons are transferred to the GaAs side until equilibrium is reached. This occurs because electron transfer raises the Fermi level on the GaAs side due to filling of the conduction band by electrons and also raises the electrostatic potential of the interface region because of the more numerous ionizer donors in the AlGaAs side. This charge transfer effects make possible an old dream of semiconductor technologist, i.e. getting conducting electrons in a high mobility, high purity semiconductor without having to introduce mobility limiting donor impurities.

Assuming that before the charge transfer the potential is flat band. After charge transfer of N_s electrons the electric field in the potential well created can be taken as constant to first order and is therefore expressed as;

$$E_o = \frac{qN_s}{\epsilon_o \epsilon_2}$$

Through the Gauss' Electrostatic potential, we get;

$$\phi(x) = -E_o x \quad \text{for} \quad x > 0$$

This now changes the Schrödinger equation for the electron envelope wave function so that it is expressed as;

$$\left[\frac{p^2}{2m^*} - q\phi(x)\right]\psi(x) = E\psi(x)$$

Therefore, the energy level in infinite triangular potential well at the ground state, (n = 0) is given by;

$$E_1 = \left(\frac{\hbar^2}{2m_n^*}\right)^{\frac{1}{3}} \left(\frac{\frac{9}{8}\pi q^3 N_s}{\epsilon_o \epsilon_2}\right)^{\frac{2}{3}}$$

This can then be summarized as equal to E_1 as;

$$= \gamma N_s^{2/3}$$

Where the constant γ is usually determined experimentally to be equal to;

$$\gamma = \left(\frac{\hbar^2}{2m_n^*}\right)\left[\frac{\frac{9}{8}\pi q^3}{\epsilon_o \epsilon_2}\right]^{\frac{2}{3}}.$$

As charge transfer increases, potential created by transferred electrons also increases. This leads to the lowering of the bottom of the Conduction and thus a 2DEG is formed when the Conduction band goes below the Fermi level and thus we get;

$$P_{2D} = \frac{m_n^*}{\pi \hbar^2}$$

This means that the energy in channel is thus given by;

$$E_1 + \frac{\pi \hbar^2 N_s}{m_n^*}$$

The above observation does not take place in all materials but in AlGaAs Fermi energy level is pushed downwards by electrostatic potential voltage V_{dep} built up at the interface is given by;

$$V_{dep} = \int_{x=0}^{x=W} E \, dx$$

Where W is the optimum width of depletion region (V_{dep}) in AlGaAs material hence V_{dep} can also be expressed as;

$$V_{dep} = \int \frac{qN_D x}{\epsilon_s \epsilon_1} \, dx = \frac{qN_D W^2}{2\epsilon_o \epsilon_1}$$

In the above expression, N_D is the donor on AlGaAs and therefore calculating the energies from the bottom of Conduction Band, we get;

$$\Delta E_c = \epsilon_1 + \frac{\pi \hbar^2 N_a}{m^*} + E_d + qV_{dep}$$

and thus ΔE_F will be given as;

$$\Delta E_F = \epsilon_1 + \frac{\pi \hbar^2 N_a}{m^*}$$

In AlGaAs E_d is the donor binding energy in which the donor concentration, N_D is equal to N_s. this is also equal to the number of electron transferred and hence we have;

$$\Delta E_c - \left(\frac{\hbar^2}{2m^*}\right)^{\frac{1}{3}} \left(\frac{8}{8}\pi q^2 \frac{N_a}{\epsilon_o \epsilon_s}\right) + \frac{\pi \hbar^2 N_a}{m_n^*} + \epsilon_d + \frac{q^2 N_s^2}{2\epsilon_o \epsilon_1} \frac{1}{N_D} = 0$$

From the above equations N_s can be calculated if the other parameters are known. In most cases, the Fermi level is determined empirically by the model given by;

$$\Delta E_F = \Delta E_{F_c}(T) + N_s$$

In practice un-doped AlGaAs spacer layer of thickness, W_{ap} is used to separate Donor atom from channel electrons to prevent coulomb interactions resulting in an increased mobility, N_s is decreased. The charge in the conduction band is therefore given as;

$$\Delta E_c = qV_{dep} + qV_{sp} + \in_d + E_1 + \frac{\pi h^2 N_\alpha}{M^*}$$

Where the voltage V_{sp} is given by;

$$V_{sp} = \frac{qW_{sp}N_s}{\in_o \in_R}$$

Using the above expression, we can calculate the 2DEG charge density N_s and it can then be related to the gate voltage, V_g as;

$$qN_s = C_s(V_g - V_m)$$

$$V_{DS} = 0$$

Where, C_s is the 2DEG capacitance per unit area which can also be expressed as;

$$C_s = \frac{\in_o \in_r}{\Delta d + W + W_{sp}}$$

$$E_1 = \left(\frac{\hbar^2}{2m_n^*}\right)^{\frac{1}{3}} \left(\frac{9}{8} \frac{\pi q^3 N_s}{\in_o \in_2}\right)^{\frac{2}{3}}$$

Which then gives E_1 as earlier obtained as equal to;

$$= \gamma N_s^{2/3}$$

Where γ is usually determined experimentally and obtained as;

$$\gamma = \left(\frac{\hbar^2}{2m_n^*}\right) \left[\frac{9}{8} \frac{\pi q^3}{\in_o \in_2}\right]^{\frac{2}{3}}$$

As charge transfer increases, potential created by transferred electrons also increases, leading to the lowering of the bottom of the Conduction. A 2DEG is formed when the Conduction band goes below the Fermi level, hence we get

$$P_{2D} = \frac{m_n^*}{\pi h^2}$$

Thus the energy in channel will also be given by;

$$E_1 + \frac{\pi \hbar^2 N_s}{m_n^*}$$

In AlGaAs Fermi energy level is pushed downwards by electrostatic potential voltage V_{dep} built up at the interface, where it will be given by;

$$V_{dep} = \int_{x=0}^{x=W} E\, dx$$

Where W is the width of depletion region (V_{dep}) in AlGaAs and is given by;

$$V_{dep} = \int \frac{qN_D x}{\epsilon_s \epsilon_1}\, dx = \frac{qN_D W^2}{2\epsilon_o \epsilon_1}$$

Where, N_D is the donor on AlGaAs and by calculating energies from the bottom of Conduction Band, we get;

$$\Delta E_c = \epsilon_1 + \frac{\pi \hbar^2 N_s}{m^*} + E_d + qV_{dep}$$

Where, Δd is the distance of the centroid of 2DEG from $x = 0$ and is usually in the range of;

$$\Delta d \sim 80 A^\circ \quad \text{for} \quad N_s = 10^{12}/cm^2$$

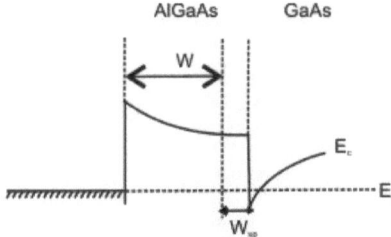

Again the threshold voltage or pinch off voltage is given by;

$$V_{th} = \phi_b - \frac{qN_D W^2}{2\epsilon_o \epsilon_1} - q(\Delta E_c - \Delta E_{Fo})$$

where, Φ_b is the Schottky barrier height on the donor layer as shown.

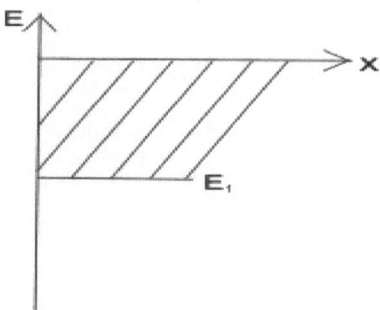

At room temperature, V_g also modulates the bound carrier density in Donor layer and free electron in Donor layer. For a simplest model V_{DS} is so large that all electrons in 2DEG channel move with V_{sat} independent of V_{DS} and therefore;

$$I_{DS} = n_s Zq V_{sat}$$

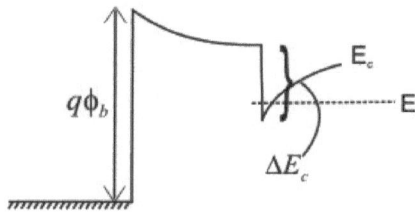

In the above expression, n_s is the electron density per surface area and is given by;

$$n_s = \frac{N_g}{ZL_g}$$

Where, z is the gate width and L_g is the gate length and thus this now causes transconductance to be expressed as;

$$g_{m0} = \frac{\partial I_{DS}}{\partial V_s} I_{sat} \cong \frac{g_{mo}}{2E_c L_g}(V_g - V_{Th})^2 = ZqV_{sat}\frac{\partial n_s}{\partial V_S} = \frac{CV_{sat}}{L_g}$$

At this stage we have the simplest model in which we have;

$$\frac{g_{m0}}{c} = \tau = \frac{V_{sat}}{L_g}$$

Therefore with $E < E_c$ we have, $v = \mu E$ and with $E > E_c$ we have $v = v_{sat}$ then the voltage in the channel will be given by;

$$V_{cb} = V(x)$$

and the current is given by;

$$I = Zqh_s(y)v(y) = \frac{c}{L_g}[V_g - V_{Th} - V(y)]\mu\frac{dV}{dy} \quad \text{for} \quad E = \frac{dV}{dy} < E_c$$

Hence current is given by;

$$I(y-y_s) = \mu\frac{c}{L_g}\int_{V(y)}^{V}[V_g - V_{Th} - V']dV'$$

Where y_s is the position of entrance to channel on the source side and therefore the saturation current is the current for which field on the drain side at $y_s + L_g$ just reaches $V_{D,sat}$ and can be expressed as;

$$I_{sat} = g_{mo}\left\{\sqrt{\frac{(V_g - V_{Th} - R_s I_{sat})^2}{+E_c^2 L_g^2}} - E_c L_g\right\}$$

Where, R_s is the source resistance. Thus for

$$E_c L_g \ll V_g - V_{Th} - R_s I_{sat}$$

We express current I_{sat} as;

$$I_{sat} \cong \frac{g_{mo}}{1 + R_s g_{mo}}(V_g - V_{Th} - E_c L_g)$$

This forms a linear behavior for a short highly conductive channel in which the I-V characteristic are shown.

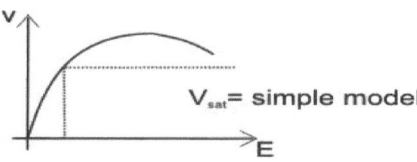

CHAPTER NINE
SWITCHING IN HIGH SPEED BIPOLAR JUNCTIONS

Semiconductors and junctions

A semiconductor has been defined as a material that has a conduction character that is lower than metals but higher than insulators at room temperature and this is based on the fact that a semiconductor always has at least an intrinsic carrier concentration as n_i. Since there exists two types of charged carriers in all semiconductors, they both aid to carry current. These are the electrons that have a negative charge and the holes that are assumed to have a positive charge as;

$$q = \pm e = \pm 1.6 x\, 10^{-19}\, C = \pm 1eV$$

It should be noted that only "free" charges carry currents where the term "Free" means that they are not bonded to any atoms. Increase or decrease of current or simply the conduction in a semiconductor can be changed through doping. Doping is defined as the introduction of foreign atoms such as B or As in Si in a material in small quantities. Doping can be done with donors N_D and thus forms donor concentration that result into an n-type semiconductor. In such a case, the density of free electrons n_n becomes larger than the density of free holes p_n present in the same n-type material. When doping is done to introduce acceptors, N_A the acceptor concentration result into a p-type material and therefore the density of free holes p_p present is larger than the density of free electrons n_p.

Therefore the intrinsic carrier concentration in intrinsic semiconductors is always given by:

$$n_i = p_i$$

This shows that the same density of free electrons and holes are found in an intrinsic semiconductor. On the other hand, in the extrinsic n-type semiconductor we have the relations:

$$\begin{cases} n_n = N_D \\ p_n = \dfrac{n_i^2}{N_D} \end{cases}$$

It is also agreed that in the extrinsic p-type semiconductor, we have the relations;

$$\begin{cases} n_p = \dfrac{n_i^2}{N_A} \\ p_p = N_A \end{cases}$$

In the above relationships, n, p the electron and hole density respectively; N_D and N_A represents the donor and acceptor doping density concentrations respectively. This will result into a total conductivity of this semiconductor to be given by;

$$\sigma_{tot} = e(n\mu_n + p\mu_p)$$

where the symbols, μ_n, μ_p represent the mobility of electrons and holes respectively. In this case mobility is a function of the scattering processes that involve effective mass of the carriers and it is thus given as;

$$\mu = \dfrac{q\tau}{m}$$

with q the charge of the electrons or holes, τ the average time between collisions during scattering, m the effective mass. It should also be noted here that the hole and electron mobility are not necessarily the same and mobility only indicates on how fast a device will work. Mobility is therefore the proportionality constant in the drift velocity-electric field relationship. For very small values of electric field, it relates to velocity as;

$$v = \mu E$$

where v is the drift velocity while E the electric field inside the material. From this expression, it can be said that electrons and holes have opposite velocities.

Drift currents in semiconductors

Drift current are carried by both electrons and holes and especially in a homogeneously doped semiconductor with a constant carrier density in which an external electric field is applied, the drift currents that flow is represented by;

$$I_{tot} = \sigma_{tot} AE = eA(n\mu_n + p\mu_p)E$$

In this expression, the letter A represents the cross sectional area perpendicular to the current flow and E represents the applied electric field. However, the electric field can be applied internally to the device structure.

Diffusion currents

When carrier gradients exist in a semiconductor, diffusion currents will occur. Hole density gradients cause hole diffusion currents, electron density gradients cause electron diffusion currents and gradients in both carrier types will cause diffusion of both, creating a total diffusion current of:

$$I_{tot} = eA\left(D_n \frac{dn}{dx} - D_p \frac{dp}{dx}\right)$$

With $D_{n,p}$ the diffusion constant of electrons respectively holes, x is the direction of carrier propagation. The Einstein equation gives the relationship between the diffusion constant and the mobility of the carrier:

$$\frac{D}{\mu} = \frac{kT}{e}$$

with, k the Boltzmann constant, T the temperature in Kelvin. In the general case where both concentration gradients and electric fields are present the total current is the sum of both drift and diffusion currents:

$$I_{tot} = eA\left(n\mu_n E + p\mu_p E + D_n \frac{dn}{dx} - D_p \frac{dp}{dx}\right)$$

This equation is normally referred to as the drift-diffusion equation of carriers and is the basic equation that describes carrier movement in semiconductor devices. The direction of the electron *particle flux* is opposite to the current flow. Particle flux is strictly governed by electrostatics, the current is dependent on the sign of the charge $q = \pm e$. The third important basic equation in semiconductor devices is the Poisson equation:

$$\frac{d^2V}{dx^2} = -\frac{\rho(x)}{\varepsilon} = -\frac{e}{\varepsilon}\left(p - n + N_D^+ - N_A^-\right)$$

With V the electrostatic potential, ρ the charge density as a function of x, while p and n the free hole and electron density respectively which are both a function of x as well as of V and N_D^+ & N_A^- the concentration of ionized doping atoms as a function of x.

Energy band diagrams

Carriers in a semiconductor are quasi free because they are not covalently bonded to the atoms. They thus can move but at the same time they undergo interactions with the atoms in the regular lattice and the mathematical description of the wave character of carriers in semiconductors in a region of forbidden energies exists is called the band-gap E_g and therefore in a perfect semiconductor there are no energy levels within the band-gap and this implies that no electrons nor holes can reside in the energy region defined by the band-gap.

It is also noted that free electrons have energies in an allowed energy band higher in energy than the band-gap, the conduction band E_c. Free holes have energies in an allowed energy band lower in energy than the band-gap which is the valence band E_v. The number of electrons at each energy value is determined by the density of states $g(E)$ and the Fermi-Dirac distribution function $f(E)$ as illustrated in figure below. It can be noted that $g(E)$ gives the distribution of energy levels or states as a function of energy and when $g(E) = 0$ in the band-gap when no energy levels. On the other hand $f(E)$ gives the probability of finding an electron at energy E.

Thus $1-f(E)$ gives the probability of finding a hole. In the figure below, the energy band diagram with the position of the Fermi level E_F as a function of the doping type (left). The integration of the product of the density of states g(E) and the Fermi-Dirac distribution function f(E) gives the carrier distribution over the energy range above or below the band-gap (right). Remember the relationship between the position of the Fermi level E_F and the density of carriers in the semiconductor as given by the Fermi-Dirac or Maxwell-Boltzmann distribution functions:

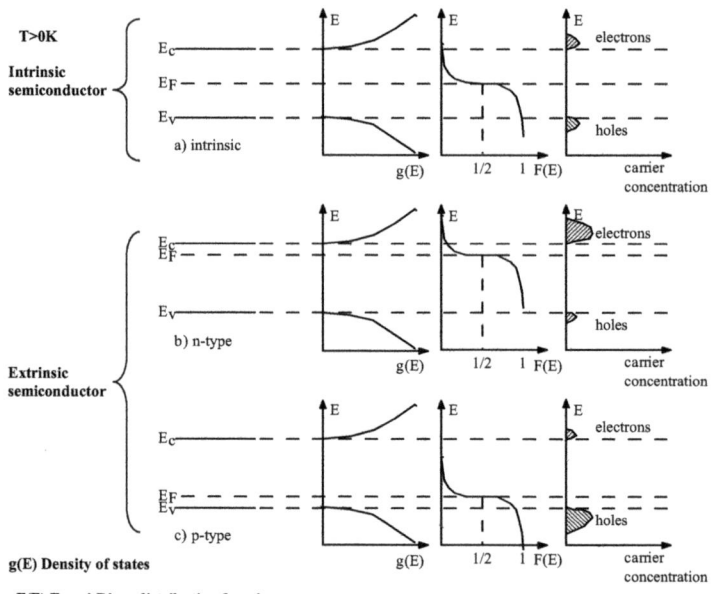

$$p = N_v e^{(E_v - E_F)/kT}$$

so that we have

$$p_i = N_v e^{(E_v - E_i)/kT}$$

$$n = N_c e^{(E_F - E_c)/kT}$$

and

$$n_i = N_c e^{(E_i - E_c)/kT}$$

with E_i the intrinsic level, N_c the effective density of states in the conduction band and Nv the effective density of states in the valence band. The intrinsic level E_i is an energy position within the forbidden gap that is determined by the intrinsic carrier concentration.

Fermi level constant

At an equilibrium state when no external bias is applied, no discontinuity or gradient can arise in the position of the Fermi level throughout the complete structure, otherwise a current or diffusion of high energy carriers to lower energetic places would flow. Thus;

$$\frac{dE_F}{dx} = 0$$

for;

$$V_{ext} = 0$$

This gives the starting point for drawing the energy band diagram since E_c-E_F and E_F-E_v is known from the doping concentration and E_g is known and remains constant in a homo-junction.

Influence of the electric field on the band diagram

Applying an electric field will result in a discontinuity in the Fermi level and will result in band bending as shown in the tilt in potential energy E_c below.

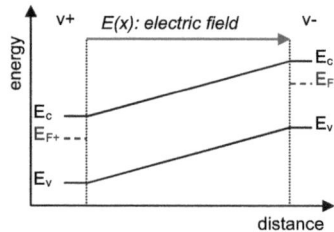

The figure above shows the relationship between the electric field, the electrostatic potential and the potential energy in which;

$$E(x) = \frac{-dV}{dx} = -\frac{1}{e}\frac{dE_c}{dx}$$

where, $E(x)$ is the Electric field, $V(x)$ is the electrostatic potential and $E_c(x)$ is the potential energy. Electric field $E(x)$ points "uphill", electrons drift "downhill". The reasoning is that electrons will flow from v- to v+ (electrostatic attraction). This means that electrons are losing potential energy since it is converted in kinetic energy, thus E_c at v- needs to be higher than E_c at v+. The electric field points from + to − and thus up the potential hill. Holes are flowing up the potential hill. You can check that this picture is completely consistent with the carrier fluxes and current directions that have been discussed before. It is also comforting to see that these directions are completely consistent with current directions used in the other courses such as Analysis of Circuits and Analogue Electronics!

High Speed Solid State Junctions

Materials that can be used for high semiconductor devices include the semiconductors of different doping types (homojunctions), different materials like Si and SiGe heterojunctions and metal-semiconductor junctions like Ohmic contact and Schottky contact. A junction is formed when two or more different materials are brought into close contact.

The work function (w_o)

To draw the energy bands of each material used in high speed electronic devices separately, we need to do it with respect of a local reference level that is defined for each material, the vacuum level E_{vac}. The position of the Fermi level with respect to the vacuum level, before contact of the material, is given by the work-function, w_o. Diffusion of electrons will appear from the material with the higher lying Fermi level to the material with the lower lying Fermi level. This process will align the Fermi levels and create band bending in the potential energy bands. The band bending is created as a result that no net current can be flowing when no external bias is applied, thus the diffusion of carriers needs to be compensated. This is done by drift of carriers which is caused by the build-up of an *internal electric field* that creates a drift current opposite to the diffusion current and allows the system to reach equilibrium.

Band diagram of junctions in high speed devices

When two different conducting materials that were initially uncharged are brought into contact, it is noted that the electrons close to the junction re-distribute themselves so that the Fermi energy is constant across the junction. This can be demonstrated with different junctions with their energy band diagrams before and after contact without external bias.

Metal-semiconductor junctions

Ohmic contact

In the cases where Ohmic contacts are used, the Energy band diagram of a metal and a p-type semiconductor are illustrated below where the left which is before contact and the right is after contact. For Ohmic contacts there should be an accumulation of majority carriers at the junction. On the right side, the current-voltage characteristic of the Ohmic contact is linear.

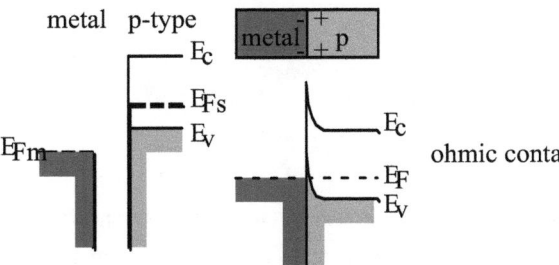

The current-voltage characteristic of this junction is linear. The slope of the curve is proportional to the conductivity of the semiconductor and the resistance of the Ohmic contact is expressed as;

$$R_{tot} = 2R_{contact} + R_{semiconducdor}$$
$$V = R_{tot}I$$

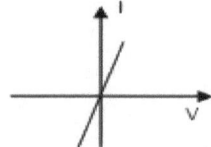

Schottky contact

In the figure below a Schottky contact on an n-type semiconductor is given. On the bottom of the diagram is the energy band diagram when no voltage is applied. The bands bend up. At the top a positive voltage is applied to the metal with respect to the semiconductor. This causes a discontinuity of the Fermi level at the junction. The electrons in the n-type material try to cross the junction towards the metal but are not allowed to go through the wide forbidden band-gap. Thus the only electrons that can contribute to the current are the ones above the potential barrier and this is only a small amount thus the current flowing at this voltage will be small.

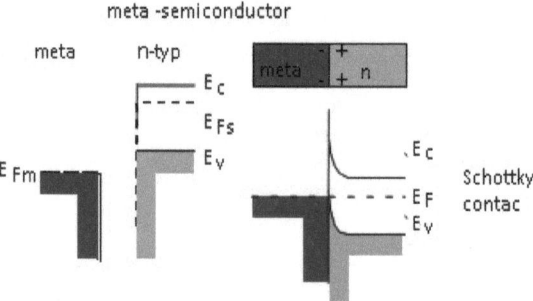

For Schottky contacts there should be depletion of majority carriers at the junction. The current-voltage characteristics of the Schottky contact are exponential. A Schottky contact can be used as a rectifier. It is a majority high speed semiconductor carrier device.

Semiconductor-semiconductor junctions

With a p-n junction and many more

The work function difference between n-Si and p-Si causes a potential barrier between n and p Si when forming the junction. Applying an external electric field will increase or decrease this potential barrier respectively inhibiting and allowing current to flow.

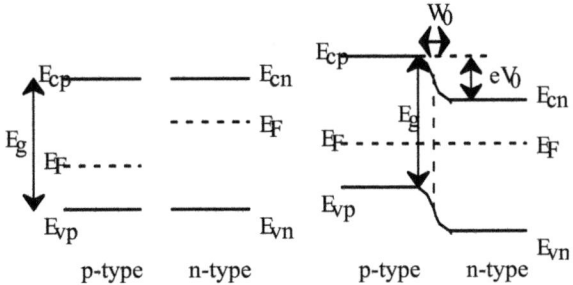

The energy band diagram of a p-n junction shown is that which has to the left and to the right after contact. Energy band diagrams are great tools to predict the behaviour of the junctions under the influence of an externally applied voltage. The figure gives a Schottky contact on an n-type semiconductor. The Schottky contact for the n-type semiconductor is characterised with a potential barrier for electrons. If a positive voltage is applied to the metal side, (V^+), the Fermi level of the metal drops with a value equal to eV with respect to the Fermi level. This decreases the potential barrier seen by the electrons in the semiconductor that are electrostatically attracted to the metal.

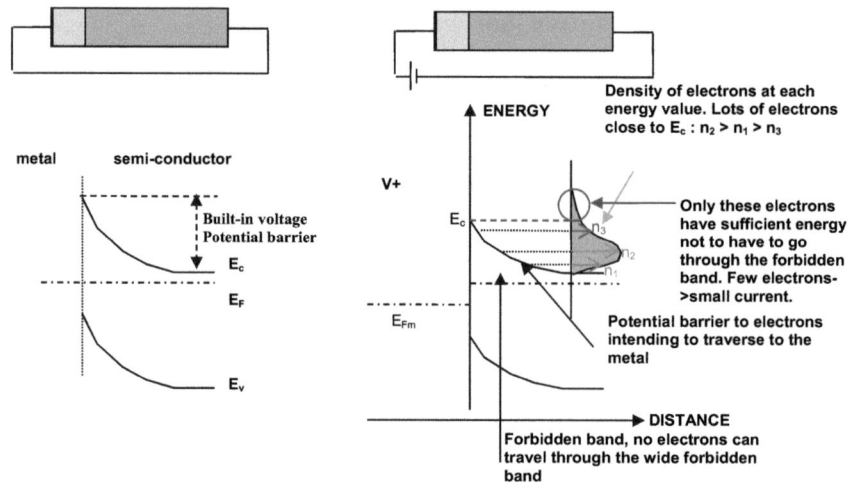

As shown in the figure above the potential barrier blocks all carrier transport of carriers with energies lower than the barrier. This is because these carriers would flow into the forbidden gap where there are no energy levels to receive them. The carriers with energy higher than the barrier will be able to "flow" across the junction and occupy free energy states at the other side of the junction. This will cause a small current to flow. If the potential barrier is reduced more by applying a large V on the metal, then the current will increase exponentially. Thus the electrostatics imposed by the +V voltage attracts the electrons, however the quantum mechanics presented by the energy band diagram shows that there is a potential barrier that blocks a certain amount of available charges.

It can be seen therefore that when semiconductors have two types of free carriers - electrons and holes; this gives the possibility for two types of currents to flow in different junctions hence drift and diffusion currents. Energy band diagrams are a handy way to predict the expected magnitude of the current in junctions under the influence of an external bias without the need for full scale device modelling.

p-n diode in high speed semiconductor devices

The short p-n diode

The definition of *short* diode is that the width of the materials that make the pn junction, are smaller or equal to the diffusion length of the minority carriers. The currents in a p-n diode are governed by *diffusion* of minority carriers in the neutral regions. As a consequence the number of minority carriers in the neutral region varies linearly, with a slope determined by;

        1) the injection of minority carriers across the junction

        2) the neutrality condition at the contacts.

However the following assumptions are made to derive the currents: the contacts are ideal and do not show any band bending; the contacts can absorb and generate the necessary carriers to keep bulk condition carrier concentrations fixed; no voltage is dropped across the neutral regions; the applied voltage drops across the depletion region only ;the amount of injected carriers across the junction due to the applied voltage is negligible compared to the majority carrier concentration determined by the doping.

Therefore under forward bias, the electrons are injected from the n-region into the p-region and holes are injected from the p-region into the n-region. This is when an externally applied potential lowers the energy barrier between the n and p type region. This injection of carriers causes a minority carrier gradient of electrons in the p-type region and a hole gradient in the n-type region. Gradients of carriers cause diffusion current. Similar gradients occur in reverse bias but they are a lot smaller because they are minority carriers rather than majority carriers that are injected across the barrier. The reverse bias leakage current is limited by the amount of holes available and/or generated in the n-type region and the amount of electrons available and/or generated in the p-type region. Since under normal circumstances there is a limited number of minority carriers available, the reverse leakage current is small.

Band diagram of unbiased high speed junction

V_0 which is also called V_{bi} is the built-in voltage and it represents the potential barrier that controls the amount of carriers that diffuse. V_0 is generated by the internal electric field that is a consequence of carrier diffusion upon contact is expressed as;

$$V_0 = \frac{kT}{e} \ln\left(\frac{N_D N_A}{n_i^2}\right)$$

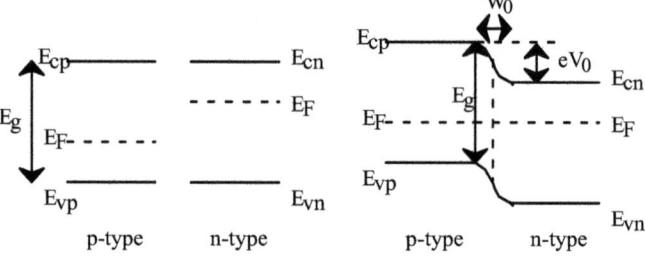

Space charge region or depletion region

In the figure below p and n material are shown before contact while at the right ***pn*** junction with the two neutral regions (that is the n and p) and the space charge region are shown where W_0 is the depletion width when no external bias is applied.

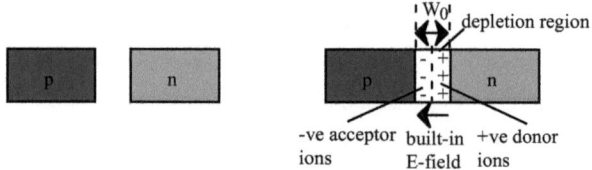

Based on the "depletion approximation approach" it is assumed that the depletion region is empty of moving carriers. It contains two sides with fixed ionized charges of opposite charge at each side. The fixed ionized charges are the ionized doping atoms. The doping atoms are fixed in the lattice and cannot move and hence cannot carry current. This region is thus more resistive than the neutral regions. The charge separation is caused by the internal electric field. The internal electric field is due to that built-in voltage V_{bi}, the built-in voltage is a result of the contact potential difference and is equal to the work function difference between n and p-type Si.

The high speed p-n junction under forward bias

In the figure below, the top figure shows forward bias condition while at the bottom show an energy band diagram of the p-n diode under forward bias. Forward bias lowers the potential barrier across the junction with a value proportional to the applied voltage, therefore more carriers will have an energy higher than the barrier and cross the junction. Thus the current is flowing and is increasing exponentially.

High speed p-n junction under reverse bias

The reverse bias adds the absolute value of the applied voltage to the built-in voltage increasing the potential barrier therefore there are no carriers with sufficient energy to travel across the barrier. There is drift of minority carriers across the junction. Since the number of minority carriers is low, the current is small and determined by the minority carrier concentration rather than the electric field. Thus the off-current in a diode is small and constant as a function of reverse bias voltage when no break-down mechanisms occur. As illustrated in the figure below, the top show the reverse bias condition while the bottom figure shows the energy band diagram of the ***pn*** diode under reverse bias.

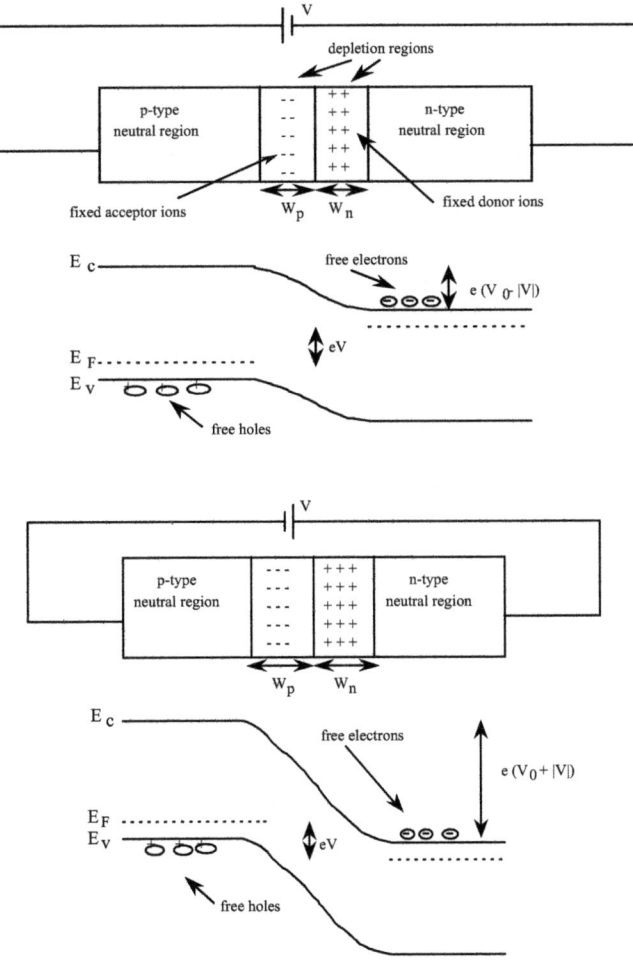

High speed minority carrier distribution under bias

Under certain bias conditions, a certain amount of excess carriers is stored in the n and p regions between the edge of the depletion region and the contact hence called neutral regions. These excess carriers will cause the delays when switching pn diodes. On the other hand under forward bias the excess minority carrier concentration at the edges of the depletion width are given as a function of the bulk minority carrier concentration and the applied voltage. The excess of carriers is defined as the surplus of minority carriers with respect to the initial minority carrier concentration that is determined by the bulk material. The excess carrier concentration at the edge of the depletion width we have at $x = w_n$ or $-w_p$) is given by:

$$\Delta n_p = n_p' - n_{p0}$$

excess electron concentration in p - type region at depletion region edge

$$\Delta p_n = p_n' - p_{n0}$$

excess hole concentration in n - type region at depletion region edge

With the injected concentrations given by:

$$\frac{n_p'}{n_{p0}} = \frac{p_n'}{p_{n0}} = e^{\left(\frac{eV}{kT}\right)}$$

where n'_p and p'_n are respectively the minority electron concentration in the p-region at the depletion region edge and the minority hole concentration in the n-region at the depletion region edge. The n_{p0} and p_{n0} are the intrinsic minority carrier concentrations.

To derive the expression for the current that flows, the graphical picture approach of the device under bias is used in that looking at forward bias only, is sufficient to derive the expression of the current in both bias regions. As shown in the figure below, the short pn diode under forward bias with exaggerated depletion width is illustrated at the top picture while at the bottom figure, an excess minority carrier distribution of short diode. Shaded regions represent the amount of stored excess minority carrier charge. The excess carrier concentration is zero and this condition is imposed by the contacts.

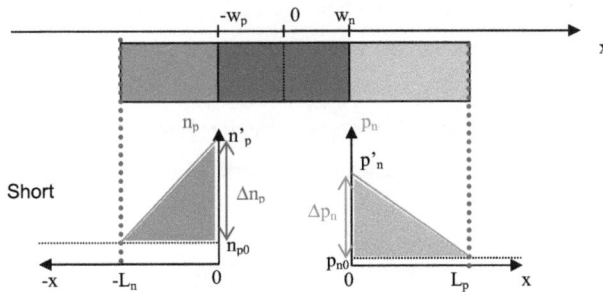

The lengths of each neutral region is smaller or equal to the minority carrier diffusion length, the minority carrier concentration is decreasing linearly in each region. The *excess* carrier concentration is given by the shaded triangles in the neutral region. The density of minority carriers is imposed by the boundary conditions in which the bulk conditions kept at Ohmic contacts. Since gradients in carrier concentration occur due to the injection process across the junction, diffusion of minority carriers will happen. Calculation of the diode currents in a short diode is easy such that the expression for the diffusion current can be expressed in several ways as shown by the expression below;

$$I_{tot} = \begin{cases} eA\left(D_n \dfrac{dn_p}{dx} - D_p \dfrac{dp_n}{dx} \right) \\ eA\left(D_n \dfrac{n'_p - n_{p0}}{0 - (-L_n)} - D_p \dfrac{p'_n - p_{n0}}{0 - L_p} \right) \\ eA\left(D_n \dfrac{n'_p - n_{p0}}{L_n} + D_p \dfrac{p'_n - p_{n0}}{L_p} \right) \\ eA\left(n_{p0} D_n \dfrac{e^{\left(\frac{eV}{kT}\right)} - 1}{L_n} + p_{n0} D_p \dfrac{e^{\left(\frac{eV}{kT}\right)} - 1}{L_p} \right) \\ eA\left(\dfrac{n_{p0} D_n}{L_n} + \dfrac{p_{n0} D_p}{L_p} \right)\left(e^{\left(\frac{eV}{kT}\right)} - 1 \right) \\ I_s\left(e^{\left(\frac{eV}{kT}\right)} - 1 \right) \end{cases}$$

There is also a majority carrier flow as the majority carriers injected across the junction have to be re-supplied. The total current is constant through the device makes calculation of the minority carrier diffusion current in each region of the pn junction sufficient for deriving the total current in the pn junction. Thus, the magnitude of the diffusion current of the electrons in the p-type region is equal to the magnitude of the electron drift current in the n-type region and this is similar for holes. The contribution of the minority and majority carrier current to the total current through the pn junction and thus the majority carriers injected across the junction are re-supplied via drift through the contact and diffuse to the other contact through the minority carrier gradient region. The minority carrier gradients determine the total current completely. The current remains constant in the space charge region because we assume at the moment that there is no recombination no generation of carriers.

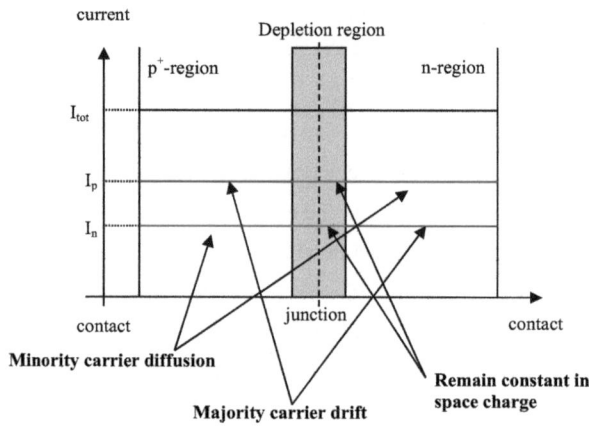

Long p-n diodes

In real diodes available commercially, the length of the materials forming the junction is not necessarily smaller or equal to the diffusion length of the minority carriers. However in the case of long diodes recombination of minority carriers occur happen accompanied whilst with diffusion at the contact. As a result, the number of minority carriers varies non-linearly in the neutral regions of the device.

Generation and recombination

Due to thermal energy, electron-hole pairs are created. This electrons and holes are freed from the atoms and become free carriers that can be involved in the conduction processes. This process is called generation of carriers and is happening continuously as long at the temperature is sufficiently high to break the valence electron-atom bonds. Of course other forms of energy can do the same. This can be illustrated using light: shining light on a semiconductor, with the correct energy, larger than the energy gap, can also release electron-hole pairs. The opposite process is called *recombination*. In this case an electron and hole will recombine and thus get fixed to the atom and become unavailable for conduction.

This is a reduction of the number of free carriers. As with generation, recombination is also continuously happening. Recombination can happen in two ways. The first is direct recombination of electron and holes which means that excess electrons in the conduction band recombine with holes in the valence band. Energy is lost by the electron and is given off as photons or light. The second type is indirect recombination. In this case there are impurity centres or lattice defects in the semiconductors, called traps – these will always be available as a 100% perfect semiconductor does not exist. The traps will cause some discrete energy levels to exist within the bandgap of the semiconductor which can have empty states to which the electrons from the conduction band can fall in order to cross the band-gap in steps. In this case the energy released by the electrons causes heating of the lattice.

The characteristic timescale involved in the recombination processes is called the *carrier lifetime* $\tau_{n,p}$. The characteristic length scales are the diffusion lengths $L_{n,p}$. These parameters give the statistically average time, respectively distance that a carrier travels before it recombines. This is a statistical average, thus means that not all carriers will recombine within that time scale, but the majority will such that at $x = L_{n,p}$ the number of carrier have been reduced by a factor Exp(1). The values are dependent on the material, the quality of the material, the doping density and the type of charged carrier. The relationship between the carrier lifetime and diffusion length is given by the diffusion constant following:

$$L_{n,p} = \sqrt{D_{n,p} \times \tau_{n,p}}$$

The intrinsic number of free electrons and holes is the same as carrier generation and recombination is in equilibrium. In a pn diode under forward bias an excess minority carrier density exists in the neutral regions of the device. When the voltage across the junction is not too high we can assume that this excess is a lot smaller than the number of majority carriers in that region. It is the gradient of the excess minority carriers that is significant. Recombination will happen between the majority carriers and the excess minority carriers. Since the density of the excess carriers is small compared to the density of the majority carriers, recombination processes can have a significant impact on the number of excess carriers, whilst being insignificant for the majority carriers in comparison with the amount available.

Recombination becomes important for devices where the neutral regions are larger than the diffusion length. If we put the *pn* diode under reverse bias then there will be a reduction of minority carriers in the neighbourhood of the depletion region edges. In this case generation of minority carriers is larger in order to maintain equilibrium conditions. Thus the rate of recombination and generation will adapt itself to the availability of excess minority carriers in the "neutral regions". When there is an excess of minority carriers, then the recombination rate will increase proportional to this excess in order to drive the semiconductor back to equilibrium. Thus the generation rate of minority carriers will then be smaller than the recombination rate. However when there is a shortage of minority carriers then the generation rate increases to try to bring the minority carrier concentration back to equilibrium conditions.

In this case the generation rate of minority carriers is larger than the recombination rate. Of course when minority carriers recombine they do so with majority carriers. As a consequence the majority carrier recombination or generation rate is equal to that of the minority carriers. We will assume that the injected minority carrier concentration is sufficiently small compared to the majority carrier concentration such that the variation of majority carriers can be neglected. The minority carrier recombination rate in the "neutral region" can be described to acceptable approximation by the simple equations:

$$U_n = \frac{\delta n_p}{\tau_n}$$

$$U_p = \frac{\delta p_n}{\tau_p}$$

Recombination in a depletion region or in those cases where the minority carrier concentration is of the same order of magnitude as the majority carrier concentration have to be described by different equations than those given above. The first step is thus to derive the expression of this variation in the case recombination happens. In order to do that we take a volume of semiconductor material through which a hole diffusion current is flowing. In the figure below, the variation of the density of holes as a function of time and place is over an infinitesimal distance Δx. This equation gives us the continuity equation for holes. It also gives the rate of change of the hole density while a current is flowing and therefore this equation that will be important to analyze the transient processes in bipolar devices under large signal switching conditions as shown in the figure;

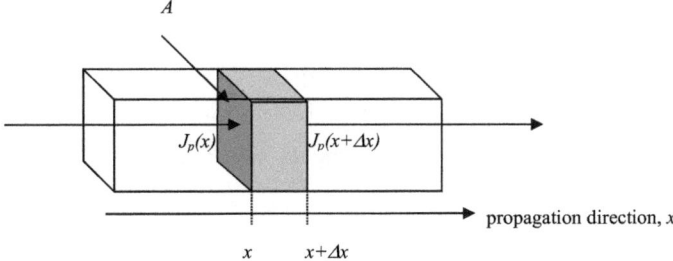

Hole current entering and leaving a volume, Δx A can be expressed as;

$$\left.\frac{\partial p_n(x,t)}{\partial t}\right|_{x \to x+\Delta x} = \frac{1}{e}\frac{J_p(x) - J_p(x+\Delta x)}{\Delta x} - \frac{\delta p_n(x,t)}{\tau_p}$$

The equation means that the rate of variation of the hole concentration is equal to the increase of the hole concentration in volume (Δx × A) per unit time minus the loss of holes due to recombination. The term Δp$_n$ is the excess minority carrier concentration. The term excess as used here means that the density of carriers larger than the minority carrier concentration defined by the doping of the bulk material only. The second term is the carrier recombination term given for Δx→0 the equation becomes a differential equation expressed as;

$$\frac{\partial p(x,t)}{\partial t} = \frac{\partial \delta p(x,t)}{\partial t} = \frac{-1}{e}\frac{\partial J_p(x,t)}{\partial x} - \frac{\delta p(x,t)}{\tau_p}$$

In a similar way this applies for electrons as:

$$\frac{\partial n(x,t)}{\partial t} = \frac{\partial \delta n(x,t)}{\partial t} = \frac{1}{e}\frac{\partial J_n(x,t)}{\partial x} - \frac{\delta n(x,t)}{\tau_n}$$

These equations are referred to as the continuity equations for the holes and the electrons respectively. If we can assume that drift is negligible, then the current is solely diffusion and the current densities can be written as:

$$J_n = eD_n \frac{\partial \delta n_p}{\partial x}$$

$$J_p = -eD_p \frac{\partial \delta p_n}{\partial x}$$

Substituting in the expressions above, it gives the continuity equations of minority carriers taking recombination into account as;

$$\frac{\partial \delta n}{\partial t} = D_n \frac{\partial^2 \delta n}{\partial x^2} - \frac{\delta n}{\tau_n}$$

$$\frac{\partial \delta p}{\partial t} = D_p \frac{\partial^2 \delta p}{\partial x^2} - \frac{\delta p}{\tau_p}$$

This equation is of utmost importance for analysing the switching characteristics of bipolar devices.

Minority carrier variations

When a steady state carrier occurs injection across a forward biased p-n diode in DC, the time derivative is zero and in steady state the continuity equations become:

$$\frac{\partial^2 \delta n}{\partial x^2} = \frac{\delta n}{D_n \tau_n} = \frac{\delta n}{L_n^2}$$

$$\frac{\partial^2 \delta p}{\partial x^2} = \frac{\delta p}{D_p \tau_p} = \frac{\delta p}{L_p^2}$$

By solving this equation with boundary conditions we obtain that;

at
$x = 0$
$\delta n = \Delta n$ & $\delta p = \Delta p$
and
at
$x = X_p, X_n$
$\delta n = 0$ & $\delta p = 0$

From the expression above, these boundary conditions mean that; $x = 0$ is the point of steady state carrier injection which only occurs at the depletion region edge due to the applied DC voltage across the junction. The second boundary condition shows that at the contacts the excess carrier concentration needs to be zero. The solution the expression used to obtain holes and which is a second order differential equation can be written as the sum of two exponentials:

$$\delta p(x) = C_1 \exp\left(\frac{x}{L_p}\right) + C_2 \exp\left(-\frac{x}{L_p}\right)$$

or using a hyperbolic sine function to express it, we can write:

$$\delta p(x) = C_1 \sinh\left(\frac{x}{L_p} + C_2\right)$$

with C_1 and C_2 integration constants that must be determined by the boundary conditions.

Mathematically, the **sinh** function can be used as possible solution in which we can find the expression of the excess hole concentration variation as a function of position as;

$$\delta p(x) = \frac{\Delta p}{\sinh\left(\frac{-X_n}{L_p}\right)} \sinh\left(\frac{x - X_n}{L_p}\right)$$

This is a rather complex equation, useful for numerical simulations only and therefore required to be simplified for the short diode approximation and the long diode approximation.

The short diode approximation

We take the limit of $X_n \ll L_p$ and for the long diode approximation we take the limit of $X_n \to \infty$. We need to calculate the current and that based on the limits of short and long diode approximations. In order to calculate the current we have to take the first derivative to give the hole current as:

$$I_p = \frac{eD_p A \Delta p}{L_p} \coth\left(\frac{X_n}{L_p}\right)$$

The following figures give the difference in excess minority carrier concentrations between the short, long and correct diode descriptions. The excess minority carrier hole concentration in the n-layer under normalized injection conditions for the different solutions of the continuity equation and for different relative differences between contact position relative to the minority carrier diffusion length. In the figure above (a), is for short X_n in the short diode approximation and describes the excess carrier concentration to excellent accuracy. (d) is the long diode approximation is excellent for long X_n. In the other two situations some error is made. When $X_n = L_p$, the excess carrier concentration is best described by the short diode approximation. When $X_n > L_p$ but of the same order of magnitude, the linear approximation becomes less accurate and the exponential equation more accurate. The error in the long diode approximation is in the non-zero excess at the contacts. The ratio of the correct current to the approximated current is given as a function of the ratio of the lengths.

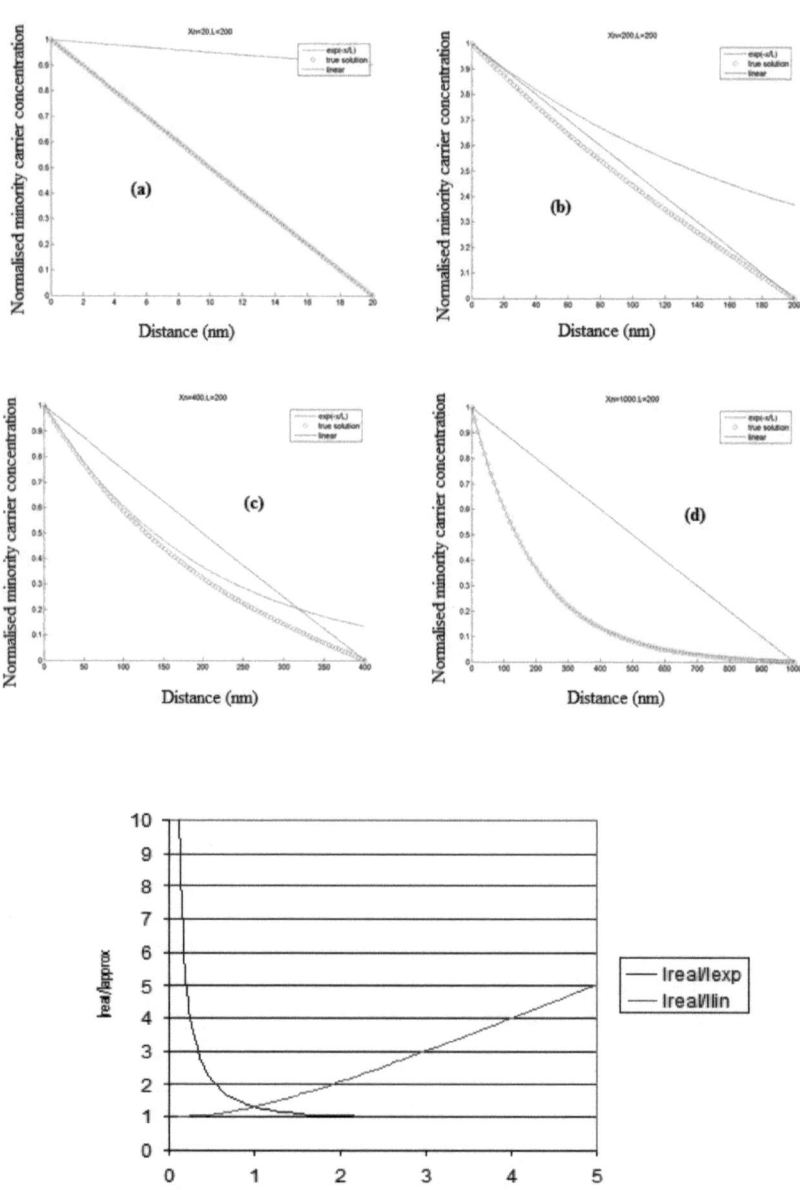

In the figure above it can be noted that:

1. the error of using the short diode approximation becomes negligible small when $X_n \ll L_p$.
2. for $X_n = L_p$ the error using the short or long diode approximation is exactly the same.
3. for $X_n > L_p$ the error made by using the long diode approximation decreases quickly and it is better to use the long rather than the short diode approximation.

The Ohmic contacts are at a distance shorter than the minority carrier diffusion length and thus $X_p < L_n$ and $X_n < L_p$, where X_p is the position of the contact of the p-layer and X_n is the position of the contact in the n-layer. In this case we neglect any recombination by setting $\tau_{n,p} = \infty$. Then under the boundary condition, the excess minority carrier concentration varies linearly. This is what is referred to as the *SHORT diode approximation*. The Ohmic contacts are at a distance larger than the minority carrier diffusion length. Thus $X_p \gg L_n$ and $X_n \gg L_p$. if we let the contacts appear at a distance equal to infinity: $X_p = X_n = \infty$. The solution becomes an exponential and is called the *LONG diode approximation*. Solving the long diode approximation we impose that the excess minority carriers have completely disappeared by recombination. The solution for the excess carrier concentration is then:

$$\delta n = \Delta n \, e^{\left(-x/L_n\right)}$$

$$\delta p = \Delta p \, e^{\left(-x/L_p\right)}$$

excess minority carrier concentration. Thus with $L = \sqrt{D\tau}$ the diffusion length, which is the average distance a carrier diffuses before recombining while τ is the average time between scattering events. Thus diffusion with recombination the excess carrier concentration is varying exponentially in a long diode in contrast to the short diode where the excess carrier concentration is varying linearly as shown in the figure below in which the variation of the excess carrier hole concentration as a function of distance due to recombination (full line). In this figure, n_0 (p_0) is the carrier concentration determined by the doping of the bulk material while $\Delta n (\Delta p)$ is the excess carrier concentration at x = 0 resulting from the carrier injection process. Likewise, the x = 0 is the edge of the depletion region and the $\delta n (\delta p)$ is the x-dependent excess carrier concentration.

f

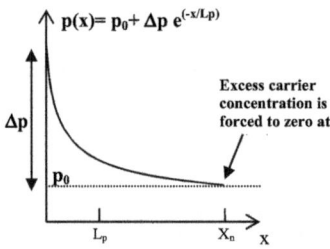

As in the short diode, the current is still determined by the diffusion of the excess minority carriers in the neutral region. The difference is that now this minority carrier excess is changing exponentially. As a consequence the minority carrier diffusion current calculated for holes is:

$$J_p(x) = q\frac{D_p}{L_p}\Delta p\, e^{\left(-x/L_p\right)}$$

Thus this current is decreasing for positive x because holes are disappearing by recombination. But the total current needs to be constant. This can be obtained by remembering that total current consists of hole and electron movement. Since we have neglected the variation of majority carriers, there must be a drift current of majority carriers that is re-supplying the recombined carriers and resupplying the carriers that are injected across the junction. The sum of this majority carrier drift current and minority carrier diffusion current is constant. The sum of the two is constant: I_{tot}. As a consequence it is the diffusion of minority carriers that limits the total current, therefore the total current must be determined by the maximum electron and hole diffusion currents that can happen under the given injection conditions. The maximum diffusion current that can flow can be calculated at the point of injection at both sides of the junction as there the gradient of the exponentials is maximu and this gives the first approach of finding the expression of the current in a long pn diode. The pn diode under forward bias with exaggerated depletion width and in the figure the bottom is the excess minority carrier distribution of long diode. The shaded regions represent the amount of stored excess charge.

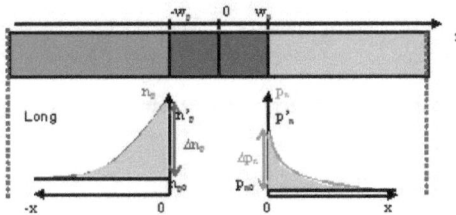

Calculation of the diode currents in a *long* diode is based on calculating the maximum diffusion current at the edge of the depletion region width. This means that we determine the excess carrier gradient at x = 0 in each region. So that this expression can also be given as;

$$I_{tot} = \begin{cases} eA\left(n_{p0}D_n \dfrac{e^{\left(\frac{eV}{kT}\right)}-1}{L_n} + p_{n0}D_p \dfrac{e^{\left(\frac{eV}{kT}\right)}-1}{L_p}\right) \\ eA\left(\dfrac{n_{p0}D_n}{L_n} + \dfrac{p_{n0}D_p}{L_p}\right)\left(e^{\left(\frac{eV}{kT}\right)}-1\right) \\ I_s\left(e^{\left(\frac{eV}{kT}\right)}-1\right) \end{cases}$$

The result obtained for the long *pn* diode current is the same as the short diode only for the short diode we take the lengths of the neutral regions to be *exactly* the same as the diffusion lengths of the minority carriers. For shorter diodes, the values of L_n and L_p will have to be changed by the actual widths of the neutral regions and the current will thus increase in shorter diodes. Each component is not constant throughout the device but the sum of electron and hole current is constant. The majority carrier component occurs as a result of recombination between minority and majority carriers, majority carriers need to be re-supplied, which causes the current. In this figure below the variation of the magnitude of the hole and electron current in a long diode and thus I_p is the minority carrier diffusion current in the n-region and a majority carrier drift current in the p-region.

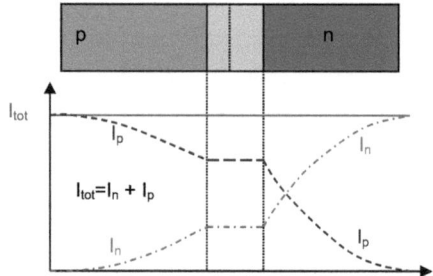

Another way of calculating currents in the *p-n* diode in the high speed switching behaviour of the p-n diodes is using the excess minority carrier charge in the neutral region. $Q_{n,p}$ is the excess minority carrier electron in the hole concentration in the p, n type region respectively. The density of the excess charge is given by the shaded area underneath the curves for Δn and Δp. To calculate the excess charge we integrate the excess minority carrier functions as expressed below:

$$Q_n = -eA \int_{-\infty}^{0} \delta n_p \, dx$$

$$Q_p = eA \int_{0}^{\infty} \delta p_n \, dx$$

The boundary conditions chosen for this calculation are taken to be infinite and using these boundary conditions represents the position of the contacts under the long channel approximation. Therefore under the assumption that the exponential variation of the excess carrier concentration is the solution of the steady state continuity equation we have to make sure that the excess carrier concentration at the contacts is zero. Current is a variation of charge as a function of time and in recombination of the excess carriers occurs we know that this recombined charge needs to be re-supplied. Every $\Delta_{n,p}$ the charge $Q_{n,p}$ disappears via recombination and this charge needs to be re-supplied in the same time scale in order to maintain steady state. The current that re-supplies these re-combined minority carriers is given by:

$$I_n = \frac{Q_n}{\tau_n} = \frac{eAL_n \Delta n_p}{\tau_n}$$

$$I_p = \frac{Q_p}{\tau_p} = \frac{eAL_p \Delta p_n}{\tau_p}$$

This current can be expressed as the stored minority charge in the neutral regions divided by the lifetime of the minority carriers in this region.

Non-ideal characteristics in pn diodes

In this section silicon based diode is discussed. The difference between an ideal and a real IV curve of a Si pn diode and when measuring a Si pn diode and plotting the current on a log scale we get a graph as in figure below.

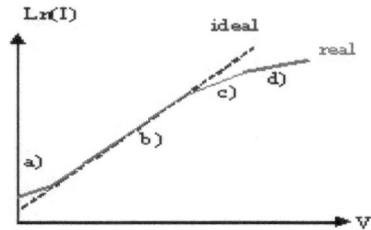

In Region (a)

The assumption in the ideal pn diode is that no recombination happens in the depletion width, thus carriers travel through unhindered. This is not completely correct, especially in pin diodes or diodes with long depletion widths. If the depletion width is larger than the carrier diffusion lengths, recombination of carriers in transit will occur. This will affect the current at low current densities. More carriers need to cross in order to obey the relationship for injected carrier concentration as a function of applied voltage.

In Region (b)

This region behaves as ideal diffusion and happens for intermediate voltages.

In Region (c)

This is the region of high level injection. A huge amount of carriers is injected across the junction and makes the number of minority carriers at the edge of the depletion region of the same order of magnitude as the majority carriers. This brings changes to the majority carrier density which we have assumed to be constant in the ideal case. In the high carrier injection regime the relationship for the injected carrier concentration that we have derived in EE1 breaks down. Its effect is a lowering of the total current flowing.

In Region (d)

At still higher voltages or for large ohmic contact resistances a large amount of the applied voltage is dropped across the ohmic contact rather than across the depletion region and thus the intrinsic diode feels less voltage applied, keeping the current lower than expected. Therefore an ideality factor *n* is introduced in the formula for the current in a pn junction that takes into account these non-ideal effects. The expression for the current then becomes:

$$I_{tot} = I_s \left(e^{\left(\frac{eV}{nkT}\right)} - 1 \right)$$

where $1 \leq n \leq 2$. n=1 in the ideal region and increases towards 2 in the recombination and high level injection region.

Switching delays in high speed pn-diodes

The p-n diodes have a switching capability in which they are all governed by the excess minority carriers. When it is under bias there exists an excess minority carrier concentration inside the neutral regions close to the depletion region. This depletion region is a function of the applied biases such that the larger the bias current flowing, the larger the excess charge produced. To change the bias current, the excess charge stored must change before the voltage across the diode junction changes. This is inherently a capacitive effect.

Switching ways

There are two ways of switching where the first method is where the diode starts from an voltage, $V_{applied} > 0V$ to form a forward bias to the off state where that $V_{applied} = 0V$. This opens the switch such that the current I = 0 Amperes. To remove the excess carriers when the diode switches from forward bias to open circuit occurs only through recombination process governed by the carrier lifetimes. A carrier lifetime is the average time an electron spends in the material before it recombines with a hole noting that τ_n and τ_p are electrons and holes minority carrier life times of respectively. It will take a time equal to μ_n and μ_p to reduce the excess carrier concentration by a factor e.

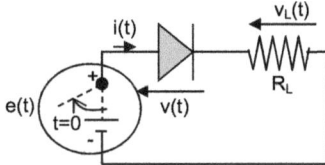

Experimental Switching circuit

As shown in the figure above a p⁺n diode switches from a forward to a zero bias through a switch that opens at t = 0. Although the current drops to zero immediately, voltage v(t) across the diode still remains till all the excess charge is completely removed. Through this process, it is ensured that there is a voltage across the load, R_L till the excess charge completely removed is such that:

$$\frac{\partial \delta p(x,t)}{\partial t} = \frac{-1}{e}\frac{\partial J_p(x,t)}{\partial x} - \frac{\delta p(x,t)}{\tau_p}$$

where the symbols have their usual meanings and if this equation is re-organized multiplying with its cross sectional area, A and integrating over all the neutral region, we obtain:

$$\int_0^{contact} \frac{-\partial J_p(x,t)}{\partial x} A = \int_0^{contact} \frac{e\delta p(x,t)}{\tau_p} A + \int_0^{contact} \frac{\partial e\delta p(x,t)}{\partial t} A$$

Taking diffusion current as zero at the contact and maximum at the edge of the depletion region (x = 0), the integral of excess minority carrier concentrations in the neutral region multiplied by unit charge, e and cross sectional area, A result into the magnitude of the total stored charge, Q_p in that region. Thus the expression for time variation of the current as a function of the stored charge in the neutral region will be:

$$i_p(t) = \frac{Q_p(t)}{\tau_p} + \frac{dQ_p(t)}{dt}$$

In other words, this expression shows that the instantaneous current $i_p(t)$ in a switching device is equal to the re-supply of the excess carrier concentration that is disappearing as a result of recombination plus the AC component that allows a build-up or depletion of charge as a function of time. We also have a special case of switching. The p⁺n diode switches from the ON state to the zero state such that when t ≤ 0, the DC forward bias imposes a current I_0 through the high speed switching device. When is at zero, t = 0-, the excess minority carrier charge is given by:

$$Q_p = I_0 \tau_p.$$

When it is at t = 0+, the current drops to zero forming an open circuit and remains at zero but as a result the charge does not follow immediately forming the boundary conditions in the neutral region such we obtain;

$$\frac{Q_p(t)}{\tau_p} = -\frac{dQ_p(t)}{dt}$$

Integration this expression over time, we obtain:

$$\int_0^t \frac{dQ_p(t)}{Q_p(t)} = -\int_0^t \frac{dt}{\tau_p}$$

$$\ln(Q_p(t)) = -\frac{t}{\tau_p} + c$$

$$Q_p(t) = C\,e^{-t/\tau_p}$$

where the letters c and C are integration constants. Since $Q_p(t = 0+) = I_0\tau_p$ the integration constant $C = I_0\tau_p$, the variation of excess minority carrier charges as a function of time is obtained by:

$$Q_p(t) = I_0\tau_p\,e^{-t/\tau_p}$$

Sometimes it is required that the readers of this book proof that the voltage remains constant across the diode. This starts by finding the relationship between the voltage across the diode and the minority carrier excess charge. This relationship is a function of the injection of carriers given by:

$$\Delta p_n = p_n' - p_{n0} = p_{n0}\left(\exp\left(\frac{e\,v(t)}{kT}\right) - 1\right)$$

where v(t) is the voltage across the diode. As long as Δp_n is non-zero according to this expression, there exists a voltage v(t) across the diode.

By ignoring the distortion of the exponential variation of the excess charge at the depletion region edge and use a pure exponential relationship of the excess carrier concentration the stored charge at any time is then given by:

$$Q_p = eAL_p\Delta p_n(t)$$

Substituting this equation in the earlier expressions will give $\Delta p_n(t)$ as an exponential function of time and equally substituting it in the above expression and extracting the voltage, v(t) will give:

$$v(t) = \frac{kT}{e} \ln\left(\frac{I_0 \tau_p}{eAL_p P_{no}} e^{-t/\tau_p} - 1 \right)$$

In practise, it is needed that the switching speed is improved and this is done by adding recombination centres to decreases carrier life time, τ_p. This can make a p^+n diode with a short p^+ region to be smaller than the re-combination length and thus we get that the excess charge dies out exponentially through recombination as a function of time and the voltage across the diode junction does not change instantaneously but instead wait till the excess charge disappears. Switching time can also be improved by switching the diode from the ON state to the OFF state by applying a reverse bias voltage during the off switching. This gives the reverse recovery transient and this current flow during the switching. The excess charges are then removed through a combined diffusion and recombination processes.

The physics associated to this process is represented in the drawings below. As shown in the figure, the magnitude of the switch-off voltage is equal to the magnitude of the applied voltage such that the time period, T is large enough to allow the diode to switch off completely in after each cycle. The second method to switch the diode is by driving the excess carriers from the neutral region through a current in which the charge will be removed through recombination and diffusion currents.

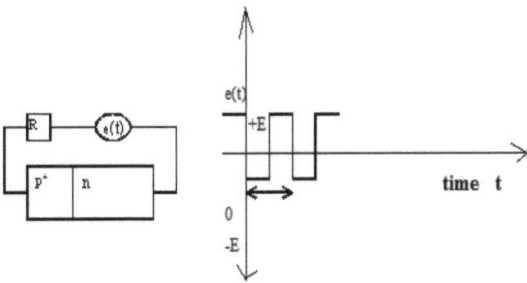

The figure below can be used to illustrate how the excess hole concentration distribution in the n-region as a function of time during the transient behaves it should be noted that the change in the gradient of the excess minority carrier density at the moment of switch-OFF is below the x-axis while that of switching ON is the current that is positive. Thus at the moment of switch OFF the current becomes negative.

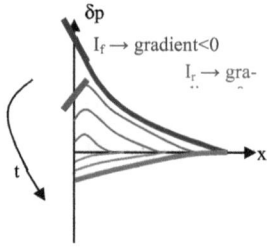

Further still analysis show that at the top as shown in the figure below, the full plotted line show the diode current-voltage characteristic while the plotted dashed line with the arrows show the current-time diode characteristic whilst switching. In the figure below, the bottom curve shows the variation of the voltage across the diode as a function of time and as such the graphs plotted at the right are for the currents and voltages as measured across the diode as a function of time. The component t_{sd} is therefore the storage delay time. From the figure at t < 0, the diode is ON and therefore a current $I_f = E/R$ flows. This implies that a forward bias current is mainly determined by the bias circuit because of the resistance of the forward biased diode that is very small and this cause a voltage drop across the diode to reduce to a negligible level when compared to that across the resistance, R.

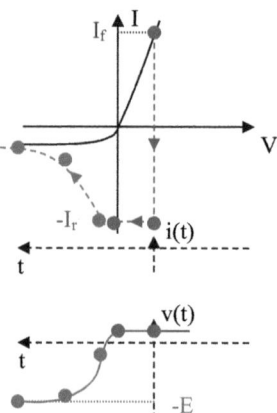

This therefore means that the load resistance becomes the limiter of the current. This current finally determines the slope of the excess carrier concentration at the depletion region edge. Similar to the electrons in the p-type region, the above slope becomes consistent with the diffusion of minority carrier holes in the n-type region as it varies from the depletion region edge to the contacts.

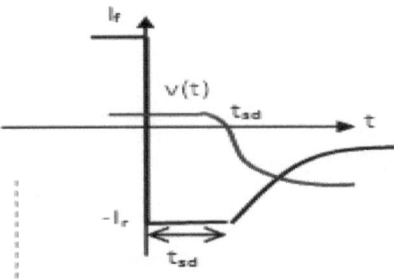

From the above graph, it is found that at t = 0, the voltage polarity is reversed though the magnitude of this voltage remains the same. Similarly a negative current $-I_r = -E/R$ flows that is many times larger than the reverse bias current and this is consistent with this applied voltage. If initially the diode was forward biased, it means that it had a narrow depletion region such that a relatively large density of stored minority carriers is only present. When the external bias is therefore reversed, the excess minority carrier charge that was previously stored in the neutral regions instantly disappears. This causes the depletion region not to increase and as a result, the voltage across the diode can also not increase immediately.

In order to deplete the neutral regions, the regions around the junction must be depleted first of the excess minority carrier charge and then later the region around the junction can be depleted of the majority carriers. This process forms a depletion region. Once this is done the external bias goes negative as the voltage drop across the diode remains consistently small and negligible when it is compared to the voltage drop across the resistance, R. Then obviously the current can be obtained by using the load resistance R and the applied voltage –E. As the negative current, $-I_r$ flows, the slope formed by the excess carrier concentration at the depletion region edge becomes exactly opposite the one in forward bias state. This is a confirmation that the minority carriers flow in the opposite way under positive bias. This behaviour show that the minority carrier excess holes diffuse from the n-type region to the p-type region while the electrons diffuse from p to n regions and this current removes any excess charge present.

The slope of Δp at the depletion region edge remains constant as long as any excess carriers' remains but when they are removed an increasing voltage dropped across the increasing depletion width is noted. The current through the circuit then increases and is only limited by the current that is allowed to flow through the diode. The final current registered is determined by the amount of minority carriers in the neutral region near the depletion region edge which is also maintained at steady state through the carrier generation processes. The leakage current becomes the OFF current through the reverse biased diode. At a time t = 0, any excess carrier concentration at the edge of the depletion region is therefore only p'$_n$. The time it takes to remove all the excess charge is called the storage delay time t$_{sd}$. As time increases to the end of the switch cycle, any excess carrier concentration present at the edge of the depletion region becomes p"$_n$ given by;

$$p'_n = p_{n0} e^{\left(\frac{eE}{kT}\right)}$$

$$p''_n = p_{n0} e^{\left(\frac{-eE}{kT}\right)}$$

To find the variation of charge as a function of time, the boundary conditions for t < 0 is applied such that the charge Q$_p$ = I$_f$Δ$_p$. In this case, at t = 0, the current becomes I = -I$_r$ and this allow the derivation of an expression for the storage delay time in which the assumption is that the condition I = -I$_r$ is for a steady state till t = t$_{sd}$. Solving using this assumption for a p$^+$n diode using Laplace transforms and re-writing the expression, we get:

$$\frac{dQ_p(t)}{dt} = i_p(t) - \frac{Q_p(t)}{\tau_p}$$

Based on this expression for the range 0 < t < t$_{sd}$ when i$_p$ remains constant at –I$_r$, we obtain the equation:

$$\frac{dQ_p(t)}{dt} = -I_r - \frac{Q_p(t)}{\tau_p}$$

Using the method of separation of variables, the expression reduces to:

$$\frac{dQ_p(t)}{Q_p(t) + \tau_p I_r} = -\frac{dt}{\tau_p}$$

This expression can be integrated and if we integrate with respect to time we get:

$$\int_{Q_p(0)}^{Q_p(t)} \frac{dQ_p(t)}{Q_p(t) + \tau_p I_r} = -\int_0^t \frac{dt}{\tau_p}$$

$$\ln\left(Q_p(t) + \tau_p I_r\right)\Big|_{\tau_p I_f}^{Q_p(t)} = -\frac{t}{\tau_p}$$

$$\ln\left(Q_p(t) + \tau_p I_r\right) - \ln\left(\tau_p I_F + \tau_p I_r\right) = -\frac{t}{\tau_p}$$

$$\ln \frac{\left(Q_p(t) + \tau_p I_r\right)}{\left(\tau_p I_F + \tau_p I_r\right)} = -\frac{t}{\tau_p}$$

$$Q_p(t) = -\tau_p I_r + \left(\tau_p I_F + \tau_p I_r\right)\exp\left(-\frac{t}{\tau_p}\right)$$

The final expression for $Q_p(t)$ can also be rewritten as

$$Q_p(t) = \tau_p \left[-I_r + (I_f + I_r)\exp\left(\frac{-t}{\tau_p}\right) \right]$$

But this expression is only valid for $0 < t < t_{sd}$ only. Based on this expression for $Q_p(t)$, the storage delay time t_{sd} can easily be extracted by the condition that at $t = t_{sd}$, then $Q_p(t_{sd}) = 0$ and this will give:

$$0 = \tau_p \left[-I_r + (I_f + I_r)\exp\left(\frac{-t_{sd}}{\tau_p}\right) \right]$$

This last expression is also valid only under the assumption of quasi steady state such that the storage delay time is finally given by:

$$t_{sd} \cong \tau_p \ln\left[1 + \frac{I_f}{I_r}\right]$$

From the derivations and discussion made above, it can be summarized that the switch from forward to reverse bias state ensures that reverse current larger than diodes' reverse saturation current while the transient current and voltage are delayed by change in stored charge. Likewise this causes a storage delay time t_{sd} during which the current is larger than the reverse bias current associated to the DC characteristics and the delays in pn diodes are a consequence of excess *minority* carrier charge. To improve the switching time in *pn* diodes one can introduce recombination centers so that the lifetime of the carriers is reduced. This makes any excess charge to disappear faster. Delay time can also be reduced by using a short device layer.

To obtain a fast rectifying devices diodes like the Schottky diodes are used in which only *majority* carriers need to be removed which is a fast process. Note that in a Schottky diode, a junction is formed by a metal-semiconductor contact. No depletion of carriers occurs in the metal and for it to switch the Schottky diode from the ON to OFF states, it is only the *majority* carriers that move back towards the junction. The use of high mobility carriers and shorter lifetime than Si (GaAs) has currently remained the only ultrahigh speed diodes in use and are specially designed *pn* diodes which strongly conduct for a short period of time in reverse bias fore step recovery and are able to switch in the pico-second range. They are used for harmonic generators.

Small signal equivalent circuit for p-n junction diodes

In most cases in circuit operation a device is functioning around a DC point and the variations around this DC point is small that the device acts only as if it is linear around. If a diode is to be used as component in a circuit, its non-linear characteristics become cumbersome especially for fast switching circuit simulation purposes. Its circuit design requires an equivalent circuit with component values dependent on the chosen DC point. Such an equivalent circuit is given in the figure below.

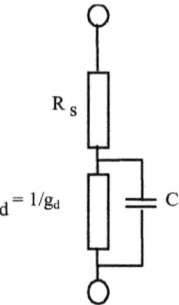

In this figure, R_s is the contact resistance, r_d is the diode's differential resistance and C is the voltage dependent capacitance. The series resistance R_S represents the non-ideal characteristic of the Ohmic contacts and the capacitance, C is due to charging and discharging effects occurring during AC operation. These results into a reverse bias where a depletion region with fixed ionized charges is built-up with an amount dependent on the applied voltage and forward bias which takes into account the variation of the built-up excess charge as a function of voltage. In the reverse bias condition, the depletion capacitance is given by;

$$C_{depl} = \frac{\varepsilon A}{W_{depl}}$$

$$C_{depl} = \varepsilon A \sqrt{\frac{eN_A N_D}{2\varepsilon(N_A + N_D)(V_0 - V)}}$$

Where the W_{depl} is the depletion width and the V_0 the built-in voltage and its variation of the depletion capacitance as a function of bias across the pn-junction can be graphically be represented as;

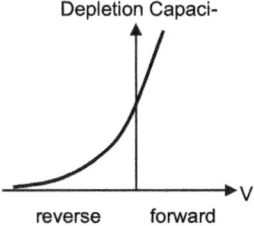

On the other hand, the forward bias which constitutes the diffusion capacitance can be obtained by;

$$C_{diff} = \frac{dQ}{dV}$$

$$C_{diff} = \frac{dQ_p}{dV} \text{ for } p^+n \text{ junction}$$

$$C_{diff} = \frac{e}{kT} I\tau$$

In this case the differential resistance, r_d relates to the slope of the curve, IV characteristic and as a result it can easily be calculated as a differential conductance expressed as:

$$\frac{1}{r_d} = g_d = \frac{dI}{dV} = \frac{e}{kT} I$$

Large signal equivalent circuit for a p-n junction diode

In the above section, we discussed on a small signal equivalent circuit that does not consist of non-linear elements. This was due to the small variation in applied voltage the IV region that was assumed to be linear. In the case of a large signal equivalent circuit, linearization is no longer possible because the whole non-linear current-voltage region is used. Thus the large signal equivalent circuit for this large signal for pn diodes is given in as shown below.

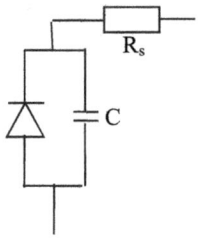

CHAPTER TEN
HIGH SPEED SWITCHING IN BIPOLAR TRANSISTORS

A Bipolar Junction Transistors (BJT) is a device which consists of three good Ohmic contacts that is composed of three different semiconductor layers. These layers are connected together. Therefore, a BJT structure consists of a series connection of a *pn* and *np* or *np* and *pn* diode structure in which the *n*-regions, respectively *p*-regions diodes overlap each other. In simple terms, these forms a *pnp* and *npn* Bipolar Junction Transistors. By having a series connection of two *pn* diodes, forms two junctions to be considered that do not have their boundaries connected to an Ohmic contact. A BJT device is therefore another device in which the minority carriers play a role similar to that of a *pn* diode. The voltage across each of its junctions determines the density of excess carriers at the edges of the depletion regions. To calculate the excess minority carrier concentration across each junction is done identically as that in the pn diode. The currents that flow in the different layers of the BJT are determined using the maximum minority carrier concentration gradients that occur near the depletion region edges. As discussed in chapter five, recombination of carriers forms the basis of determining whether a material combination results into amplification or not.

BJTs Principle of operation

In a pn diode, a reverse biased p-n junction constitutes a current given by;

$$I_{tot} = I_s \left(e^{\left(\frac{eV}{nkT}\right)} - 1 \right)$$

This expression can be reduced to become;

$$I_{tot} = -I_s$$

which then is independent of the magnitude of the reverse bias voltage V. This therefore implies that further reduction by taking into account of minority carriers gives the expression for I_s as:

$$I_s = eA \left(\frac{n_{p0} D_n}{L_n} + \frac{p_{n0} D_p}{L_p} \right).$$

The magnitude of this current I_s varies as the density of minority carriers, p_{n0} and n_{p0}, varies in each layer. In reverse bias its the minority carriers that cross the junction from the n-type material to the p-type while the majority carriers or electrons from the p-type to the n-type whose magnitude determines the current I_s as illustrated in the figure below.

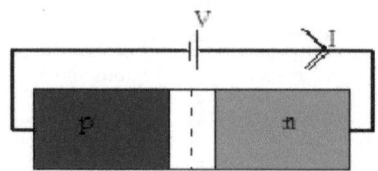

Reverse biased p-n junction

If doping density in each layer is reduced, the magnitude of the minority carriers increases and therefore the reverse bias current in the *pn* diode increases. Of course doping is not a good option for active control of the off-current. A way of controlling the current density on this diode is done by shining the correct frequency of light on the junction.

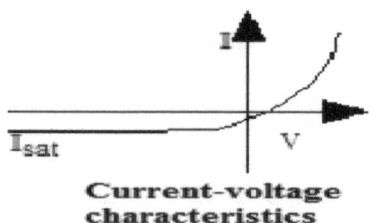

Current-voltage characteristics

This light is absorbed to create electron-hole pairs in which the minority carriers are near the depletion region edge. Any of these electron hole pairs that feel the electric field across the depletion region are swept across the depletion region increasing the off-current. In practical applications, this is done in a controlled way depending on the intensity of the light such that more intense light generates more carriers.

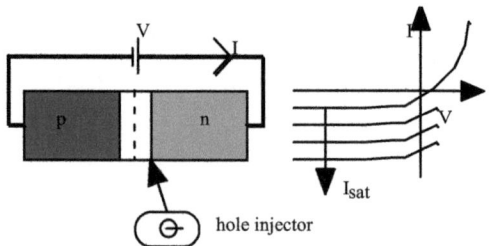

In the above figure the left side figure show holes that are injected onto the n-type region near the depletion region edge of a reversed biased pn-junction while on the right side the off-leakage current is shown to increases as the minority carrier density increases due to the hole injection process in the left side. In a practical way the active control of the off-current in a reversed biased pn diode is derived by using the forward biased p^+-n junction in which the p+ is a heavily doped p-side. A heavily doped p layer in a pn junction will cause the total current to be mainly dependent on holes.

From the gradients of the minority carriers in each region it can be illustrated that in the n-region, the gradient is given by;

$$p'_n - p_{n_0} \approx p'_n = p_{n_0} \exp\left(\frac{V_{EB}}{V_T}\right)$$

and in the p-type layer the gradient is given by;

$$n'_p - n_{p_0} \approx n'_p = n_{p_0} \exp\left(\frac{V_{EB}}{V_T}\right).$$

In both cases, $V_{EB} > 0V$ for forward bias and $V_T = \frac{kT}{e}$.

Since in most semiconductors, the values of

$$N_A \gg N_D$$

it is also true that;

$$p_{n_0} \gg n_{p_0}$$

And therefore the hole minority carrier gradient that define the hole diffusion current in the n-region is much larger than the electron minority carrier gradient in the p^+ region. Thus since;

$$I_p \gg I_n$$

we call a p^+n junction a hole injector and if this hole injector is sufficiently close to a reverse biasedi junction, the hole injector increases the amount of minority carriers that flow across the junction increasing the off-current in exactly the same way.

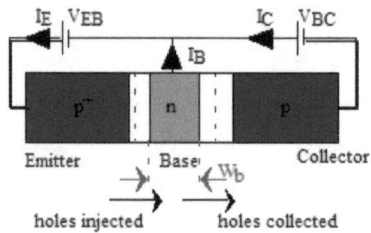

forward biased p⁺n junction, acting as a hole injector

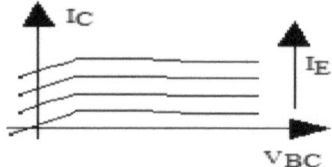

holes injected

As shown in the two figures above a combination of a forward biased p⁺n junction, acting as a hole injector and a reverse biased np junction acting as a hole collector – the BJT is shown and its graphical characteristics in which the collector current (holes collected) as a function of the emitter current is plotted. This gives the whole combination of *pn* and *np* junction forming a *pnp* bipolar junction transistor (BJT).

The heavily doped p+ layer of the injector diode is called the *emitter* (E), the layer that is in common for the two diodes is the *base* (B) and the layer that will collect all the holes that are injected by the p⁺n diode (thus injected by the EB junction) will be collected by the reverse biased np junction. The p-layer of the hole collecting diode is called the *collector* (C). If the base width is sufficiently small the collector current that will flow as a function of reverse bias voltage across the BC junction. Thus if the emitter current – which is the current across the EB junction – increases, then the collector current is increasing accordingly. The magnitude of the current is determined by the minority carriers that are injected across the EB junction rather than the reverse bias voltage across the BC junction.

If we take the approximation that the emitter current consists of a hole current I_p only, and all the injected holes get through the base, then the collector current is equal to the emitter current and there is no current amplification between I_E and I_C. The BJT operates by injection and collection of minority carriers. A good BJT is therefore an a-symmetric device in which the doping density in the emitter is higher than the doping density in the collector. The main carrier flow in a pnp BJT is holes. In normal active mode, the operation of a ***pnp*** BJT is given by:

$$V_{EB} > 0V$$

and

$$V_{BC} < 0V$$

A large amount of holes are injected by the emitter into the base and these are then collected by the collector. The magnitudes these holes is determined by the doping densities. If recombination of holes occurs in the base region then it can be expected that the collector current will decrease, as there will be less holes available for collection at the collector side. As a result there will also be electron currents in the BJT where these electron currents have a direct relationship to the base current in a pnp BJT. These can be illustrated using the figure below;

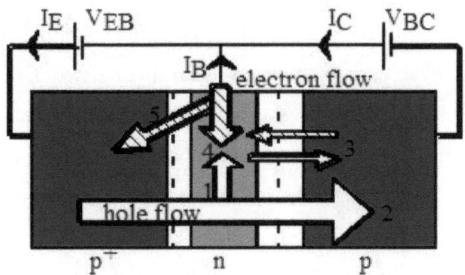

BJT in common base configuration in forward active mode

In the above figure the BJT in common base configuration is in forward active mode with the flow of both holes and electrons (flux) indicated with arrows. At the emitter-base junction, the holes are injected into the base while the electrons are injected from the base into the emitter. This implies that the electrons are disappearing out of the base region. If there would be no re-supply of these disappearing electrons then the density of majority carriers in the base would decrease.

To accommodate this decrease in majority carriers while maintaining charge neutrality, the depletion region at the EB junction has to adapt. This means that the forward biased voltage across the EB junction decreases and thus reduces the amount of injected carriers. Thus no re-supply of electrons would lower the hole current to zero. Therefore in order to maintain a collector current, the lost electrons from the base through the forward biased EB junction has to be re-supplied by an electron base current. The injected holes diffuse through the base with a probability that these holes recombine with the electrons available in the base region resulting in the base current decreasing. If the number of electrons in the base is not re-supplied the charge density relationship in the base would break down and hole current would disappear. In order to maintain the hole current I_C for the given biasing conditions, the base contact has to re-supply the electrons.

Therefore the thermally generated electrons and holes constitute the reverse saturation current across the base-collector junction and this reverse biased BC junction supplies electrons to the base. As a result the amount of electrons at the base current needs re-supply that is smaller than the re-supply current. The need to maintain a given steady state bias conditions, the base current I_B is found to consist of three terms given as:

$$I_B = I'_B + I''_B - I_{CB0}$$

Where, I''_B is the base current to re-supply and I_{CB0} is the reverse bias leakage current that injects carriers from collector into base. The base current re-supplies the majority carriers that are lost in the base however, charge neutrality is not maintained and the voltage drop across the different junctions adapt themselves. The total current, I_C that is flowing through the device is maintained. This book uses the common base configuration because it gives a direct insight into the voltage across each junction EB and BC separately. The main difference in this analysis is that the boundary conditions in the base junctions are imposed by the voltage across each junction rather than an Ohmic contact as was the case for a single pn junction. Therefore the minority carrier excess in the base at the depletion region edges is different from the bulk minority carrier concentration. They are determined by the voltage across each junction by the formulae given as;

$$\frac{n'_p}{n_{p0}} = \frac{p'_n}{p_{n0}} = e^{\left(\frac{eV}{kT}\right)}$$

The variation of the minority carrier concentration becomes dependent on the recombination process such that when recombination is taken into account, the variation of the minority carriers in the base are easily determined.

Currents in a BJT

The derivation of the currents for a BJT is very easy if the short diode approximation method is used and in such a case, both the emitter and the base layers are considered to be very thin. This does not take into account recombination. To calculate, a choice of including recombination or not is done and this choice depends on the accuracy that is required and the speed with which derivations and calculations will be done. Thus in a BJT under normal active mode or where the EB is forward biased while the BC is reverse biased the short diode approximation method ignores recombination making it sufficiently accurate. However this makes the following assumptions; No recombination in the emitter and base region, the reverse bias collector current into the base is negligible and the depletion regions are neglected.

Under normal forward active mode, the currents in the emitter base diode determine all the BJT currents completely. This is because holes diffusing through the base are collected by the collector forming the collector current. Electrons on the other hand escape from the base into the emitter and have to be re-supplied by the base current. This is why the electron current of the E-B diode define the base current. The emitter current therefore is taken as the total current in the E-B diode and is given by:

$$I_E = I_B + I_C$$

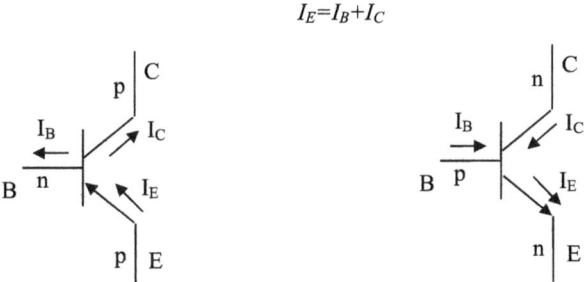

In a *pnp* and a *npn* BJT symbols shown below indicate the directions of the different currents and therefore applying nodal analysis, the sum of current into node is equal to the sum of currents out of node and thus:

For the *pnp* BJT, we have;

$$I_E = I_C + I_B$$

and for the *npn* BJT we have:

$$I_C + I_B = I_E$$

This can then be easily illustrated using the figure below.

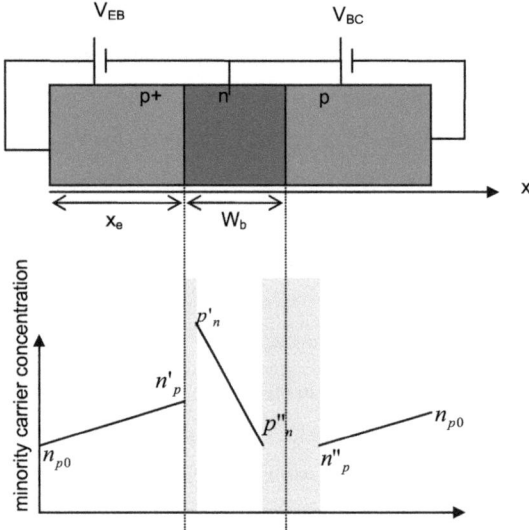

In this figure the top figure is a *pnp* BJT in forward active mod while below it is the minority carrier density variation plot in each layer for ta short BJT. The shaded regions shown are the depletion regions. Based on this concept we find that in high speed bipolar devices diffusion of minority carriers are the basis for the calculation of the current. Diffusion is also noted to occur due to a carrier gradient. To derive current in the BJT is majorly determined by the minority carrier concentration at each side of the junction and for both junctions in the device exactly the same way as in the pn diodes. The injected minority carrier concentration at the emitter EB junction where electrons come from the base is obtained by:

$$n'_p = n_{p0} e^{\left(\frac{eV_{EB}}{kT}\right)}$$

At the base where holes come from the emitter at the EB junction, the injected minority carrier concentrations are also obtained from:

$$p'_n = p_{n0} e^{\left(\frac{eV_{EB}}{kT}\right)}$$

Likewise, the BC junction the minority carrier holes are injected to the collector at the base have a reduction of the minority carrier concentration in the reverse biased collector given by:

$$p''_n = p_{n0} e^{\left(\frac{-eV_{BC}}{kT}\right)}$$

The minority carrier concentration at the Ohmic contact of the emitter and collector also imposes a bulk minority carrier concentration such that the electron current of the EB diode is given by:

$$I_n = eD_n A \frac{n_{p0} - n'_p}{0 - x_e} = eD_n A \frac{n_{p0} e^{\left(\frac{eV_{EB}}{kT}\right)} - n_{p0}}{x_e} \approx \frac{eD_n A n_{p0}}{x_e} e^{\left(\frac{eV_{EB}}{kT}\right)}$$

On this basis then, the hole current crossing the EB junction is given by a reduced expression illustrated as:

$$I_p = eD_p A \frac{p'_n - p''_n}{W_b} \approx eD_p A \frac{p_{n0} e^{\left(\frac{eV_{EB}}{kT}\right)}}{W_b} = \frac{eD_p A p_{n0}}{W_b} e^{\left(\frac{eV_{EB}}{kT}\right)}$$

With boundary conditions given by;

$$p''_n \ll p'_n$$

and

$$p_{n_0} \ll p'_n,$$

the emitter current in the pn diode current is expressed as:

$$I_E = I_n + I_p$$

Such that the base current is given by:

$$I_B = I_n$$

This expression is thus equivalent to the re-supply of the electrons that have left the base via the EB junction. This implies then that the collector current can be obtained from:

$$I_C = I_p.$$

Therefore all the injected holes by the EB junction are finally collected by the BC junction and can be determined by the short diode approximation method. The important parameters thus defining the quality of the BJT transistor are:

The emitter efficiency, γ

This parameter show how much of the total γ =1 because in that case $I_n = 0$, no electrons are lost out of the base through the E-B junction as shown from the expression given as;

$$\gamma = \frac{I_p}{I_p + I_n}$$

The current transfer ratio, α

The current transfer ratio gives the ratio of the input current to the output current and it is expressed as;

$$\alpha = \frac{I_C}{I_E}.$$

In practical devices, this ratio is preferable to be as close to 1 as possible so that no amplification between emitter and collector current occurs. When α = γ no recombination takes place. However this is not be true in the case of recombination because then I_C is lower compared to I_p injected across the E-B junction.

The current amplification factor, β

The current gain is called amplification and shows that we want I_B as small as possible for a given I_C β as large as possible as shown by the expression;

$$\beta = \frac{I_C}{I_B}$$

In case recombination is taken into account, then three major assumptions are used. These assumptions are that the recombination occurs at he emitter, base and collector region, the reverse bias collector current into the base is negligible is zero $I_{CB0} = 0A$) and that the depletion regions are always ignored or neglected in all calculations.

When we have recombination, the first observation made is that the minority carrier concentrations at the junctions are assumed to be the same and can be determined by the voltage across the junctions. This implies that at the emitter and the collector regions we can expect the variation of the minority carrier density to be the same since the boundary condition imposed by the contacts ensures that excess minority carrier concentration goes to zero at the contacts. At this stage an exponential approximation to the minority carrier concentration is applied for the emitter and collector while at the base, boundary conditions are imposed by the junction voltages. This ensures that exact solutions to the continuity equations apply and the sum of any two exponentials forming a solution of the second order differential equations can be used to give the continuity equations represented as:

$$\frac{\partial^2 \delta n}{\partial x^2} - \frac{\delta n}{L_n^2} = 0$$

$$\frac{\partial^2 \delta p}{\partial x^2} - \frac{\delta p}{L_p^2} = 0$$

If the figure below is used to illustrate the boundary requirements, then the boundary conditions for the base region as shown are surrounded by two junctions in forward active mode such that;

$$V_{EB} > 0V$$

and

$$V_{BC} < 0V$$

for a *pnp* BJT. This simply implies that;

at
$x = 0$;

$$\delta p_n(0) = \Delta p_{n_E} = p'_n - p_{n_0} \approx p_{n_0} \exp\left(\frac{V_{EB}}{V_T}\right)$$

$x = W_b$;

$$\delta p(W_b) = \Delta p_{n_C} = p''_n - p_{n_0} \approx -p_{n_0}$$

This takes x = 0 as the starting point of the base region where W_b is the width of the base.

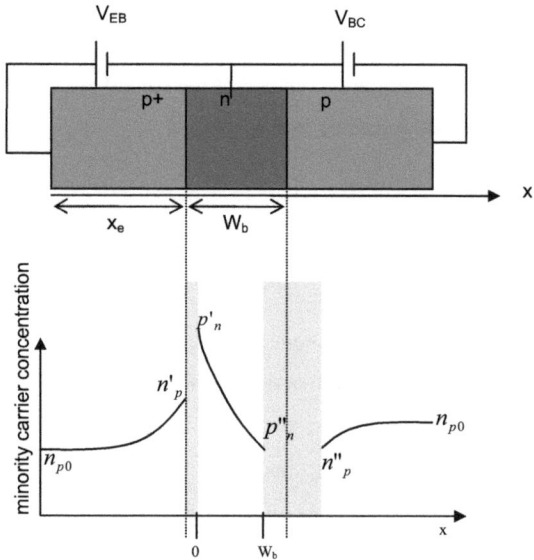

The solutions to these differential equations having these boundary conditions no longer obey a decaying exponential and thus their general solution is a second order differential equation. For the holes we obtain the solution as:

$$\delta p_n(x) = C_1 \exp\left(\frac{x}{L_p}\right) + C_2 \exp\left(\frac{-x}{L_p}\right)$$

Where C_1 and C_2 are constants that are determined by the boundary conditions such that:

$$\delta p_n(0) = p'_n = C_1 + C_2$$
$$\delta p_n(W_b) = -p_{n_0} = C_1 \exp\left(\frac{W_b}{L_p}\right) + C_2 \exp\left(\frac{-W_b}{L_p}\right)$$

When these boundary conditions are substituted in the expressions above, we obtain the expression for the minority carrier variation in the base when recombination is taken into account. What remains now is to calculate the currents in cases where recombination is taken into account. The gradients of the minority carrier concentration must be calculated first at the edges of the depletion regions for each junction. These are points that carry the maximum minority carrier gradients.

Applying the above expression to the emitter, the emitter current is obtained to be expressed as:

$$I_E = -eAD_p \frac{d\delta p_n}{dx}\bigg|_{x=0} + eAD_n \frac{d\delta n_p}{dx}\bigg|_{x=0}$$

The result also is equivalent to the sum of the hole and the electron current as determined by the gradient of the minority carriers at the EB junction. Similarly the collector current can also be obtained from:

$$I_C = -eAD_p \frac{d\delta p_n}{dx}\bigg|_{x=W_b}$$

This expression is also the gradient of the hole minority carrier concentration in the base at the B-C junction. The base current through this expression now consist of two major components as given by;

$$I_B = I'_B + I''_B$$

where we have now assumed that $I_{CB0} = 0A$. It also consists of the electron current lost across the E-B junction given by;

$$I'_B = eAD_n \frac{d\delta n_p}{dx}\bigg|_{x=0}$$

This component of current needs re-supply from the recombined electron population in the base. As a consequence, the number of electrons that recombine in the base is always equal to the number of holes recombined. Though he number of holes that recombine in the base form the hole minority carrier gradient between the emitter and the collector side, the current-difference that follows from the gradient difference at the junction at both sides of the base is the current that needs to be re-supplied through the base contact to compensate the electron charge loss in the base. Therefore this current can be given as:

$$I''_B = -eAD_p \frac{d\delta p_n}{dx}\bigg|_{x=0} + eAD_p \frac{d\delta p_n}{dx}\bigg|_{x=W_b}$$

where A is the cross sectional area of the depletion surface.

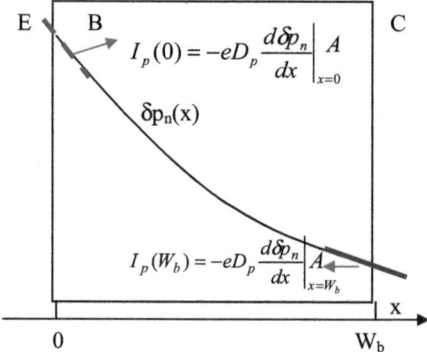

When the base region of a ***pnp*** BJT is plotted with the variation of excess minority carrier concentration under forward active mode when recombination is taken into account then the dashed line (red) in the figure below gives the gradient of the minority carrier concentration at the emitter side. As a result then, the full bold line (blue) forms the gradient of the minority carrier concentration at the collector side. Due to the injection process of holes at the E-B junction into the base and the collection process at the collector, there will be a hole charge Q_p in the base that is consistent with this injection/collection process. The holes are diffusing through the base with a characteristic time constant τ_t, called the transit time. The transit time determined by the diffusion process and the base width given as;

$$\tau_t = \frac{W_b^2}{2D_p}.$$

The steady state hole current formed is therefore obtained as;

$$I_p = \frac{Q_p}{\tau_t}.$$

This hole current is equal to the collector current I_C such that the holes that are diffusing in the base recombine with a characteristic time τ_p, which is the hole lifetime in the base. A charge, Q_p disappears per τ_p and is re-supplied by the E-B current. However, hole charge, Q_p recombines with the same amount of electron charges in the base and therefore the amount of charge disappearing in the base is;

$$Q_n = Q_p.$$

It is thus noted that an amount Q_n disappears per τ_p which is re-supplied by the base current to keep $Q_n = Q_p$ by an amount given by;

$$I_n = \frac{Q_n}{\tau_p} = \frac{Q_p}{\tau_p}.$$

When no loss of electrons from the base into the emitter is experienced, $I'_B = 0A$ and current I_n is equal to the base current $I_B = I''_B$ so the currents and current amplification factor are given in function of charge and time constants as:

$$I_C = \frac{Q_p}{\tau_t}$$
$$I''_B = \frac{Q_p}{\tau_p} \Rightarrow \beta = \frac{I_C}{I''_B} = \frac{\tau_p}{\tau_t}$$

High Speed Switching of the BJT

The switching delay in a BJT is well understood when a biasing circuit is used and a square wave voltage is applied to the base that drives the pnp BJT from off state to an on state. The resulting variation of the collector current is then measured as a function of time and the formed waveforms of the applied base voltage and the collector current are recorded on oscilloscope as shown in the figure below.

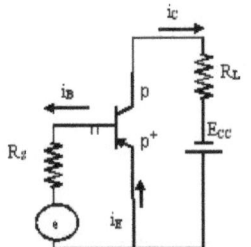

Simple switching circuit for the pnp BJT in common emitter forward active mode

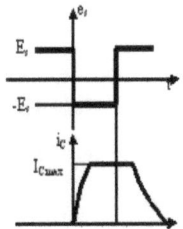

Output current i_c as a function of time during switching

Switching cycle

The BJT starts by switching from the OFF state when both junctions are reversed biased to the ON state through a base current driving the BJT into oversaturation in both junctions as shown below.

373

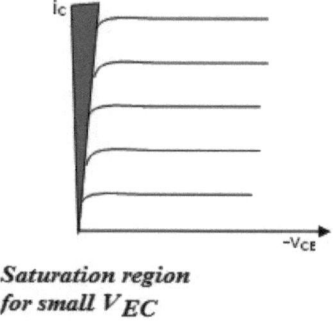

Saturation region for small V_{EC}

In the second stage of switching, the BJT is driven the OFF state from oversaturation into reverse bias across both junctions. The hole current and hole flux are thus in the same direction while the electron current and electron flux are in the opposite directions due to the negative charge of the electrons.

The Switch-ON state

The circuit illustration below, the switch on can be solved by using the load line technique. Current - voltage characteristics of an ideal transistor in a given configuration on the left part for different values of the control current result into the plot below. As shown in the figure above, the base controls the collector current with a control base current given by I_B. in the figure each single line in the graph is for one value of I_B such that the lowest characteristic curve is the one with the smallest base current. As I_B increases, I_C also increases and thus the constant current value of I_C for an active BJT mode is related as;

$$I_C = \beta I_B$$

As shown in the figure, the shaded region is the saturation region and in this region, both emitter-base and base-collector junction are forward biased and this causes the minority carrier charge to built-up in the base under saturation and over-saturation conditions.

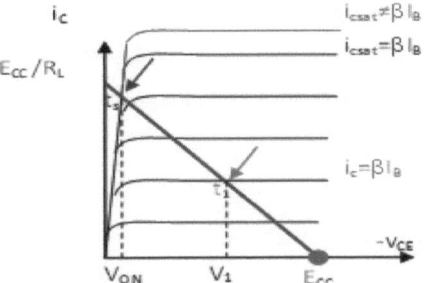

Switch ON from OFF to saturation

The expression for the load line can then be represented by the expression given by:

$$i_c = \frac{-1}{R_L}(-v_{ce}) + \frac{E_{CC}}{R_L}$$

When the BJT transistor is in the OFF state, the voltage across it is $-V_{CE} = E_{CC}$ and it has leakage current through the reverse biased diodes while when it is on the ON state, the voltage across goes to zero. Current is therefore limited by the load line only. When switching from OFF state to the ON state, the collector current needs to follow the load line such that each cross-section of the characteristics curve and the load line gives the solution. As a result, the base current is increases and the output current, i_C increases until the BJT goes into saturation at a time t_s. Finally, the voltage across the base-collector junction depends on the cross-point between the load line and BJT characteristics. As such the base-collector voltage varies from the reverse bias condition into forward bias or saturation condition. The maximum collector current that can flow through the circuit is approximated to:

$$i_c = \frac{E_{CC}}{R_L}$$

On the other hand when the base current is further increased by using the control circuit, the collector current stagnates from increasing and the relationship $I_C = \beta I_B$ breaks down. But when the base-emitter junction of the pnp BJT is switched from reverse to forward bias, holes are injected from E into B and an excess hole charge is built up in the base. The amount of excess charge in the base is determined by the magnitude of the base current. The collector current in a BJT, under the assumptions that $I_{BC0}=0$, is determined by the minority carrier concentration variation in the base, thus by the gradient, at the BC junction. As a result the minority carrier variation in the base for the pnp BJT in OFF state is illustrated by the curve below.

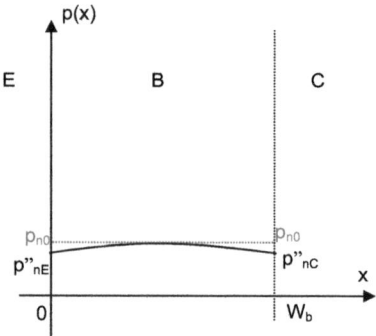

In the curve, there is a gradient of minority carriers near the junctions and this gradient is non-linear when recombination is taken into account. When the base current drives the BJT into saturation, the minority carrier variation in the base for the pnp BJT at the starting point of saturation can also be illustrated as shown in the figure below.

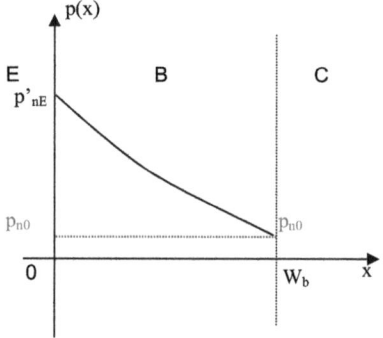

At a saturation point, the voltage condition shown in the curve across each junction is given by:

$$v_{eb} > 0 \ \& \ v_{bc} = 0$$

After the BJT saturates, The collector current remains constant after saturation because the load line limits the maximum collector current such the charge in the base is called the saturation charge Q_s. If the emitter-base bias is such that the base current drives the transistor into over-saturation, then the associated charge build-up in the base is larger than the saturation charge such that:

$$Q_p > Q_s$$

where

$$Q_p = \tau_p I_B,$$

Then;

$$I_C \neq \beta I_B.$$

At saturation level, we have $Q_b = Q_s$ while beyond saturation we have $Q_p > Q_s$. The minority carrier variation in the base for the pnp BJT in forward active mode is thus oversaturation as illustrated in the figure below.

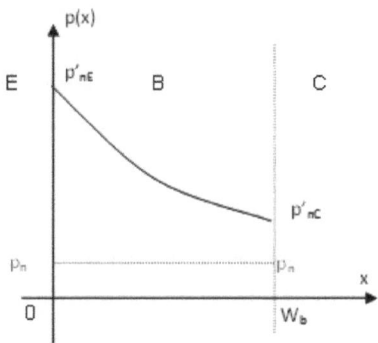

Driving the transistor ON state using over-saturation ensures fast switch-on times, however this will create a large density of store minority carrier charge in the base that will need to be removed before the BJT can switch OFF state.

The Switch-OFF

Now the pnp BJT will be switched OFF by applying a reverse bias voltage on the EB junction. This removes charge out of the base by a reverse pn diode current, the EB junction remains forward biased and the BC junction too until all excess charge larger than the saturation charge is removed. Thus the collector current remains constant until the excess charge $Q = Q_p - Q_s$ is removed. The base-collector voltage can only return to reverse bias when this excess charge is removed. When the charge in the base has reached $Q=Q_s$ the BC junction first goes to zero bias, then with further reduction of the charge in the base, the BC junction goes into reverse bias and the collector current can decrease following the load line. The base-emitter junction is still forward biased until all remaining minority carrier charge is removed from the base, after which the EB junction can go into reverse bias. The exponential reduction of the collector current is caused by the base-emitter diode switch-off.

In order to calculate the switching dynamics in a p^+np BJT, the following differential equation needs to be solved as;

$$i_p(t) = \frac{Q_p(t)}{\tau_p} + \frac{dQ_p(t)}{dt}$$

This differential equations can be solved using Laplace transforms, but in the bipolar devices' switching experiments an easier way exists because in large signal switching, steady state conditions occur before and after switching on the current that controls the switching dynamics. To turn ON from 0 bias to ON state the switching happens at $t = 0s$ such the boundary conditions are for $t < 0$ $i_p(t = 0) = 0$. This then is switching from zero bias in current and voltage. This give $Q_p(t=0)=0$ while for the time, $t \geq 0$ $i_p(t)=I_B$ and the EB bias circuit then determines the current that flows. The potential drop across the diode becomes much smaller causing the depletion region to be removed. Thus $I_B = E/R_s$.

Under certain given boundary conditions, a steady state current condition is reached after switching and the switching expression for the time variation of $Q_p(t)$ after $t = 0$ can be The final expression illustrates the variation of charge in the base as a function of time when the EB junction is switched from 0 to forward bias I_B. If at any time switching happens from the reverse bias to the forward bias, then a delay of time t_r occurs before the charge builds up in the base. This delay t_r is found to be related to the removal of the depletion region before $t = 0s$ modified for steady state condition to take the form of:

$$I_B = \frac{Q_p(t)}{\tau_p} + \frac{dQ_p(t)}{dt}$$

$$\frac{dQ_p(t)}{Q_p(t) - I_B \tau_p} = \frac{-dt}{\tau_p}$$

$$\int_{Q_p(t=0)}^{Q_p(t)} \frac{dQ_p(t)}{Q_p(t) - I_B \tau_p} = \int_0^t \frac{-dt}{\tau_p}$$

$$\ln \left(Q_p(t) - I_B \tau_p \right)_{Q_p(t)} - \ln \left(Q_p(t) - I_B \tau_p \right)_{Q_p(0)} = \frac{-t}{\tau_p}$$

$$\ln \left(\frac{Q_p(t) - I_B \tau_p}{-I_B \tau_p} \right) = \frac{-t}{\tau_p}$$

$$Q_p(t) - I_B \tau_p = -I_B \tau_p \exp\left(\frac{-t}{\tau_p}\right)$$

$$Q_p(t) = I_B \tau_p \left(1 - \exp\left(\frac{-t}{\tau_p}\right)\right)$$

If the EB drive current I_B is smaller than that necessary to drive the BJT into saturation, then the collector current would be given by:

$$i_C(t) = \frac{Q_p(t)}{\tau_t} = \frac{I_B \tau_p}{\tau_t} \left(1 - \exp\left(\frac{-t}{\tau_p}\right)\right)$$

with τ_t being the base transit time. When the EB drive current I_B is larger used to drive the BJT into saturation states, the load line limits the current that flow into the output. The charge in the base is thus equal to the saturation charge Q_B^{sat} at $t = t_s$ so that the maximum collector current $i_c(t)$ is then obtained from:

$$i_C(t) \approx \frac{E_{CC}}{R_L}$$

Thus allow the extraction of the time delay to saturation because at $t = t_s$ to be equated as;

$$\frac{I_B \tau_p}{\tau_t}\left(1-\exp\left(\frac{-t_s}{\tau_p}\right)\right) = I_C^{\max} = \frac{E_{CC}}{R_L}$$

$$t_s = \tau_p \ln\left(\frac{1}{1-\dfrac{I_C^{\max} \tau_t}{I_B \tau_p}}\right)$$

t_s here is small for smaller values of τ_p and for I_C small compared to βI_B, where is equal to;

$$\beta = \frac{\tau_p}{\tau_t}.$$

The turn OFF state from ON state to the OFF state can also happen at $t = 0$s. The turn-off time in this case is based on two factors which are the removal of the charge in excess of the saturation charge. This is when the collector current remains constant and this time is called the storage delay time in the BJT and the removal of the excess charge that prevents the EB junction to turn into reverse bias. This is similar to the **pn** diode drive-off delay time. Two different techniques can be applied to turn the BJT off. One is to switch to 0 bias and current conditions and the other one is by driving the EB diode into reverse bias. The second one will give much faster switching times as we have seen for the pn diode switching. The boundary conditions for the ON bias state to the OFF or zero state can be taken as for $t < 0$ $i_p(t=0)=I_B$ when the BJT is ON gives $Q_p(t=0) = I_B \tau_p$ while $t \geq 0$ $i_p(t) = 0$. The calculation method is the same as for the ON switching but the t = 0 boundary conditions have changed. After switching the steady state control current is zero.

$$0 = \frac{Q_p(t)}{\tau_p} + \frac{dQ_p(t)}{dt}$$

$$Q_p(t) = I_B \tau_p \exp\left(\frac{-t}{\tau_p}\right)$$

Although the base charge is decreasing exponentially, the collector current first remains constant. This is because first the excess saturation charge needs to be removed before the collector current can change its value and follow the variation of the base charge. Thus initially $i_c(t)$ is determined by the load line for $t < t_{sd}$ we have;

$$i_C(t) \approx \frac{E_{CC}}{R_L}$$

And thus gives collector current for, $t \geq t_{sd}$ as;

$$i_c(t) = \frac{Q_p(t)}{\tau_t} = \frac{I_B \tau_p}{\tau_t} \exp\left(\frac{-t}{\tau_p}\right)$$

The storage delay time associated to the removal of the excess base charge above saturation t_{sd} can then be found at $t = t_{sd}$. The storage delay time is then:

$$t_{sd} = \tau_p \ln\left(\frac{\beta I_B}{I_C}\right)$$

Thus t_{sd} is small for small τ_p and for βI_B small compared to I_C. At boundary conditions for the ON state at saturation bias to OFF at reverse bias for $t < 0$ is such that $i_p(t=0)=I_B$ at the BJT is ON state resulting into $Q_p(t=0)= I_B \tau_p$. For, $t \geq 0$ $i_p(t)=-I_B$ ($-I_B=-E/R_s$ imposed by bias circuit we obtain;

$$-I_B = \frac{Q_p(t)}{\tau_p} + \frac{dQ_p(t)}{dt}$$

$$Q_p(t) = I_B \tau_p \left(2\exp\left(\frac{-t}{\tau_p}\right) - 1\right)$$

And for $t < t_{sd}$ we get that the collector current ic is given by;

$$i_C(t) \approx \frac{E_{CC}}{R_L}$$

$$i_c(t) = \frac{Q_p(t)}{\tau_t} = \frac{I_B \tau_p}{\tau_t}\left(2\exp\left(\frac{-t}{\tau_p}\right) - 1\right) \quad \text{also for } t \geq t_{sd}$$

The resulting storage delay time is thus obtained from:

$$t_{sd} = \tau_p \left[\frac{I_B \tau_p}{I_{Csat} \tau_t \left(\frac{1}{2} + \frac{1}{2}\frac{I_B \tau_p}{I_{Csat} \tau_t}\right)}\right]$$

In this last expression, the additional term in the denominator decrease the storage delay time compared. i.e

$$\left(\frac{1}{2} + \frac{1}{2}\frac{I_B \tau_p}{I_{Csat} \tau_t}\right)$$

BJT Schottky diode clamp

In order to be able to switch-on fast and also switch-off fast without being hindered by the excess oversaturation charge in the base, a clever technique has been invented to prevent the charge build-up of the base region after saturation; this is done by introducing a Schottky diode clamp. A Schottky diode is a majority carrier device that does not store minority carriers and thus switches faster than a pn junction. The on-voltage of the Schottky diode is determined by the metal-semiconductor work function difference. The on-voltage of a pn diode is determined by the built-in voltage as shown in the figure below.

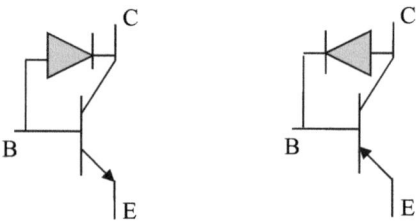

Introducing a Schottky diode (metal-semiconductor junction) keeps the charge in the base region at the value given at time t_s. This is a result of the fact that the Schottky diode switches on (goes into conduction mode) at lower voltages than a pn junction made of the same material. Therefore the Schottky diode will fix the base-collector voltage at the Schottky on-voltage $\ll 0.7V$.

Equivalent BJT circuit

The figure below gives the complete intrinsic small signal equivalent circuit. The EB diode is represented by the differential resistance in parallel with the depletion and diffusion capacitance. The depletion capacitance can be neglected as for amplification the BJT is used with EB forward biased.

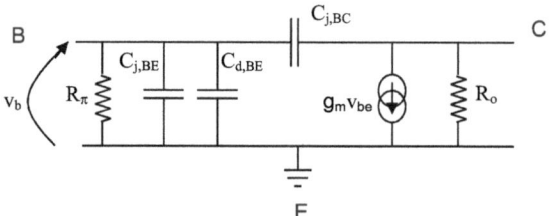

Between base and collector we find the depletion capacitance of the reverse biased BC junction. This capacitance is also referred to as the Miller capacitance and introduces an unwanted feedback network from the output C to the input B. g_m is the transconductance, it is the derivative of the output current i_C to the control voltage v_{be}. R_o is the output resistance. This resistance is associated to base with modulation which imposes an increase of the collector current as a result of the reverse bias BC junction depletion width extending into the base. This growing depletion width with increasing reverse bias reduces the effective base width. This is associated to the early voltage V_A.

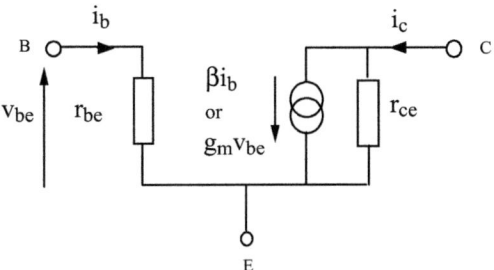

Small-signal current generator h_{fe} is given by;

$$h_{fe} = \frac{dI_C}{dI_B} = \frac{i_c}{i_b} = \frac{\tau_p}{\tau_t}$$

For a short pnp BJT we have:

$$\beta = \frac{I_p}{I_n} = \frac{D_p P_{n0} x_e}{D_n n_{p0} W_b} = \frac{D_p N_A x_e}{D_n N_D W_b}$$

for high gain h_{fe} we require a small base width W_b and a large emitter doping, compared to base doping such that transconductance is given by the expression;

$$g_m \stackrel{\Delta}{=} \frac{\partial I_C}{\partial V_{BE}} = \frac{I_C}{nV_t}$$

So that its input resistance, $r_{be} = r_\pi$ is given by;

$$r_\pi \stackrel{\Delta}{=} \frac{\partial V_{BE}}{\partial I_B} = \beta \frac{\partial V_{BE}}{\partial I_C} = \frac{\beta}{g_m} = \frac{nV_t}{I_B}$$

While its output resistance, $r_{ce}\, r_o$ is obtained from;

$$r_o \stackrel{\Delta}{=} \frac{\partial V_{CE}}{\partial I_C} \cong \frac{\partial V_{CB}}{\partial I_C} = \frac{|V_A|}{I_C}$$

The base-emitter capacitance is due to the diffusion and depletion capacitance of the base-emitter diode. The base-collector capacitance is due to the base-collector depletion capacitance. The cut-off frequency of the BJT can be derived from the small signal current gain:

$$h_{fe} = \frac{i_C}{i_B} = \frac{g_m}{i_B} v = \frac{\beta}{1 + j\omega(C_{j,BE} + C_{d,BE})r_\pi}$$

$$\left|\frac{i_C}{i_B}\right| = 1 \cong \frac{\beta}{2\pi f_T (C_{j,BE} + C_{d,BE})r_\pi}$$

$$f_T = \frac{1}{2\pi \tau}$$

The transit time of the carriers from emitter to collector:

$$\tau = \frac{C_{j,BE} \, nV_t}{I_E} + \frac{w_B^2}{2D_{n,B}} = \tau_E + \tau_B$$

APPENDIX

Interactions of Photons with Matter

Photons are electromagnetic radiation with zero mass, **zero charge**, and a velocity that is always c, the speed of light.

Because they are electrically neutral, they **do not steadily lose energy** through coulombic interactions with atomic electrons, as do charged particles.

Photons travel some considerable distance before undergoing a more

"catastrophic" interaction leading to **partial or total transfer** of thephoton energy to **electron energy**.

These electrons will ultimately deposit their energy in the medium.

Photons are far **more penetrating** than charged particles of similar energy.

Energy Loss Mechanisms

photoelectric effect

Compton scattering

pair production

Interaction probability

linear attenuation coefficient, μ,

The probability of an interaction per unit distance traveled

Dimensions of inverse length (eg. Cm^{-1}).

$$N = N_0 e^{-\mu x}$$

The coefficient μ depends on photon energy and on the material being traversed.

mass attenuation coefficient,

$$\frac{\mu}{\rho}$$

The probability of an interaction per g cm^{-2} of material traversed - Units of $cm^2 \, g^{-1}$

$$N = N_0 \, e^{-\left(\frac{\mu}{\rho}\right)(\rho x)}$$

Mechanisms of Energy Loss: Photoelectric Effect

In the photoelectric absorption process, a photon undergoes an interaction with an absorber atom in which the **photon completely disappears**.

In its place, an energetic **photoelectron** is ejected from one of the bound shells of the atom.

For gamma rays of sufficient energy, the most probable origin of the photoelectron is the most tightly bound or K shell of the atom.

The photoelectron appears with an energy given by

$$E_{e-} = h\nu - E_b$$

(E_b represents the binding energy of the photoelectron in its original shell)

Thus for gamma-ray energies of more than a few hundred keV, the photoelectron carries off the majority of the original photon energy

Filling of the inner shell vacancy can produce **fluorescence radiation**, or x ray photon(s).

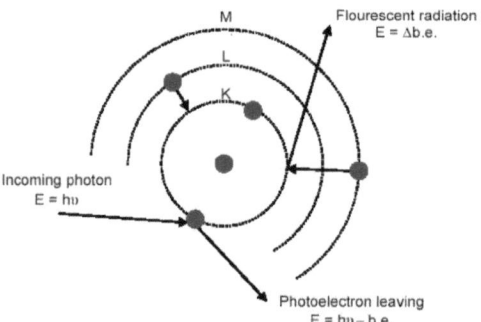

The photoelectric process is the predominant mode of photon interaction at

o relatively low photon energies
o high atomic number Z

The probability of photoelectric absorption, symbolized τ (tau), is roughly proportional to where the exponent n varies between 3 and 4 over the gamma-ray energy region of interest.

This severe dependence of the photoelectric absorption probability on the atomic number of the absorber is a primary reason for the preponderance of **high-Z materials (such as lead) in gamma-ray shields**. The photoelectric interaction is most likely to occur if the energy of the incident photon is **just greater than the binding energy** of the electron with which it interacts.

Compton Scattering

Compton scattering takes place between the incident gamma-ray photon and an electron in the absorbing material.

It is most often the predominant interaction mechanism for gamma-rayenergies typical of radioisotope sources.

It is the most **dominant interaction mechanism in tissue**

In Compton scattering, the incoming gamma-ray photon is **deflected** through an angle θ with respect to its original direction.

The photon transfers a portion of its energy to the electron, which is then known as a *recoil electron,* or a ***Compton electron***.

All angles of scattering are possible.

The energy transferred to the electron can vary from zero to a large fraction of the gamma-ray energy.

The Compton process is **most important** for energy absorption for **soft tissues** in the range from 100 keV to 10MeV.

The Compton scattering probability is is symbolized σ (sigma):

almost *independent of atomic number Z*;

decreases as the photon energy increases

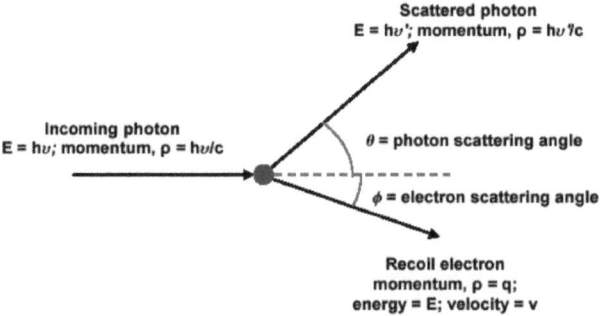

Events in the Compton (incoherent scattering process)

directly proportional to the number of electrons per gram, which only varies by 20% from the lightest to the heaviest elements (except for hydrogen).

Compton Scattering Energetics
Maximum energy transfer to recoil electron:

angle of electron recoil is forward at $0°$, $\psi = 0°$,

the scattered photon will be scattered straight back, $\theta = 180°$

With $\theta = 180°$, $\cos \theta = -1$ the expressions above simplify to:

The Table below illustrates how the amount of energy transferred to the electron varies with photon energy. Energy transfer is not large until the incident photon is in excess of approximately 100 keV.

Photon Energy, 5.11 keV	Photon Energy, 5.11 MeV
$\alpha = \dfrac{5.11 \text{ keV}}{0.511 \text{ MeV}} = 0.010$	$\alpha = \dfrac{5.11 \text{ MeV}}{0.511 \text{ MeV}} = 10$
$E_{e(max)} = 5.11 \text{ keV} \times \left(2 \times \dfrac{0.01}{1.02}\right)$	$E_{e(max)} = 5.11 \text{ MeV} \times \left(2 \times \dfrac{10}{21}\right)$
$= 0.10 \text{ keV}$	$= 4.87 \text{ MeV}$
$h\nu' \text{ (min)} = 5.11 \text{ keV} \times \dfrac{1}{1.02}$	$h\nu' \text{ (min)} = 5.11 \text{ MeV} \times \dfrac{1}{21}$
$= 5.01 \text{ keV}$	$= 0.24 \text{ MeV}$
Energy transferred: 2%	Energy transferred: 95%

For low-energy photons, when the scattering interaction takes place, little energy is transferred, regardless of the probability of such an interaction.

As the energy increases, the fractional transfer increases, approaching 1.0 for photons at energies above 10 to 20 MeV.

Pair Production

If a photon enters matter with an energy in excess of 1.022 MeV, it may interact by a process called *pair production*.

The photon, passing near the nucleus of an atom, is subjected to strong field effects from the nucleus and may disappear as a photon and reappear as a positive and negative electron pair.

The two electrons produced, e- and e+, are **not scattered orbital electrons**, but are created, *de novo*, in the energy/mass conversion of the disappearing photon.

Pair Production Energetics

The kinetic energy of the electrons produced will be the difference between the energy of the incoming photon and the energy equivalent of two electron masses (2 x 0.511, or 1.022 MeV).

$$E_{e+} + E_{e-} = h\nu - 1.022 \text{ (MeV)}$$

Pair production probability, symbolized κ (kappa),

Increases with **increasing photon energy**

Increases with atomic number approximately as Z^2

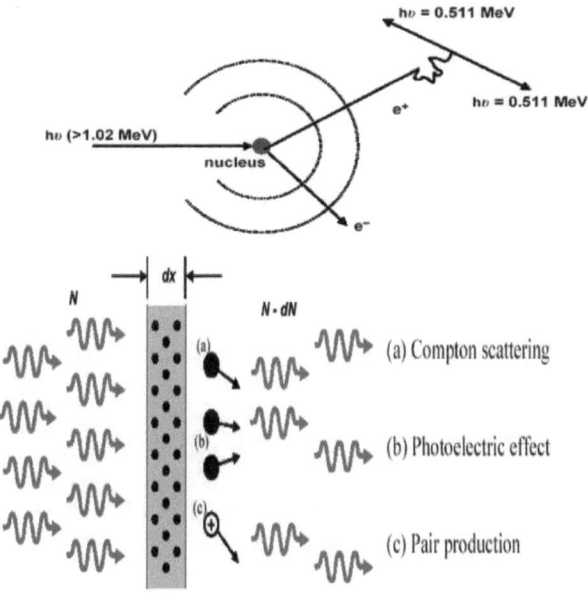

(a) Compton scattering

(b) Photoelectric effect

(c) Pair production

Photoelectric effect: produces a scattered photon and an electron, varies as ~ Z_4/E_3

Compton effect: produces an electron, varies as ~ Z

Pair production: produces an electron and a positron, varies as ~Z_2

Bulk Behavior of Photons in an Absorber

Attenuation Coefficients

Linear attenuation coefficient μ:

The probability of an interaction per unit distance traveled. μ has the dimensions of inverse length (eg. Cm^{-1}).

$N = N_0 e^{-\mu x}$

The coefficient μ depends on photon energy and on the material being traversed.

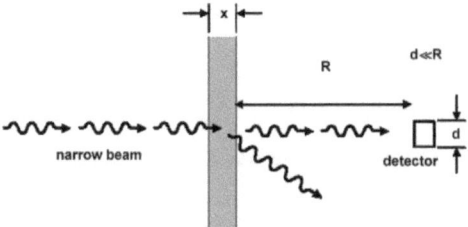

The interaction probability μ is actually the sum of the three possible photon interaction mechanisms:

$$\mu = \tau + \sigma + \kappa$$

τ is the photoelectric effect interaction probability

σ is the Compton scattering interaction probability

κ is the pair production interaction probability

Flux = photons/cm^2 sec or Intensity = energy x flux

$$Intensity(I) = \left(\frac{h\nu}{photon}\right)\left(\frac{\#\, photons}{Area \bullet time}\right)$$

The **mass attenuation coefficient,** μ/ρ, is obtained by dividing μ by the density ρ of the material, usually expressed in $cm^2 g^{-1}$.

Linear Attenuation, Energy Transfer and Energy Absorption

Not all of the energy of the incoming photons that interact in the material is necessarily absorbed there. Energy absorbed = Energy transferred − Energy "lost". *Some energy may be lost from the absorber region due to fluorescence or bremsstrahlung.*

Linear Attenuation: probability of an interaction

Energy Transfer: energy transferred to short-range electrons

Energy absorbed: energy transferred to short-range electrons *minus* the bremsstrahlung photons

A photon interaction will, in general, result in *transfer* of energy to a short range particle (mostly electrons).

The energy transferred to the electron may be *absorbed* within the material, dx, or it may leave the region of interest.

Each of the energy transfer mechanisms has an energy transfer attenuation coefficient and an energy absorption attenuation coefficient.

These coefficients each depend differently on the incoming photon energy and the Z of the absorber. Energy transferred "locally" i.e., to electrons.

$$\tau_{tr} = \tau\left(1 + \frac{\delta}{h\nu}\right)$$ - photoelectric effect: δ is the average energy emitted as fluorescence

$$\sigma_{tr} = \sigma \frac{E_{avg}}{h\nu}$$ - Compton scattering: $E_{avg}/h\nu$ is the average photon energy converted into electron energy

$$\kappa_{tr} = \kappa\left(\frac{h\nu - m_0 c^2}{h\nu}\right)$$ - pair production: $(h\nu - mc^2)/h\nu$ is the fraction of energy converted to photons by $\beta^+\beta^-$ annihilation.

Energy Transferred and Energy Absorbed for Incident Photons of Various Energy (for Carbon)

Photon energy, E_{tot} (MeV)	Average energy transferred, E_{tr} (MeV)	Average energy absorbed, E_{ab} (MeV)
0.01	0.00865	0.00865
0.10	0.0141	0.0141
1.0	0.440	0.440
10.0	7.30	7.04
100.0	95.63	71.90

$$\mu_{en} = \mu_{tr}(1-g)$$

$$\mu_{tr} = \tau_{tr} + \sigma_{tr} + \kappa_{tr}$$

µtr is the total kinetic energy of all electrons produced by photons

µtr takes no account of subsequent bremsstrahlung

$$\mu_{en} = \tau_{tr}(1-g)$$

g is the average fraction of initial photon energy, hv, transferred to electrons and subsequently emitted as bremsstrahlung. g is largest for high-Z absorbers;

$$I = I_0 e^{-\mu x}$$

$$E_{ab} = E_{tr}(1-g)$$